Ernst Heinrich Hirschel (Ed.)

Finite Approximations in Fluid Mechanics

Notes on Numerical Fluid Mechanics
Volume 14

Ernst Heinrich Hirschel (Ed.)

Finite Approximations in Fluid Mechanics

DFG Priority Research Programme
Results 1983–1985

V

Friedr. Vieweg & Sohn Braunschweig/Wiesbaden

CIP-Kurztitelaufnahme der Deutschen Bibliothek

Finite approximations in fluid mechanics:
DFG priority research programme results 1983-1985 /
Ernst Heinrich Hirschel (Ed.). – Braunschweig;
Wiesbaden: Vieweg, 1986.
(Notes on numerical fluid mechanics; Vol. 14)
ISBN 3-528-08088-4

NE: Hirschel, Ernst Heinrich [Hrsg.]; Deutsche
Forschungsgemeinschaft; GT

Manuscripts should have well over 100 pages. As they will be reproduced photomechanically they should be typed with utmost care on special stationary which will be supplied on request. In print, the size will be reduced linearly to approximately 75 %. Figures and diagrams should be lettered accordingly so as to produce letters not smaller than 2 mm in print. The same is valid for handwritten formulae. Manuscripts (in English) or proposals should be sent to the general editor Prof. Dr. E. H. Hirschel, Herzog-Heinrich-Weg 6, D-8011 Zorneding.

Produced by Lengericher Handelsdruckerei, Lengerich
Printed in Germany

ISSN 0179-9614

ISBN 3-528-08088-4

This volume is dedicated to

Prof.Dr. E.A. Müller, Göttingen

on the occasion of this sixtieth birthday

Foreword

This volume contains twenty reports on work which was conducted between 1983 and 1985 in the Priority Research Programme "Finite Approximations in Fluid Mechanics" of the Deutsche Forschungsgemeinschaft (DFG, German Research Society).

The main purpose of this publication is to give an insight into the scope of the programme, to demonstrate the state of the art, to encourage cooperation in the future, and to point to new problems. All papers are contributed voluntarily, and the authors remain fully responsible for the contents.

E.H. Hirschel
Editor

Preface

The Deutsche Forschungsgemeinschaft sponsors the development of numerical methods in fluid mechanics since 1983 in a Priority Research Programme "Finite Approximations in Fluid Mechanics". One of the major aims of the programme is to construct numerical solutions of the conservation equations for inviscid and viscous flows with complex boundaries, for boundary layers, wakes, and jets, for internal flows, for transitional and turbulent flows, and for diffusion processes. In a number of investigations cooperation between mathematicians and engineers could be initiated in the frame of the programme. Improvement of the rate of convergence, construction of suitable grids, development of discretization techniques, and careful analysis of the solution properties broadened the road to some of the nonlinear, time-dependent and three-dimensional problems mentioned earlier, at least to some extent.

The undersigned take this opportunity to thank the Deutsche Forschungsgemeinschaft for initiating and supporting the programme over the past three years. Grateful acknowledgement is also due to the reviewers of the programme: to Prof.Dr. E.Becker, who passed away much too early in 1984. He shaped the work of the programme substantially, in particular in the phase of its initiation, through his always encouraging and constructive criticism: to Profs. Dr. R.Ansorge, Dr. G.Böhme, and Dr. R.Rannacher, whose invaluable expertise helped to steer the programme safely through the annual reviews. We also thank Prof.Dr. H.G.Dohmen and Dr. W.Lachenmeier for administering the programme in various councils and committees of the DFG.

Finally we wish to thank the Vieweg Verlag for publishing the results of our programme in the Notes on Numerical Fluid Mechanics, and Prof.Dr. E.H. Hirschel for editing this volume.

Bonn-Bad Godesberg, December 1985

E. Krause E.A. Müller

CONTENTS

CONTENTS (continued)

MULTIGRID METHODS FOR PROBLEMS FROM FLUID DYNAMICS - DEVELOPMENT OF A 2D TRANSONIC POTENTIAL FLOW SOLVER

K. Becker *)
Institut für Methodische Grundlagen
GMD, St. Augustin

Introduction

Multigrid methods have a more complex structure than other "old fashioned" iterative methods for the solution of large linear or nonlinear systems like e.g. Gauß-Seidel, SOR, Jacobi, ADI etc. Each of these methods can be a part of a multigrid algorithm. So "multigrid" is on a higher level. It is an idea or a strategy which leads to special algorithms for special problems of application. Therefore - in contrast to most of the methods mentioned above - users remain with the problem of multigrid design or development.

Special MG-problems arise in the calculation of transonic two-dimensional full potential flow. A first hint on the difficulties is given by the observation of increasing convergence factors if the maximum speed of the flow is increased (e.g. by increasing the free stream Mach number or the angle of attack). This dependency of the convergence factors on some problem-dependent parameter is not very nice and should be eliminated.

The first reason for this dependency seems to be a lack of smoothing in the relaxation algorithm originally proposed for the hyperbolic equations. This can be overcome by introducing a latest value type of relaxation which on the other hand unfortunately is divergent for some Fourier components. A combination of both gives fairly good results - as good as in subsonic flow.

But as the maximum Mach number is increased further the convergence properties change abruptly. A pulsing of the supersonic region (i.e. the shock location) can be observed which results in a "convergence"

*) The author is now with TE212, MBB-UT, Hünefeldstr. 1-5, 2800 Bremen 1

factor of about 1. The fine grid smoothing and the coarse grid correction processes seem to fight against each other instead of working together. The reason is to be seen in the FAS-treatment of the shock position which in our opinion cannot be overcome by changing only parts of the algorithm. The discretization has to be adapted to allow shocks on grid lines as well as in between.

1. Essentials of the Multigrid Idea

The idea of multigrid results from a close look at the convergence behaviour of iterative methods, i.e. relaxation methods. Most of them damp high frequency Fourier components of the error between the current approximation and the final solution very fast, but they are very slow for low frequency Fourier components [3], [6], [13]. This means that the error is smooth after a few relaxation sweeps and that the overall slow convergence is caused by low frequency Fourier components.

Taking into account that the defintion of high and low frequency depends on the grid size, i.e. a low frequency component of a grid G_h may be a high frequency component of the next coarser grid G_{2h} (see Figure 1), we end up with the idea of multigrid: mainly separate treatment of the different Fourier parts of the error on different grids. Instead of working on a problem only on one grid a sequence of fine and coarse grids is used. This sequence is run through in a prescribed manner applying relaxation sweeps and grid transfer operations (see Figure 2).

To avoid a misunderstanding this look at Fourier components is only a theoretical one. In practical computations one never has to do something like Fourier analysis.

2. Examples of Applications for Multigrid Ideas

Starting point for the application of multigrid methods are elliptic differential equations and elliptic systems [3]. Fourier analysis is a common thing for this type of partial differential equations, e.g. in the definition of ellipticity. Besides this some elliptic model problems allow exact proofs of the convergence of multigrid algorithms. Realistic estimates of the very high speed of convergence have been achieved, and the typical and very important result is that the speed of convergence is independent of the number of unknowns of the discrete system [6] , [13] .

Today multigrid methods are applied to the solution of any kind of elliptic boundary value problem , with all kinds of boundary conditions and on nearly arbitrary domains of integration. We can mention e.g.

- problems with non-constant coefficients,
- nonlinear problems (with solution-dependent coefficients),
- problems with mixed or/and first order derivatives,
- problems with anisotropic operators or singular perturbations,
- systems of differential equations.

The application of multigrid is by no ways restricted to elliptic problems. Parabolic and hyperbolic problems with some kind of a stationary solution have been treated by multigrid methods since some time [11] . At least ideas of the multigrid concept have a strong influence on the development of solution algorithms for the Euler equations (e.g. Jameson [9] , Ni [12] , Jespersen [10] ,...). The main idea is to extend the multi-grids to the dimension of time.

Additionally there are a lot of multigrid algorithms for mixed type problems like the full potential equation [5] , [14] or for hybrid type problems like the stationary Euler equations [7] .

Theoretical investigations have been done for most of the problems mentioned above by Brandt [3] , Hackbusch [6] , Stüben/Trottenberg [13] and many others. From their results one can extract some guidelines for the development of concrete multigrid algorithms.

3. Multigrid treatment of the full potential equation

The development of multigrid algorithms for the calculation of two-dimensional compressible potential flows around airfoils gives rise to several problems: the treatment of the mixed type equation, the unboundedness of the domain, the jump of the potential across the Kutta-cut and the Neumann boundary condition on the profile. All these points can be handled without much computational effort [1], [2].

But much more severe problems arise in the treatment of shocks. At certain flow conditions we observed an instability of the convergence behaviour of our algorithm. This instability seems to be due to the location of the shock with respect to the three finest grids used for discretization [2].

3.1. The Continuous Problem

The perturbation potential ϕ of a flow around airfoils at subsonic free stream velocity suffices the full potential equation

$$(a^2 - u^2)\phi_{xx} - 2uv\phi_{xy} + (a^2 - v^2)\phi_{yy} = 0 \tag{1}$$

where

$$a^2 = \frac{1}{M_\infty^2} - \frac{\gamma - 1}{2}(q^2 - 1)$$

$$q^2 = u^2 + v^2$$

$$u = \phi_x + \cos\alpha$$

$$v = \phi_y + \sin\alpha .$$

M_∞ is the free stream Mach number, γ the ratio of specific heats (γ = 1.4 for air), α is the angle of attack.

As the flow is along the surface of the airfoil, the potential has to fulfill the Neumann boundary condition

$$\frac{\partial\phi}{\partial n} = -\frac{\partial\tilde{\Phi}}{\partial n} \tag{2}$$

where $\tilde{\Phi} = x\cos\alpha + y\sin\alpha$ is the potential of the unperturbed flow and

n is the outer normal vector on the profile.

Far away from the airfoil we have the usual far field condition

$$\phi = \frac{\Gamma}{2\pi} \arctan \left(\sqrt{1 - M_\infty^2} \tan(\Theta - \alpha) \right) \quad (3)$$

where Γ is the circulation of the flow and Θ is the polar angle.

Besides these conditions the potential has to suffice a so called Kutta condition which can be expressed as forcing continuity of velocity at the trailing edge of the profile.

3.2. Discretization

The quasilinear full potential equation is discretized on a cartesian grid like described in the paper of Carlson [4] . In the region of subsonic flow where equation (1) is elliptic central differencing is used whereas in the supersonic zone appearing in transonic flows - where equation (1) is hyperbolic - Jameson's rotated difference scheme is used. The velocity components u and v are always evaluated from central difference formulas. Some care has to be taken to the discretization of the Neumann boundary condition (2). A rather straightforward Taylor expansion leads to a 7-point difference molecule involving gridpoints inside the profile, two cartesian components of the boundary normal vector and the distance of a boundary point to the grid. Figure 3 shows a typical example.

The discretization has to be chosen adaptively by the solution algorithm because the location and the size of the supersonic zone are not known in advance. Shocks which mathematically appear as kinks in the potential function are not treated by special shock operators. So they always turn out to be in the discrete sense situated on grid points or lines. Their width is three mesh cells.

3.3 Multigrid Strategy

Within the MG-context a local grid refinement is used near the profile (see Figure 4) that on the one hand allows very high resolution on the surface (i.e. up to a meshsize of 1/200 on a normalized profile). On the other hand it gives the possibility to reach a far field distance of about 10 chords without any grid stretching or similar coordinate transformations. Coarsest grids with a meshsize of 2 - 3 times the chord length of the profile can be used.

Our algorithm is a typical FMG-FAS-Multigrid algorithm based on a mixed W-V-Cycle which is designed to efficiently reduce the work on coarse grids. (Due to the refinement strategy coarse grids may have the same number of unknowns as fine grids.)

Full residual weighting and bilinear interpolation of corrections are used as intergrid transfer operators.

The relaxation procedure is a successive line relaxation where the lines are chosen as columns, i.e. nearly perpendicular to the main flow direction.

The solution on the coarsest grid is approximately calculated by 4 - 5 relaxation sweeps.

A special technique is used for the update of the circulation and the far field boundary values which is relevant for flows with lift. Both the circulation and the far field boundary values are changed only on the coarsest grid. The fine grid information which naturally has to be taken into account is brought over by the residual transfer for the circulation equation. This kind of treatment seems to be the best with respect to the global character of the circulation variable.

4. Convergence Analysis

With the algorithm outlined above we observed the convergence factors shown in Table 1 for subsonic flows at different free stream Mach numbers. The mean values for the reduction of maximum residuals per cycle are taken over 8 iterations and show out to be smaller than 0.1 independent of the free stream conditions*).

For transonic flows we have analyzed two different relaxation procedures for the hyperbolic equations. They differ by the weighting between "old" and "new" values entering the correction of the potential at a certain point.

The relaxation procedure proposed by Jameson [8] leads to the convergence factors shown in Table 2. These factors clearly show an increasing behaviour with respect to an increasing Mach number M_∞. This behaviour is predicted by the local mode analysis [2] : In the elliptic region there is a smoothing factor of about 0.447 which gives an estimation of 0.089 for a (2,1)-Cycle. So we expect a convergence factor below 0.1 in the subsonic case which is really found. In the hyperbolic region the smoothing factor is about 0.707 which gives an estimate of 0.353 for the convergence factor.

As the supersonic zone increases we expect an increase of the convergence factors from below 0.1 to about 0.3, and this is clearly shown in Table 2. The result of this observation is that the convergence behaviour of this MG-algorithm in transonic flows is dominated by the slow convergence in the hyperbolic region.

To overcome this remedy we took a latest value type of relaxation in the hyperbolic region which is characterized by mostly using "new" values. (For a clear description we refer to our previous work [2] .) For this type of relaxation we obtain an average smoothing factor in the hyperbolic region which is far below 0.447. So we expect the overall convergence to be dominated by the convergence behaviour in the elliptic region. This is clearly demonstrated in Table 3: Allmost all convergence factors are below 0.1.

At $M_\infty = 0.79$ a problem occurs that gets harder when the free stream Mach number is increased: The convergence becomes unstable or unreliable at some distinct values of M_∞.

---

*) We omit a flow with lift here, but the same result is also true for that case [1] .

If we take for example the case $M_\infty = 0.81$ we get the graph of maximum residuals after each iteration as shown in Figure 5. The behaviour is typical for all cases called non-convergent in Table 4. The mean value of residual reduction is about 1, so that we speak of non-convergence rather than divergence. For the purpose of comparison also the convergence history of the convergent case $M_\infty = 0.84$ is shown.

In [2] we have reported upon a lot of possible changes in the algorithm to overcome this convergence instability, but unfortunately we did not find the right way to do it. The following observation can give a first hint to a solution of the stability problem.

As long as the shock is aligned with the 3 finest grids we get very fast convergence. But as far as the shock is in between coarse grid points (or lines), the coarse grid correction step of the multigrid algorithm produces wrong corrections in the neighbourhood of the shock. This is because the discretization does not allow shocks in between grid points. So the shock moves even if it is already at the right location on the fine grid. Figure 6 shows typical distributions of the correction potential parallel to the x-axis close to the profile. In the convergent case there is no visible difference between the potential before and after the coarse grid correction though the correction function shows a strong oszillation and steep gradients. In the non-convergent case there is a similar behaviour of the correction function, and the potential itself exhibits only small changes in the region upstream of the shock. But obviously the shock has moved. This causes a large correction everywhere downstream the shock and gives a totaly wrong prediction for the new iterate.

To overcome this difficulty the discretization has to be changed in such a way that it allows shocks in between grid lines. We think this is a necessary condition for the convergence of the whole algorithm. Work is now going on to achieve a uniform convergence behaviour, i.e. a high speed of convergence which is - within the natural bounds for the physical model of the full potential equation - independent of the flow situation.

References

[1] Becker, K.: Ein Mehrgitterverfahren zur Berechnung subsonischer Potentialströmungen um Tragflächenprofile, GMD-Bericht Nr. 152, Gesellschaft für Mathematik und Datenverarbeitung mbH, Bonn, 1985.

[2] Becker, K.: Numerische Berechnung transonischer Potentialströmungen um Tragflächenprofile - Untersuchung eines Mehrgitterverfahrens, GMD-Arbeitsbericht, Gesellschaft für Mathematik und Datenverarbeitung mbH, Bonn, 1985.

[3] Brandt, A.: Multigrid Techniques: 1984 Guide with Applications to Fluid Dynamics, GMD-Studie 86, Gesellschaft für Mathematik und Datenverarbeitung mbH, Bonn, 1984.

[4] Carlson, L.A.: Transonic Airfoil Flowfield Analysis Using Cartesian Coordinates, NASA CR-2577, Washington, DC, 1975.

[5] Caughey, D.A.: Multigrid Calculation of Three-Dimensional Transonic Potential Flows, Appl. Math. Comp. 13 (1983), 241-260.

[6] Hackbusch, W.: Multi-grid Convergence Theory, Lecture Notes in Mathematics 960 (1982), 177-219.

[7] Hemker, P.W.: Defect Correction and Higher Order Schemes for the MG-Solution of the Steady Euler Equations, Proc. 2nd Europ. Conf. Multigrid Methods, Köln, 1.-4. Oktober 1985.

[8] Jameson, A.: Transonic Flow Calculations, Numerical Methods in Fluid Dynamics (H.J. Wirz, J.J. Smolderen, Eds.) Hemisphere Publishing Corp., Washington, London, 1978.

[9] Jameson, A.: Numerical Solution of the Euler Equation for Compressible Inviscid Fluids, Princeton University, Report MAE 1643, 1984.

[10] Jespersen, D.C.: Recent Developments in Multigrid Methods for the Steady Euler Equations, Lecture Series on Computational Fluid Dynamics, von Kármán Institute for Fluid Dynamics (1984).

[11] Kroll, N.: Direkte Anwendungen von Mehrgittertechniken auf parabolische Anfangsrandwertaufgaben, Diplomarbeit, Institut für Angewandte Mathematik, Universität Bonn, Bonn, 1981.

[12] Ni, R.H.: A Multiple Grid Scheme for Solving the Euler Equations, Proc. AIAA 5th Computational Fluid Dynamics Conference, Palo Alto, 1981, 257-264.

[13] Stüben, K., Trottenberg, U.: Multigrid Methods: Fundamental Algorithms, Model Problem Analysis and Applications, Lecture Notes in Mathematics 960 (1982), 1-176.

[14] Van der Wees, A.J.: Robust Calculation of 3D Transonic Potential Flow Based on the Non-Linear FAS Multi-Grid Method and a Mixed ILU/SIP-Algorithm, NLR MP 84003 U (1984).

Figures

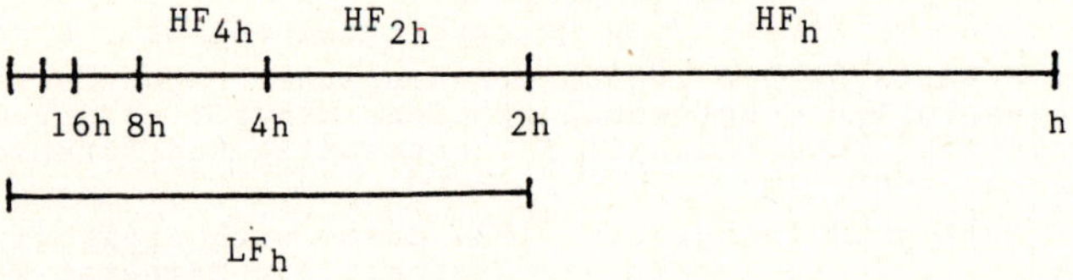

Figure 1: High Frequency and Low Frequency ranges depending on the grid size

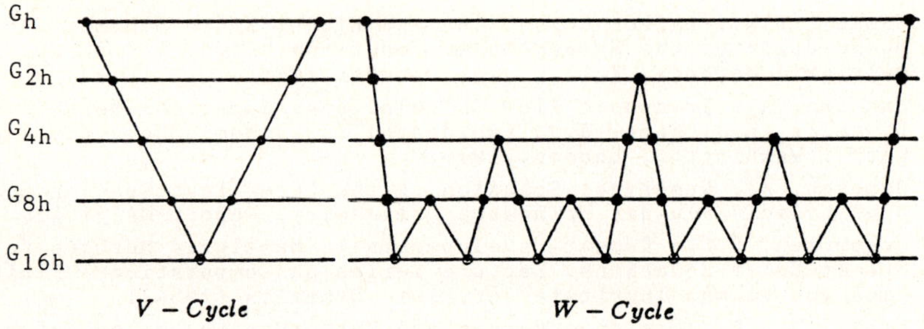

Figure 2: Different possibilities of multigrid cycling

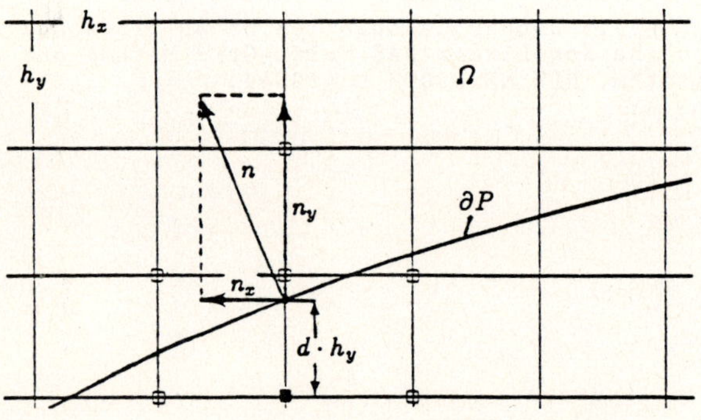

Figure 3: Discretization of Neumann boundary condition

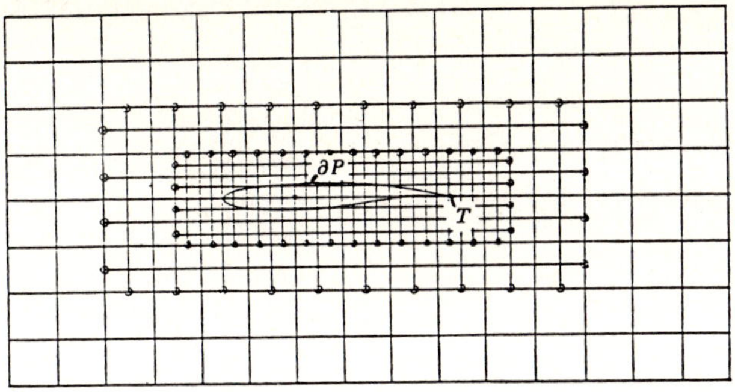

Figure 4: Local grid refinement near the profile

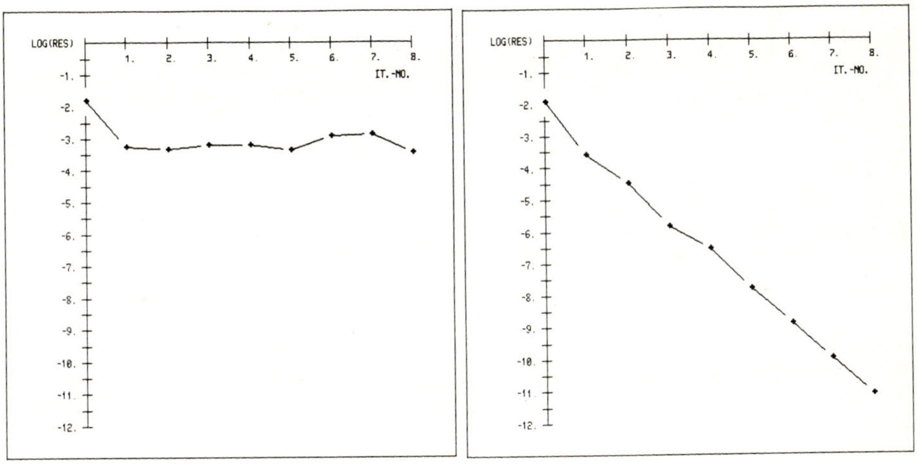

non-convergent case, $M_\infty = 0.81$ convergent case, $M_\infty = 0.84$

Figure 5: Convergence histories for two different flow situations, NACA0012, $\alpha = 0°$, $h_x = h_y \approx 1/46$ (8 Grids)

potential correction

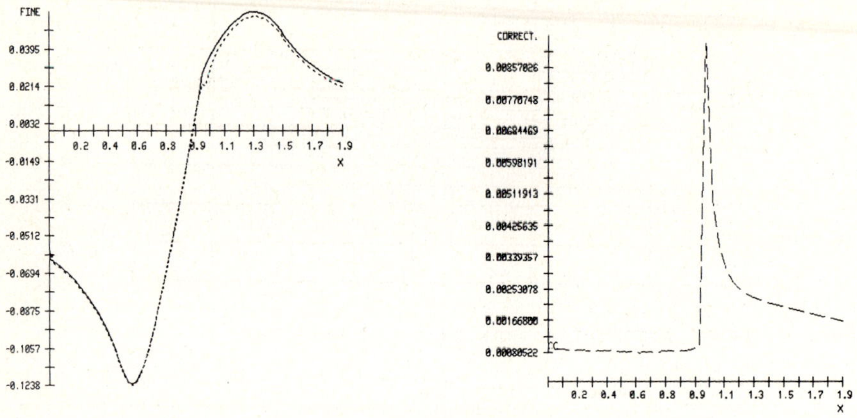

non-convergent case, NACA0012, $M_\infty = 0.81$, $\alpha = 0°$

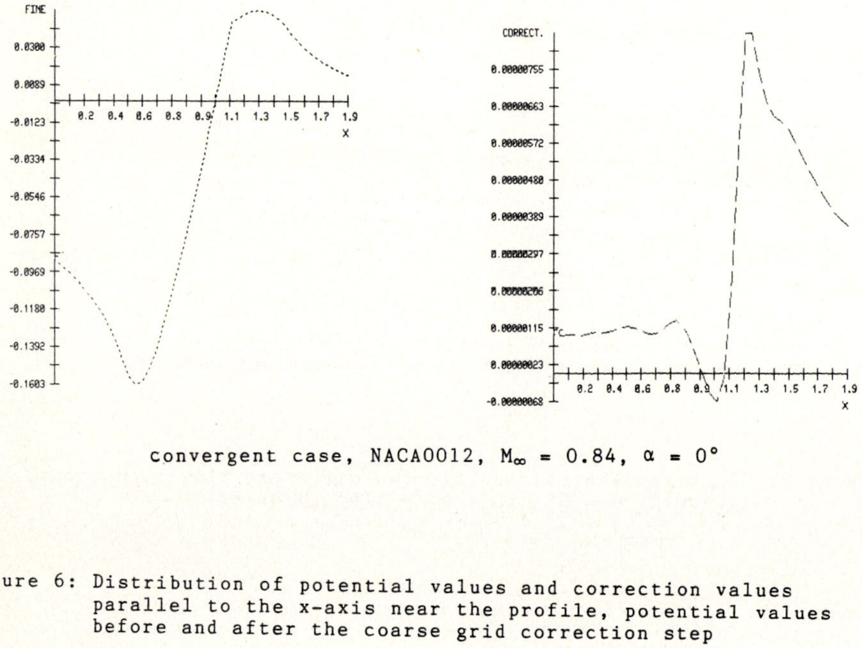

convergent case, NACA0012, $M_\infty = 0.84$, $\alpha = 0°$

Figure 6: Distribution of potential values and correction values parallel to the x-axis near the profile, potential values before and after the coarse grid correction step

Tables

M_∞	0.10	0.20	0.30	0.40	0.50	0.60	0.70
C_h/CY	.073	.071	.068	.064	.058	.053	.048

Table 1: Convergence factors per MG-Cycle, subsonic flows, NACA0012, $\alpha = 0°$, $h_x = h_y \simeq 1/183$ (10 Grids)

M_∞	0.72	0.73	0.74	0.75	0.76	0.77	0.78	0.79	0.80
C_h/CY	.059	.054	.051	.059	.115	.228	.241	.338	.305

Table 2: Convergence factors per MG-Cycle using Jameson type of relaxation (transonic flows), NACA0012, $\alpha = 0°$, $h_x = h_y \simeq 1/46$ (8 Grids)

M_∞	0.73	0.74	0.75	0.76	0.77	0.78	0.79	0.80
C_h/CY	.054	.051	.053	.053	.102	.090	.237	.084

Table 3: Convergence factors per MG-Cycle using latest value type of relaxation, NACA0012, $\alpha = 0°$, $h_x = h_y \simeq 1/46$ (8 Grids)

M_∞	0.78	0.79	0.80	0.81	0.82	0.83	0.84	0.85
C_h/CY	.117	.321	.074	1.16	.093	.923	.081	.581

Table 4: Convergence factors per MG-Cycle using mixed type of relaxation (Jameson + latest value), non-convergent cases at distinct values of M_∞, NACA0012, $\alpha = 0°$, $h_x = h_y \simeq 1/46$ (8 Grids)

A BOUNDARY ELEMENT SPECTRAL METHOD FOR NONSTATIONARY VISCOUS FLOWS IN 3 DIMENSIONS

W. Borchers, F.K. Hebeker, R. Rautmann
Fachbereich Mathematik-Informatik der Universität
Warburger Straße 100
D-4790 Paderborn, Germany

1.1. INTRODUCTION

More than 50 years after J. Leray's pioneering work and more than 30 years after E. Hopf's fundamental results the Navier-Stokes equations remain a challenge as well for pure as for numerical mathematicians. Namely in the theory, as a central point, the well-posedness of the initial-boundary value problem of these equations could not be settled for the general 3-dimensional case, inspite of tremendous mathematical efforts. On the other side the numerous new theoretical approaches are opening many new ways for numerical approximations.

This is of special practical interest, since in the contrast to their individual merits, even the best known numerical approaches show serious disadvantages in practice: The method of eigenfunction expansion, which usually offers the most natural approximation scheme, led to many theoretical insights (concerning existence, regularity, asymptotic behaviour, error estimates and optimum degree of approximation [15,26,30]). However, complete systems of Stokes eigenfunctions have been discovered recently and for special domains only [2,34]. Therefore this method could not be used for numerical purposes up till now. On the other side, finite difference - or finite element methods are restricted by the huge practical difficulties in 3-dimensional domains of a complicated geometry.

Hence *boundary* element methods have been developed since a decade, for they are convenient to handle and time-saving in comparison with finite element methods. They have been successfully applied to a large variety of problems arising in the engineering sciences (see e.g. [3],[38]). Additional fields of applications may be opened also in viscous fluid dynamics, if boundary element methods are combined with fast spectral methods to overcome the crucial problem of the nonhomogeneous terms in the differential equations.

Therefore the aim of our studies, which partly will be reviewed in the following, is to develop an efficient mixed numerical method for nonstationary viscous incompressible flows. The "boundary element spectral method" described below combines functional analytic tools with boundary element methods [10] and a fast spectral method [2]. The latter uses a complete system of Stokes eigenfunctions in the cube [2,34] in order to handle the nonhomogeneous terms in the differential equations.

1.2. THE INITIAL-BOUNDARY VALUE PROBLEMS. TIME-DISCRETIZATION. NOTATIONS.

We consider the nonstationary interior Stokes problem

$$\frac{\partial}{\partial t} u - \Delta u + \nabla p = f, \quad \text{div } u = 0 \quad \text{in } \Omega , \tag{1.1}$$

$$u_{|\partial\Omega} = g \quad \text{on the boundary } \partial\Omega ,$$

$$u(0,x) = u_o(x) \text{ at the initial time } t = 0$$

in a smoothly bounded domain Ω of 3-space $\mathbb{R}^3$. The vector $u(t,x) = (u_1, u_2, u_3)$ or the positive value $p(t,x)$ denotes the velocity (at time t at the point $x \in \bar{\Omega}$), or the pressure divided by the mass density, respectively, of a viscous incompressible flow in Ω. If additionally a conservative exterior force is prescribed, its potential can be added to p. We assume the viscosity and the mass density of the flow to be constant, both being equal to 1.

In addition we consider the exterior nonstationary Stokes problem

$$\frac{\partial}{\partial t} u - \Delta u + \nabla p = f, \quad \text{div } u = 0 \quad \text{in } \mathbb{R}^3 \setminus \bar{\Omega} \tag{1.2}$$

$$u_{|\partial\Omega} = g, \quad u_{|\infty} = 0 \quad \text{at infinity,}$$

$$u(o,x) = u_o(x) \ .$$

The linear Stokes problems with prescribed f and g play a central rôle in the development of efficient numerical algorithms for the viscous fluid dynamics, since the Navier-Stokes initial-boundary value problem has the form (1.1) or (1.2) with the right hand side

$$f = - u\nabla u \equiv - \sum_{j=1}^{3} u_j \frac{\partial}{\partial x_j} u \ .$$

This *nonlinear* problem is commonly reduced to the linear one by means of iterative or time-stepping procedures, the convergence of which can be guarantied however only locally in time or under smallness assumptions.

For the numerical treatment of (1.1) or (1.2) we introduce the time grid $t_k = k \cdot h$ with a step length $h > o$ and $k = o, 1, \ldots$. By f_k, g_k we denote the restrictions of f, g to $t = t_k$. Substituting the difference quotient $h^{-1} \cdot (u_k - u_{k-1})$ in (1.1), (1.2) instead of $\frac{\partial}{\partial t} u$, we find the resolvent equation

$$(\lambda - \Delta)\, u_k + \nabla p_k = f_k + \lambda u_{k-1} , \quad \text{div } u_k = 0 \text{ in } \Omega \tag{1.3}$$

$$u_{k|_{\partial\Omega}} = g_k$$

for the Stokes problem. Since the resolvent parameter $\lambda = 1/h > 0$ belongs to the resolvent set of the Stokes operator, the boundary value problem (1.3) is uniquely solvable if f_k, g_k, u_k and $\lambda > 0$ are given in a suitable way. Evidently (1.3) holds with $u_k = u(t_k, x)$, $p_k = p(t_k, x)$ approximately (with an error term of order $O(h)$) for any sufficiently smooth solution u, p of (1.1) or (1.2). With $k = 1, 2, \ldots$, (1.3) represents the Rothe-scheme or backwards Euler-scheme for the successive approximate computation of $u(t,x)$ and $p(t,x)$ starting with the initial value u_o of u. For the interior problem stability and convergence of the difference scheme (1.3) follow in the framework of the general theory [5,29].

The fundamental matrix of the elliptic system (1.3) (in the sense of Douglis-Nirenberg) is known and gives a representation of the unique solution u_k, p_k in terms of g_k and the volume potential of $f_k + \lambda u_{k-1}$. This allows the computation uf u_k, p_k by the boundary element method [10] (see section 3 below) in an efficient and time saving way. Most of the computing time would be needed for the evalution of the volume potential, if a numerical integration scheme would be used. Therefore, an efficient spectral method has been developed for this task [2] (see section 4 below).

1.3. A REMARK ON THE NONLINEAR CASE

In the nonlinear case (which will not be treated here in detail) with a suitable approximation for the term $f = - u\nabla u$ [27], we can approximate solutions of the 3-dimensional Navier-Stokes problem by means of the procedures described in section 3 and 4 in the linear case. Proofs of stability and convergence of these Navier-Stokes approximations are available locally in time (or globally for "small" initial values) in the framework of more general approaches [32,35]. The convergence of the approximations and their spatial partial derivatives up to the second order *on the boundary* $\partial\Omega$ of the flow region is of central importance for the computation of flow separation. Schemes for high-order approximations of this type have been developed in [31,32,35].

2. THE BOUNDARY INTEGRAL EQUATIONS

The resolvent equations form a generalized elliptic systems in the sense of Douglis and Nirenberg. However, a simple fundamental matrix

$$\Gamma_{ij}(x) = - \frac{\delta_{ij}}{4\pi|x|} e^{-\sqrt{\lambda}|x|} + \frac{1}{4\pi\lambda}\Big[(\frac{\delta_{ij}}{|x|^3} - 3\frac{x_i x_j}{|x|^5})(1-e^{-\sqrt{\lambda}|x|} - \sqrt{\lambda}|x|e^{-\sqrt{\lambda}|x|}) + \frac{\lambda x_i x_j}{|x|^3} e^{-\sqrt{\lambda}|x|}\Big] \quad (2.1)$$

exists, and the ith column $\Gamma^{(i)}$ of Γ solves these equations with pressure function

$$P_i(x) = -x_i/(4\pi|x|^3) + \text{const.} \quad (2.2)$$

A hydrodynamical potential theory, fundamental to any boundary element method, is available (see Varnhorn [35]). In particular we are given a Green's representation formula for any regular solution of the differential equations as a sum of a (hydrodynamical) volume -, simple layer -, and double layer potential.

A special attention should be paid to the volume potential. In [12] a useful formula of numerical cubature is developed. But, in any case, the numerical evaluation of volume potentials is an extremely expensive procedure (see [35]). Hence to circumvent this task the way of using fast spectral solvers has independently been found and subsequently elaborated by W. Borchers and F.K. Hebeker. At first we suppress this task by putting $f = 0$ in (1.1) and (1.2), we will return to the general case in Sec. 4.

The interior problem (1.1) is solvable if the data satisfy (see [20])

$$\int_{\partial\Omega} (g\cdot n)do = 0 . \tag{2.3}$$

In this case the ansatz

$$u = W\psi, \quad p \text{ analogous} \tag{2.4}$$

in terms of a hydrodynamical double layer potential

$$(W\phi)_i(x) = \int_{\partial\Omega} t'\ (\Gamma^{(i)}(x-z))\cdot\phi(z)do_z, \tag{2.5}$$

where

$$t'(u,p) = pn + \Sigma(\nabla u_j)n_j + \frac{\partial u}{\partial n} \tag{2.6}$$

denotes the (adjoint) stress vector and n the normal vector (exterior w.r.t. Ω), leads (by means of the jump relations) to the boundary integral equations system of the second kind (cf. Borchers [1])

$$(I + 2W + 2N)\psi = 2g \quad \text{on } \partial\Omega \tag{2.7}$$

to determine the unknown surface source vector ψ. Here denotes I the identity matrix and N the boundary operator

$$N\phi(x) = n(x) \int_{\partial\Omega} (n\cdot\phi)do. \tag{2.8}$$

It turns out [1] that (2.8) has a unique solution ψ (when g satisfies (2.3)), and consequently the corresponding potential (2.4) solves the problem (2.1).

The exterior problem (1.2) is unconditionally solvable (if $f = 0$), and the ansatz

$$u = W\psi , \quad p \text{ analogous} \tag{2.9}$$

in terms of a hydrodynamical double layer potential produces the system

$$(I - 2W)\psi = -2g \quad \text{on } \partial\Omega. \tag{2.10}$$

It has been previously shown [35], that (2.10) has a unique solution ψ. Then the corresponding potential (2.9) solves the problem (1.2).

The surface potentials in (2.7) and (2.10) form weakly singular integrals on the boundary, hence these integral equations are well fitted for numerical purposes. For the evaluation of the potentials (2.4) and (2.9) in the space the Gaussian integral [20] should be utilized.

3. THE BOUNDARY ELEMENT METHOD

The most simple and efficient way to discretize the boundary integral equations seems to be a collocation-type boundary element method. Let us describe it for the typical case of a body "similar to a ball", i.e. represented by (normalized) polar coordinates:

$$x \in \partial\Omega : x = F(\theta,\varphi), \ (\theta,\varphi) \in [0,1]^2. \tag{3.1}$$

Note that a large class of bodies including even those with corners and edges are representable in this way. The following method has been developed in [10], [12].

The method consists of decomposing the parameter space $[0,1]^2$ of the boundary into small quadratic elements Q of mesh size h. As trial functions we use globally continuous and piecewise bilinear polynomials on the parameter space $[0,1]^2$, subordinated to the quadrangulation. Then we are looking for an approximate surface source ψ_h of this kind so that the unknown degrees of freedom are computed from the collocation equations

$$(I_h + 2W_h + 2N_h)\psi_h = 2g \quad \text{in } (\theta_i,\varphi_j) \tag{3.2}$$

for the interior problem, or

$$(I_h - 2W_h)\psi_h = -2g \quad \text{in } (\theta_i,\varphi_j). \tag{3.3}$$

for the exterior problem. These equations have to be satisfied at the collocation nodes (θ_i,φ_j) only, namely at all of the grid points of the mesh.

Hence we are led to a linear algebraic system containing a nonsparse and nonsymmetric, but relatively small and compact system matrix. For details see [12], for a fast multigrid solver see [9], [12] in case of Stokes equations ($\lambda=0$).

An important point is the chosen way of numerical quadrature of the surface integrals in (3.2) and (3.3). By choice of the trial functions they are reduced to integrals over the small elements Q, and consequently they are evaluated numerically by the

a) simple midpoint rule;

b) 2x2 Gaussian rule.

Note that, in any case, the quadrature nodes and the collocation nodes form disjoint sets, hence a special tratment of the weakly singular integrals is not always required.

Finally, we notify a convergence result when applying the simple midpoint rule and neglecting any special treatment of the singularity. The following estimates have been proved in [12]:

$$\sup_{\partial\Omega} |\psi_h - \psi| = O(h \cdot \log\tfrac{1}{h}), \tag{3.6}$$

and for the corresponding potentials

$$\sup_{G} |u_h - u| = O(h \cdot \log\tfrac{1}{h}) \tag{3.7}$$

holds, uniformly in any domain G with positive distance from $\partial\Omega$. The factor $\log\frac{1}{h}$ is originating from potential theoretic considerations.

4. THE CASE OF A NONHOMOGENEITY

Let us return now to the general case $f \neq 0$ in the problems (1.1) and (1.2). The usual treatment by means of hydrodynamic volume potentials proves to be extremely expensive, if a sufficiently high accuracy is required. For an impressive numerical test computation see below Sec. 5.

In [25] it is shown, that the spectral method is a very efficient way to solve problems of fluid dynamics in simple domains numerically. On the other hand boundary element methods are known to work well in domains with complicated boundaries. And so as an alternative we propose here a "boundary element spectral method", which seems to be very promising as preliminary numerical results have shown (see Sec. 5).

Consider the interior problem (1.1). We split this linear problem:

$$u = u_1 + u_2, \quad p = p_1 + p_2, \tag{4.1}$$

where (u_1, p_1) is any solution of the nonhomogeneous Stokes differential equations, but (u_2, p_2) solves (1.1) with $f = 0$ and the boundary conditions

$$u_{2|\partial\Omega} = g - u_{1|\partial\Omega} \quad . \tag{4.2}$$

The latter is numerically obtained by the boundary element method described in Sec. 2 and 3. Hence we merely have to determine any particular solution (u_1, p_1) of the nonhomogeneous Stokes equations.

Extend f (smoothly) to a cube containing Ω so that $f_{|\partial C} = 0$. A complete set of eigensolutions of Stokes equations in a cube with periodicity condition has been constructed recently by Borchers [2]. There eigensolutions $(e_{k,\alpha})$ (with constant pressure!) are suitably composed trigonometric functions, hence they allow to use decisively the FFT for the numerical enforcement. Furthermore, in [2] an explicit series representation for the orthogonal projection of the spaces under consideration onto their divergence-free subspaces is given.

Let us assume, that the extended f is in a certain complex Sobolev space of periodic functions H^s_π (C) over the cube $C = (0,a)^3$. Then the above solution (u_1, p_1) of the problem

$$(\lambda-\Delta)u_1 + \nabla p_1 = f \ , \quad \nabla\cdot u_1 = 0 \ , \quad u_1 \in H^{s+2}_\pi(C) \ , \tag{4.3}$$

has the representation

$$u_1 = \frac{f_o}{\lambda} + \sum_{k\in Z^3\setminus\{0\}} (\lambda+|2\pi a^{-1}k|^2)^{-1}\left(f_k - \frac{f_k\cdot k}{|k|^2}\right) e^{2\pi a^{-1}ik\cdot x} \ , \tag{4.4}$$

$$p_1 = a(2\pi i)^{-1} \sum_{k\in Z^3\setminus\{0\}} |k|^{-2} f_k\cdot k \; e^{2\pi a^{-1}ik\cdot x} \ , \tag{4.5}$$

as long as $-\lambda \notin \{|2\pi a^{-1}k|^2 \ / \ k \in Z^3\}$. Here the coeffients $f_k \in \mathbb{C}^3$ are given by the integrals

$$f_k := a^{-3} \int_C f(x)\, e^{-2\pi a^{-1}ik\cdot x} dx \quad . \tag{4.6}$$

If we denote by $P_{n-1}u_1$ the projection onto the space H_n ($n \in \mathbb{N}$) spanned by the set $\{e^{2\pi a^{-1}ik\cdot x} / \ |k| < n\}$, then we can show, that the rate of decay of the truncation error $(1-P_n)u_1$ depends on the smoothness of the solution and is more than algebraic, whenever the solution is infinitely smooth. Moreover, when the 3 D-product trapezoidal rule is used, this result holds even then, when the discretization error associated with the numerical computation of (4.6) is included.

To be precise, we have the following result:

For any $f \in H_\pi^s(C)$ *the solution* u_1 *of (4.3) belongs to* H_π^{s+2}. *If* $\rho, r \geq 0$ *with* $\rho + r = s+2$, *then*

$$\| u_1 - P_{n-1} u_1 \|_\rho \leq c\, n^{-r} \| f - P_n f \|_s , \tag{4.7}$$

where the constant depends only on a, λ.

As an approximation of (4.6) we define the complex vectors

$$T_n(f) := a^3 n^{-3} \sum_{k \in [o,n-1]^3} f(x_k) , \quad n \in \mathbb{N},\ n \geq 2 \tag{4.8}$$

with $[o, n-1]^3 := \{0,1,\ldots, n-1\}^3$ and $x_k := a\, n^{-1} k$. Let us now replace the Fourier coeffients f_k in the truncated series $P_{n-1} u_1$ by

$$\beta_k := a^{-3}\, T_{2n}\, (f \cdot e^{-2\pi a^{-1} i k \cdot x}) \tag{4.9}$$

and denote by $u_1^{(n)}$ the resulting function. Since the formula (4.8) is exact for any $f \in H_n$, the following error estimates are valid:

Suppose $f \in H_\pi^s(C)$ *with* $s > 3/2$, $\lambda > 0$. *Then the solution* u_1 *and its approximation satisfy*

$$\| u_1 - u_1^{(n)} \|_\rho \leq c\, n^{-\theta} \| f - P_n f \|_s , \tag{4.10}$$

where $\rho, \theta \geq 0$, $\rho + \theta = s + 2$. *The constant depends on* $a, \lambda, \theta, \rho, s$.

From (4.10) and the Sobolev imbedding theorems it is clear, how to derive error estimates in the topology of uniform convergence.

The method indicated in this section easily extends to related problems of the applications.

5. SOME COMPUTATIONAL RESULTS

Many numerical results in 3-D have been obtained by the second author to test the proposed boundary element methods [8], [9], [10], [11], [12], [13], [14]. Subsequently we will give three test examples for Stokes differential equations (case $\lambda = 0$). Further some encouraging numerical examples corresponding to the problem (4.4) of the spectral part are obtained by the first author and reported below. Currently the combination of both software packages to test the full boundary element spectral method is nearly completed.

The first example to our boundary element method concerns the test against Stokes' wellknown formula for the viscous drag of a sphere [33] which has been obtained by analytical means. Let the uniform onflow $v_\infty = (1,0,0)$ be parallel to the x-axis. Then the exact value of the viscous drag of a ball of the radius $R = 1$ is

$$D = (6\pi, 0, 0).$$

Our numerical computations have been carried out (without using any symmetry properties of the sphere) on a finite element mesh of the mesh size h = 1/16. The nonsparse system matrix here is of size 867 x 867, and the numerical computations are carried out with simple precision on a PRIME 750 computer. The approximate drag has been computed from formula (3.3). Then by applying the different formulae of numerical quadrature a), b) the following comparison is obtained in [13] (only the x-component of the drag is given):

Tab. 1

	computed drag	exact drag	relative error
quadrature a)	19.45	18.85	3.2 %
quadrature b)	18.99	18.85	0.7 %

In the second example we compute the flow around an irregular body, represented in polar coordinates by

$$r \leq 1 + \varepsilon.\sin^3 2\pi\theta.\cos 6\pi\phi + \varepsilon.\sin 2\pi\theta$$

using normalized polar coordinates $o \leq \theta,\phi \leq 1$. In case of $\varepsilon = o$ the body degenerates to a ball. When ε increases, some conical corners turn out at the "poles" $\theta = o$ or $\theta = 1$, and the body looses all of its symmetry properties. The body is illustrated in Fig. 1, where its cross-section in the x,z-plane is plotted in case of $\varepsilon = o.3$.

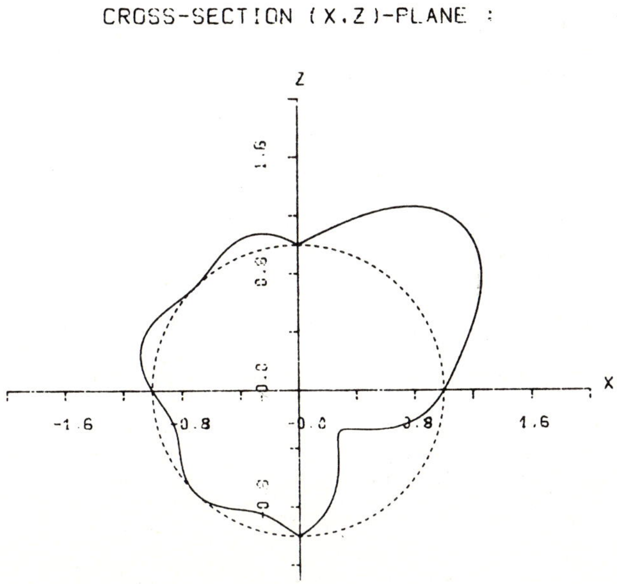

FIG. 1: Irregular Body, Cross-Section x,z-Plane.

Moreover, in case of $\varepsilon = o.5$ a spine arises ($\theta = o.75$ in Fig. 1), hence any potential theory must fail there. In this case our boundary element method is tested against an artificial solution of the problem (1.2). We use those parameters of our first example and apply the formula a) of numerical quadrature. Here the mean relative error of the computed velocity field is given as a function of ε (see [14]):

Tab. 2

ε	0.0	0.1	0.2	0.3	0.4	0.5
relative error %	5.5	5.6	6.1	7.5	9.2	16.9

Particularly, the spine arising at $\varepsilon = 0.5$ proves to destroy the numerical procedure, hence confirming the presently available mathematical theory (see Wendland [36],[37] in case of Laplace and Helmholtz equations , Hebeker [14] for Stokes equations).

In the third example our method is applied to solve the initial boundary value problem of the nonstationary Navier Stokes equations of homogeneous fluids, where we consider the flow included in a ball of radius $R = 1$. By time differencing methods (implicit for the viscous term, but explicit for the convective term) we obtain for each time step a stationary problem similar to (1.1), with the differential equations replaced by (time step δt)

$$\frac{1}{\delta t} u - \Delta u + \nabla p = f \quad , \quad \operatorname{div} u = o \quad \text{in } \Omega .$$

A simple fundamental solution corresponding to this is available, and a tested numerical algorithm is due to Varnhorn (see [35]), who extended our computer programs. The results are given in Fig. 2, where the parameters $h = 1/12$ and $\delta t = 1/10$ have been used. Here the computational work took many hours of computing time, mainly due to the frequent evaluation of the volume potentials. The spectral part (Sec. 4) of our method serves to circumvent this.

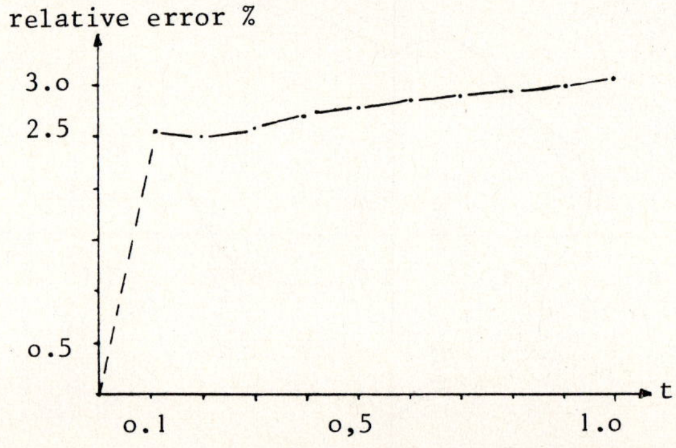

FIG. 2: Nonstationary Navier Stokes Equations, Moderate Reynolds Number.

In the following we will give some impressive numerical experiments with the spectral part, carried out by the first author, which show excellent agreement with the above given theoretical convergence rates. Consider the function

$$u(x) = e^{z}(4+e^{z})^{-2}\ (0,1,-1) \qquad (z = i(x_1+x_2+x_3))$$

defined in the cube $C = [0,2\pi]^3$. It is a solution of problem (4.3) if f is given by

$$f(x) = (\lambda-\Delta)\, u\,(x) + \nabla p\,(x)$$

$$p\,(x) = \gamma \cdot \prod_{\alpha=1}^{3} (2\pi-x_\alpha)^{\mu} x_\alpha^{\mu} \quad , \quad \gamma, \mu = \text{const.}$$

Thus, μ indicates the degree of smoothness and for various μ the constant γ is chosen such that $\delta := \max |\nabla p(x)|/\max|u(x)|$ has a prescribed value. The integer n in the tables below corresponds to the truncated series

$$P_n\, u(x) = \sum_{|k|<n} u_k\, e^{ik\cdot x} \quad .$$

Tab. 3: Pure Spectral Approximation with $\delta \approx 10^3$

n	λ	μ	relative error %	CPU (sec)
8	1	2	9,70	~ 80
		3	8,85	
		4	0,81	
		5	0,68	
		6	0,14	
	10^3	2	0,01	
16	1	2	0,65	~ 500
		3	0,61	
		4	0,01	
		5	0,01	
		6	0,004	

These results show, that the spectral method allows to achieve high accuracy at a very low number of frequencies even in the presence of a strong pressure variation.

In practice the nonhomogeneity f is not given in the whole cube. In order to simulate the process of extension we consider the solution $u(x) = \text{rot}\ (\rho(|x|) \cdot e \times x)$, where e is a constant vector and the cutoff function ρ is three times continuously differentiable and $= 1$ in the ball $|x - \pi| \leq 1$, $= 0$ if $|x - \pi| \geq 2$, and otherwise it is equal to a polynomial of seventh degree in $r = |x|$. Hence $f = (\lambda-\Delta)u$ is only Lipschitz continuous. The corresponding pressure is constant. We have the following results:

Tab.4 : Pure Spectral Approximation with Nonsmooth Nonhomogeneity

n	λ	relative error (%)	CPU (sec)
8	0	2,65	~ 80
	1	1,91	
	10	0,74	
16	0	0,40	~ 500
	1	0,29	
	10	0,09	

In the last example we take as a solution of the problem (1.1) with $g \equiv 0$, $\lambda = 0$, and $\Omega = \{x / |x| \leq 1\}$, the function

$$u(x) = (e^{-|x|^2} - e^{-1})\,(x_3 - x_2,\ x_1 - x_3,\ x_2 - x_1).$$

The following computations have been carried out with the coupled boundary element spectral program. According to the indices $m = 1$ resp. $m = 2$ in the next table the corresponding nonhomogeneity $f = \Delta u$ is smoothly extended to the whole cube resp. extended by zero.

Tab. 5: Boundary Element Spectral Approximation

n	m	relative error (%)	CPU (sec)
8	1	1,31	$\sim 2 \cdot 10^4$
	2	2,53	

In order to analyze primarily the effect of the spectral part the boundary element part has been carried out with a very fine grid.

Recently, also a plot program has been created in order to visualize the flow past a ball. Any projection of the velocity field onto a plane can be plotted. A first result is shown in Fig. 3, where the uniform onflow is parallel to the (positive) x-axis, and the plane is fixed by one of its points (o,o,o.24) and its normal vector n = (o,o,1) . And the flow configuration is observed from the point of view (5,-5,5) (see [14]). This plot program will serve to visualize the Navier Stokes flow when the computer programs are so far developed.

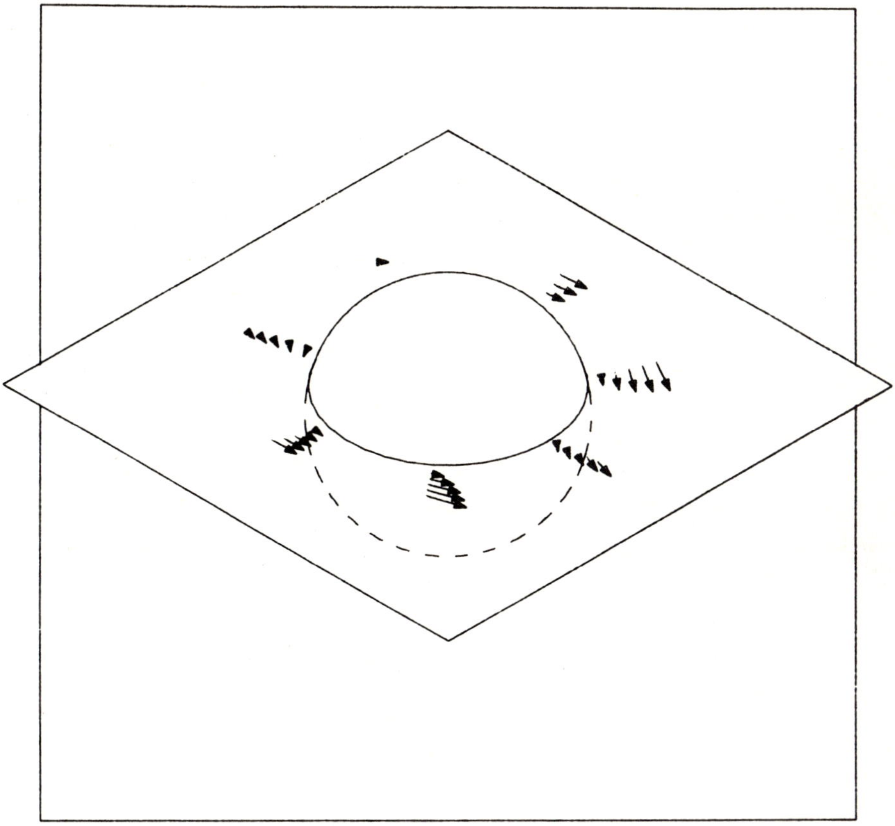

FIG. 3 : Visualization of 3-D Flow, Case of a Ball.

REFERENCES

[1] BORCHERS, W.: "Über das Anfangs-Randwertproblem der instationären Stokes-Gleichung", Z.A.M.M. 65 (1985), T 329.

[2] BORCHERS, W.: "Eine Fourier-Spektralmethode für das Stokes-Resolventenproblem". Submitted for publication, Univ. of Paderborn 1985, 13 pp.

[3] BREBBIA, C.A. and MAIER, G. (eds.): "*Boundary Elements VII*", 2 vols., Berlin 1985.

[4] FISCHER, T.M. and ROSENBERGER, R.: "A Boundary Integral Method for the Numerical Computation of the Forces Exerted on a Sphere in Viscous Incompressible Flows Near a Plane Wall". Preprint, Univ. of Darmstadt 1985, 30 pp.

[5] FUJITA, H.: "On the Semi-Discrete Finite Element Approximation of the Evolution Equation $u_t + A(t)\, u = 0$ of Parabolic Type, Topics in Num. Analysis, John J.H. Miller (ed.), Acad. Press New York 1977.

[6] HACKBUSCH, W.: "Die schnelle Auflösung der Fredholm'schen Integralgleichung zweiter Art". *Beitr. Numer. Math.* 9 (1982), 47-62.

[7] HEBEKER, F.K.: "A Theorem of Faxén and the Boundary Integral Method for 3-D Viscous Incompressible Flows". Techn. Rep., Univ. of Paderborn 1982, 14 pp.

[8] HEBEKER, F.K.: "A Boundary Integral Approach to Compute the 3-D Oseen Flow Past a Moving Body". M. Pandolfi and R. Piva (eds.), *Numerical Methods in Fluid Dynamics*. Braunschweig 1984, 124-130.

[9] HEBEKER, F.K.: "On a Multigrid Method for the Integral Equations of 3-D Stokes Flow". W. Hackbusch (ed.), *Efficient Solution of Elliptic Systems*. Braunschweig 1984, 67-73.

[10] HEBEKER, F.K.: "Zur Randelemente-Methode in der 3-D viskosen Strömungsmechanik". Habilitation Thesis, Univ. of Paderborn 1984, 127 pp.

[11] HEBEKER, F.K.: "A Boundary Element Method for Stokes Equations in 3-D Exterior Domains". J.R. Whiteman (ed.), *The Mathematics of Finite Elements and Applications*. London 1985, 257-263.

[12] HEBEKER, F.K.: "Efficient Boundary Element Methods for 3-D Exterior Viscous Flows". Submitted for publication, Univ. of Paderborn 1985, 35 pp.

[13] HEBEKER, F.K.: "Efficient Boundary Element Methods for 3-D Viscous Flows". C.A. Brebbia and G. Maier (eds.), *Boundary Elements VII*. Berlin 1985, 9:37-44.

[14] HEBEKER, F.K.: "On the Numerical Treatment of Viscous Flows against Bodies with Corners and Edges by Boundary Element and Multigrid Methods". Manuscript in preparation, ca. 30 pp.

[15] HEYWOOD, J.C.: "The Navier-Stokes Equations: On the Existence, Regularity and Decay of Solutions", Indiana Univ. Math. J. 29 (1980) 639-681.

[16] HEYWOOD, J.G., RANNACHER, R.: "Finite Element Approximation of the Nonstationary Navier-Stokes Problem I, Regularity of Solutions and Second Order Error Estimates for Spatial Discretization", Siam J. Numer. Anal. 19 (1982) 275-311.

[17] HSIAO. G.C. and KOPP, P. and WENDLAND, W.L.: "A Collocation Method for some Integral Equations of the First Kind". *Computing* 25 (1980), 89-130.

[18] HSIAO, G.C. and KOPP, P. and WENDLAND, W.L.: "Some Applications of a Galerkin Collocation Method for Boundary Integral Equations of the First Kind". *Math. Meth. Appl. Sci.* 6 (1984), 280-325.

[19] HSIAO, G.C. and KRESS, R.: "On an Integral Equation for the 2-D Exterior Stokes Problem". *Appl. Numer. Math.* 1 (1985), 77-93.

[20] LADYZHENSKAJA, O.A.: "*The Mathematical Theory of Viscous Incompressible Flow*". New York 1969.

[21] MC CRACKEN, M.: "The Resolvent Problem for the Stokes Equations on Halfspace in L_p", Siam J. Math. Anal. Vol. 12, No. 2, 201-228 (1981).

[22] MADAY, Y., QUARTERONI, A.: "Spectral and Pseudospectral Approximations of the Navier Stokes Equations". *Siam J. Numer. Anal.* 19 (1982), 761-780.

[23] NEDELEC, J.C.: "Approximation des Equations Integrales en Mechanique et en Physique". Ecole Polytechnique Palaiseau 1977, 127 pp.

[24] NOVAK, Z.: "Use of the Multigrid Method for Laplacean Problems in 3-D". W. Hackbusch and U. Trottenberg (eds.), *Multigrid Methods*. Berlin 1982, 576-598.

[25] ORSZAG, S.A.: "Numerical Simulation of Incompressible Flows Within Simple Boundaries". I. Galerkin (Spectral) Representations. Stud. Appl. Math. 2 (1971), 293-326.

[26] RAUTMANN, R.: "On the Convergence Rate of Nonstationary Navier-Stokes Approximations", Springer Lecture Notes in Mathematics 771 (1980) 425-449.

[27] RAUTMANN, R.: "On the Navier-Stokes Initial Value Problem" (Sir G.I. Taylor Memorial Lecture), Proc. Int. Symp. "Problems of Nonlinear Continuum Mechanics" Kharagpur 1980, Indian Soc. Theor. Appl. Mech. Kharagpur, 1-16 (1982).

[28] RAUTMANN, R.: "A Semigroup Approach to Error Estimates for Nonstationary Navier-Stokes Approximations", Meth. Verf. der Math. Physik 27 (1983) 63-77.

[29] RAUTMANN, R.: "Zur Konvergenz des Rothe-Verfahrens für instationäre Stokes-Probleme in dreidimensionalen Gebieten", ZAMM 64 (1984) T 387-388.

[30] RAUTMANN, R.: "On Optimum Regularity of Navier-Stokes Solutions at Time t = o" Math. Z. 184 (1983) 141-149.

[31] RAUTMANN, R.: "Three Dimensional Flows: Models and Problems", Proc. German-Italian Symp. "Applications of Mathematics in Technology" Rome 1984, Teubner Stuttgart (1984) 58-78.

[32] RAUTMANN, R., VARNHORN, W.: "Die Navier-Stokessche Randwertaufgabe mit einer Differenzennäherung", ZAMM 65 (1985), T 360-362.

[33] SCHLICHTING, H.: "*Grenzschicht Theorie*". 8th ed., Karlsruhe 1982.

[34] TEMAM, R.: "*Navier Stokes Equations and Nonlinear Functional Analysis*". Reg. Conf. Ser. Appl. Math., Siam, Philadelphia 1983.

[35] VARNHORN, W.: "Zur Numerik der Gleichungen von Navier-Stokes". Doctoral Thesis, Univ. of Paderborn, to appear.

[36] WENDLAND, W.L.: "Die Behandlung von Randwertaufgaben im $\mathbb{R}^3$ mit Hilfe von Einfach- und Doppelschichtpotentialen". *Numer. Math.* 11 (1968), 380-404.

[37] WENDLAND, W.L.: "Boundary Element Methods and Their Asymptotic Convergence". P. Filippi (ed.), *Theoretical Acoustics and Numerical Techniques*. CISM Courses and Lectures 277, Wien 1983, 135-216.

[38] WENDLAND, W.L.: "On some Mathematical Aspects of Boundary Element Methods for Elliptic Problems". J.R. Whiteman (ed.), *The Mathematics of Finite Elements and Applications V*. London 1985, 193-227.

[39] ZHU, H.: "A Boundary Integral Equation Method for the Stationary Stokes Problem in 3-D". C.A. Brebbia (ed.), *Boundary Element Methods*. Berlin 1983, 283-292.

Supported by "Deutsche Forschungsgemeinschaft".

NAVIER-STOKES COMPUTATIONS OF TWODIMENSIONAL LAMINAR FLOWS IN A CHANNEL WITH A BACKWARD FACING STEP

U. Brockmeier, N.K. Mitra, M. Fiebig
Institut für Thermo- und Fluiddynamik
Ruhr-Universität Bochum, 4630 Bochum 1, FRG

ABSTRACT

Flowfields in a channel with a backward facing step have been computed in the Reynolds number range of 50 to 150 from full Navier - Stokes equations by a SOLA-algorithm. Computed lengths of the separation zone, wall shear and velocities at chosen points lie in the middle of the scattered results of computations of the same problem reported in a GAMM - workshop. In flowfields computed for a Reynolds number 1389, besides the main separation behind the step two more separation zones occur. The computed separation lengths behind the step and at the bottom wall are in good agreement with avaible experimental results.

INTRODUCTION

The computation of the flowfield in a twodimensional channel with a backward-facing step (Fig. 1) has become a model problem to test the accuracy and efficiency of computational schemes of the Navier-Stokes equations. The report [1] of a recent GAMM workshop presents a comparison of 22 computations of the flowfield in a channel with a backward facing step from different authors who (with one exception) solved the full Navier-Stokes equations using different finite difference and finite element schemes. The range of Reynolds numbers in these computations lie within 50 and 150 where the Reynolds number is defined with U_{max} at the channel inlet and the step height (H-h). The length of the channel after the step is 38(H-h) and the velocity profile at the channel inlet is parabolic. The computational and experimental results (see Ref. [1]) show one separation bubble behind the step. The dimensionless length of the separation bubble from Navier-Stokes computations lie between 2.5 and 3.4 where the experiment gives a value of 3.. Computed results of wall shear stresses show even larger disagreement.

Recently Osswald and Ghia et. al. [2] published computational results for flows in longer channels with Reynolds numbers larger than in the GAMM-workshop. The results have been obtained from the solution of unsteady Navier-Stokes equations in a bodyfitted coordinate system. These results show the existence of two more separation bubbles on the top and bottom wall besides the main separation bubble downstream of the step. At large Reynolds number (Re =2000, based on the volume flow rate) a periodic flow with a vortex street has been observed by Osswald et. al.[2]. Such separation structures with three bubbles have also been experimentally observed by Armaly and Durst [3]. Osswald et. al.[2] use an implicit time dependent scheme for the solution of vorticity and stream function equations. The computed velocity profile does not compare well with experiments of Armaly and Durst [3].

The purpose of the present work is to compute the flow in a channel with a backward facing step with a modified version of the SOLA-algorithm based on a Marker and Cell(MAC) technique and compare the results with those from the GAMM-workshop and from Osswald et. al.[2] and Armaly et. al.[3]. Our objective is to test how the results from MAC, which is one of the oldest numerical techniques for the solution of Navier-Stokes equations, compare with more modern finite differences and finite elements techniques as used in GAMM-workshop and by Osswald et. al. [2] . For the latter test case we will implement a multigrid technique in the SOLA-algorithm [8] in order to accelerate the convergence rate. In all of our computations we will use simple cartesian coordinates.

MATHEMATICAL FORMULATION

Two-dimensional unsteady incompressible flow is governed by the following continuity and momentum equations which are written in dimensionless form in cartesian coordinates :

$$\frac{\partial u}{\partial x} + \frac{\partial v}{\partial y} = 0 \tag{1}$$

$$\frac{\partial u}{\partial t} + \frac{\partial u^2}{\partial x} + \frac{\partial uv}{\partial y} = - \frac{\partial p}{\partial y} + \frac{1}{Re}\left(\frac{\partial^2 u}{\partial x^2} + \frac{\partial^2 u}{\partial y^2}\right) \tag{2a}$$

$$\frac{\partial v}{\partial t} + \frac{\partial uv}{\partial x} + \frac{\partial v^2}{\partial y} = - \frac{\partial p}{\partial y} + \frac{1}{Re}\left(\frac{\partial^2 v}{\partial y^2} + \frac{\partial^2 v}{\partial x^2}\right) . \tag{2b}$$

These equations are solved numerically with the semi-implicit hydrodynamic code SOLA[4]. The procedure incoporated in SOLA is the correction of the explicitly advanced velocity field by an iterative procedure applied to the implicit continuity equation. This procedure is mathematically equivalent to the application of the artificial compressibility method at each time step [5], and avoids the need of pressure boundary conditions as associated with the solution of a Poisson equation for the pressure[6].

NUMERICAL SCHEME

The governing equations are discretized on a staggered grid, where the pressure p is defined at cell centers and the velocities u and v are respectively defined at the cell faces in x- and y-direction. The momentum equations are discretized at the cell faces as

$$u^{n+1}_{x+h/2,y} + {}^{\Delta t}/_{h} \; (p^{n+1}_{x+h,y} - p^{n+1}_{x,y}) = u^{n}_{x+h/2,y} + A^{n}_{x+h/2,y} \tag{3a}$$

$$v^{n+1}_{x,y+h/2} + {}^{\Delta t}/_{h} \; (p^{n+1}_{x,y+h} - p^{n+1}_{x,y}) = u^{n}_{x,y+h/2} + B^{n}_{x,y+h/2} \tag{3b}$$

where n denotes the time level and $h=(\Delta x,\Delta y)$ is the mesh size. The convective and viscous terms are calculated explicitly and are combined in the terms A and B of eq.'s (3). The discretization of these terms employs linear interpolation to determine the velocities at points, where they are not defined, and is given in detail in [4].
In the first step of the new time level, the velocity field is advanced by eq.'s(3), using the old pressure solution p^n as a first guess for p^{n+1}. The new velocity field is then corrected by the following formulas, which are appelied successivly to each cell :

$$D^k_{x,y} = 1/\Delta x(u^k_{x+h/2,y} - u^k_{x-h/2,y}) + 1/\Delta y (v^k_{x,y+h/2} - v^k_{x,y-h/2}) \quad (4)$$

$$\tilde{u}^k_{x+h/2,y} = u^k_{x+h/2,y} + \frac{\Delta t}{\Delta x} (\omega/\tau \; D^k_{x,y})$$

$$u^{k+1}_{x-h/2,y} = \tilde{u}^k_{x-h/2,y} - \frac{\Delta t}{\Delta x} (\omega/\tau \; D^k_{x,y})$$

$$(5)$$

$$\tilde{v}^k_{x,y+h/2} = v^k_{x,y+h/2} + \frac{\Delta t}{\Delta y} (\omega/\tau \; D^k_{x,y})$$

$$v^{k+1}_{x,y-h/2} = \tilde{v}^k_{x,y-h/2} - \frac{\Delta t}{\Delta y} (\omega/\tau \; D^k_{x,y})$$

$$\Delta p^k_{x,y} = -\omega/\tau \; D^k_{x,y}$$

$$\text{with} \quad \tau = 2\Delta t \; (1/\Delta x^2 + 1/\Delta y^2)$$

$$p^{k+1}_{x,y} = p^k_{x,y} + \Delta p^k_{x,y}$$

where k denotes the iteration steps and the tildes represent partially advanced quantities from cells already corrected. ω is an overrelaxation parameter. The iteration is carried out until the divergence of the velocity becomes neglegibly small in each cell. The above iteration scheme has been derived by applying Newton's method to the implicit continuity equation [7].
One important feature of this work is concerned with acceleration of the convergence rate of the iterative procedure by application of a multigrid technique to the computational sequence (eq.'s 4 and 5). The multigrid approach results in a vast increase of the convergence rate and considerably reduces the CPU time needed to obtain fully converged solutions. The multigrid technique will not be further discussed in this paper, since it has been presented elsewhere [8].

BOUNDARY AND INITIAL CONDITIONS

The boundary condition at the channel inlet is given by :

$$V_{\Gamma,i} = \begin{pmatrix} u(y) \\ 0 \end{pmatrix} \quad ;\Gamma := \text{boundary} \; ; \; V = \begin{pmatrix} u(x,y) \\ v(x,y) \end{pmatrix}$$

where u(y) represents a parabolic profile.
At the channel walls the inpermeability and no-slip conditions are used:

$$V_{\Gamma w} = \begin{pmatrix} 0 \\ 0 \end{pmatrix} .$$

On the boundary the so called reflection technique of original SOLA is replaced by a second order aproximation in order to maintain consistency of formulation of the second order derivatives [6].
At the channel exit the second derivative of V is set to zero.

$$\frac{\partial^2 V_{\Gamma,e}}{\partial x^2} = 0 \quad ; \quad \frac{\partial^2 V_{\Gamma,e}}{\partial y^2} = 0 .$$

At time t=0 the fluid in the channel is at rest.

RESULTS AMD DISCUSSION

In the present study five test cases have been considered. The geometries and Reynolds numbers are given in Fig. 1 and Table 1.

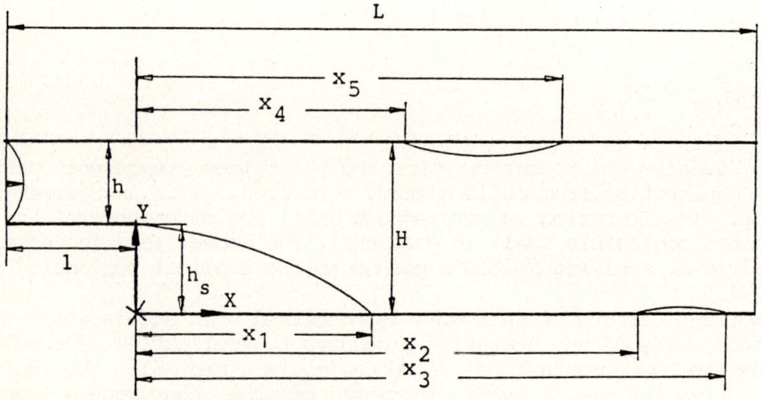

Fig. 1 : Geometry of the channel.

Tab. 1 : Geometrical and flow parameters for the considered test cases.
Inlet profile : parabolic
U_{max} : maximum value of the velocity at channel inlet
U_{Av} : mean outflow velocity

Case	L	l	H	h	$Re_s = U_{max}(H-h)/\nu$	$Re = U_{Av} \cdot H / \nu$
1	22	3	1.5	1.	50	–
2	22	3	1.	0.5	50	–
3	22	3	1.5	1.	150	–
4	22	3	1.	0.5	150	–
5	22	3	1.	0.5149	1389	737

Figures 2 and 3 show the streamlines in the channel and in the recirculation zone respectivly. The u-velocity profile at several locations downstream of the step and the shear stresses on bottom wall are given in Figures 4 and 5 respectively.

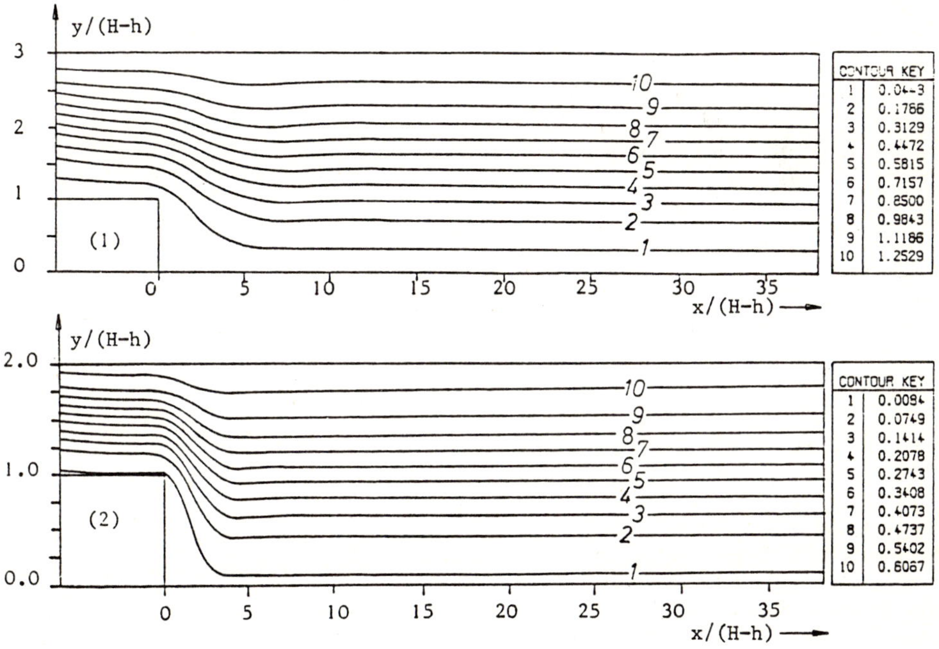

Fig. 2a : Streamline plots in the channel for cases (1-2).

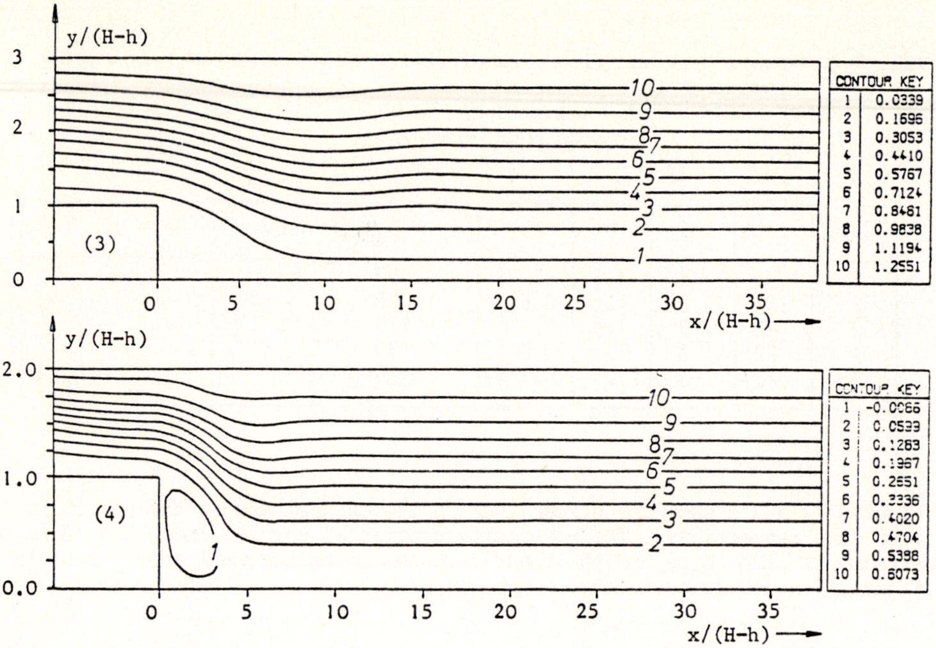

Fig. 2b : Streamline plots in the channel for cases (3-4).

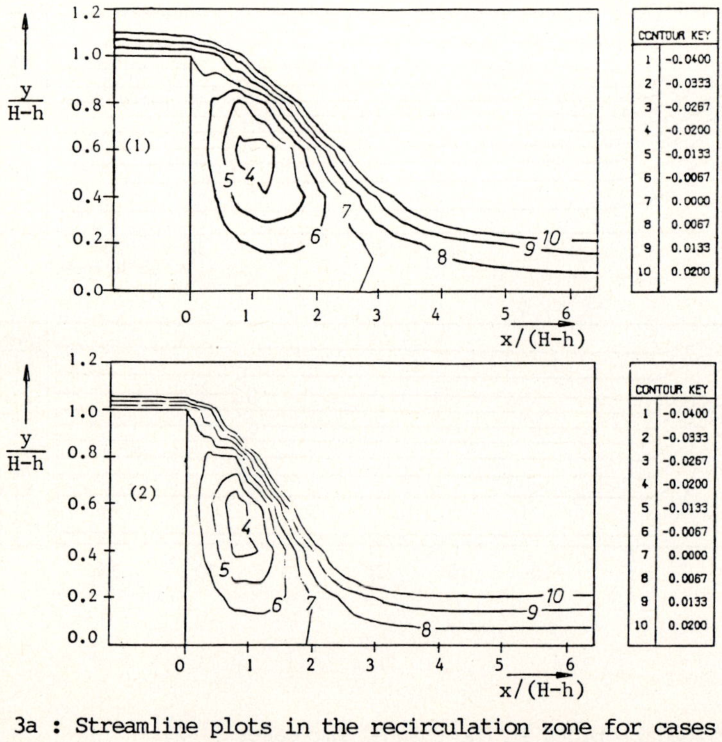

Fig. 3a : Streamline plots in the recirculation zone for cases (1-2).

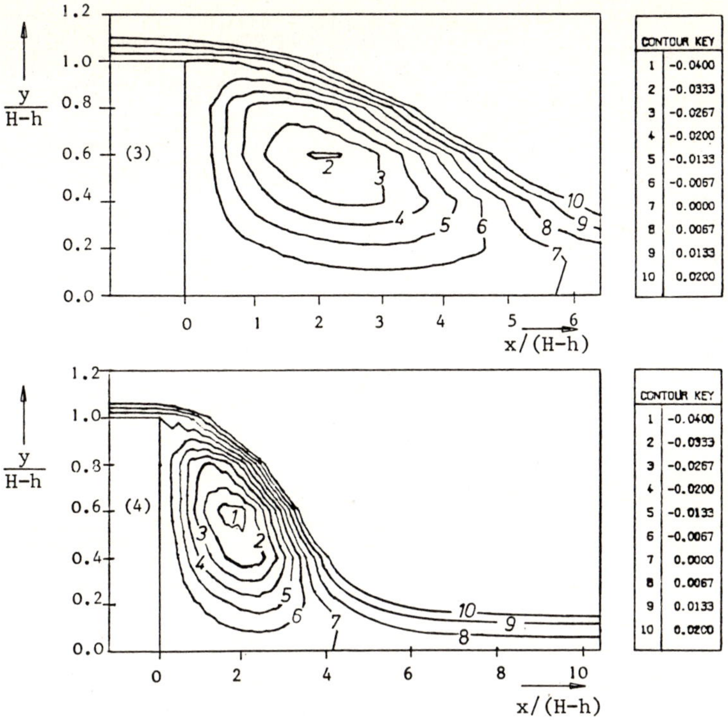

Fig. 3b : Streamline plots in the recirculation zone for cases (3-4).

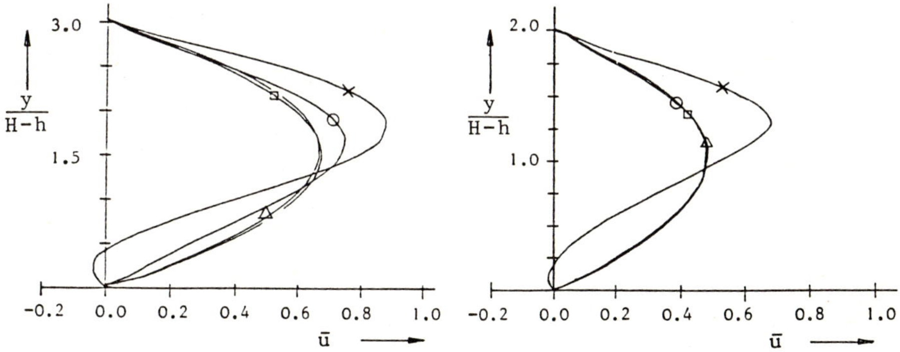

Fig. 4a : Velocity profiles $\bar{u} = u/U_{max}$ for cases (1-2) at position x/(H-h):

×: 1.6; ○: 4.0; □: 8.0; Δ: 16.0.

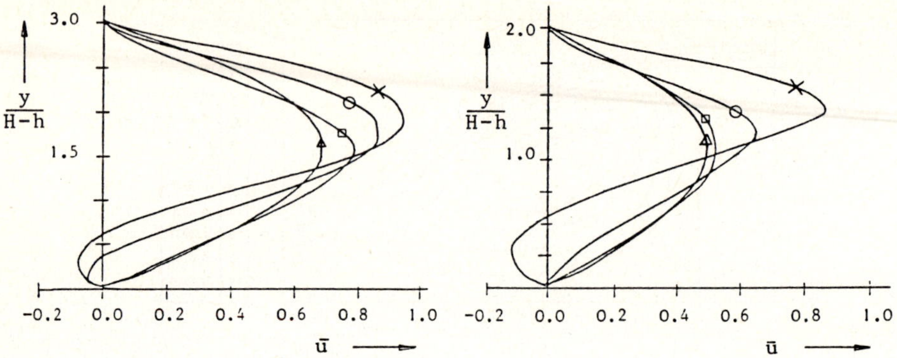

Fig. 4b : Velocity profiles $\bar{u} = u/U_{max}$ for cases (3-4) at position x/(H-h):

×: 1.6; o: 4.0; □: 8.0; Δ: 16.0.

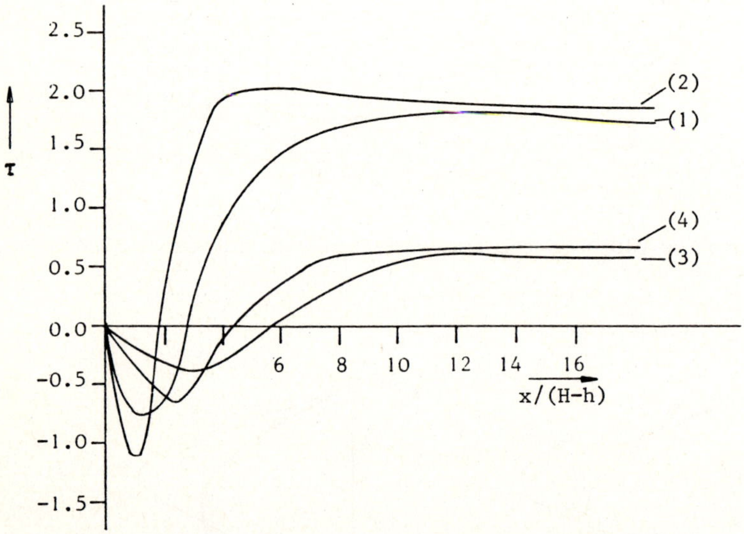

Fig. 5 : Shear stress distribution at the botton wall;

$$\tau = \frac{H-h}{Re\, \bar{U}_{max}} \frac{\partial u}{\partial y} 10^{-2} .$$

Numerical values for the shear stress distribution (Fig. 5) at several points in the channel and points of reattachment are compared in Table 2 and Table 3 respectivly with experiments of Kueny et. al. [9] and the averaged value from all other contributors of the GAMM-workshop [1].

Tab. 2 : Shear stress $\tau = \frac{H-h}{Re\, U_{max}} \frac{\partial u}{\partial y} 10^{-2}$.

Exp: Experimental results of Kueny et. al. [9].
Av : Averaged value from contributed results in GAMM-workshop[1].
Cal: present calculations.

x/(H-h) \ Case		1	2	3	4
1.6	Exp	-0.53	-	-0.21	-0.31
	Av	-0.75	-0.97	-0.28	-0.40
	Cal	-0.75	-0.58	-0.29	-0.40
4	Exp	0.80	-	-0.26	-0.04
	Av	0.92	1.96	-0.31	-0.35
	Cal	0.88	1.93	-0.30	-0.02
8	Exp	-	-	0.23	0.57
	Av	1.83	1.96	0.39	0.59
	Cal	1.69	1.94	0.34	0.59
16	Exp	-	-	-	-
	Av	1.78	1.98	0.51	0.64
	Cal	1.75	1.87	0.51	0.63

Tab. 3 : Reattachment points x/(H-h) for the separation behind the step.

Case	1	2	3	4
Exp	3.	-	6.	4.5
Av	2.76	1.96	5.85	4.72
Cal	2.75	1.94	5.71	4.32

The computed and the experimental results compared in Table (2-4) are in good agreement concerning the position of the reattachment point, the shear stresses and the axial velocities (see Tab. 4) at all locations in the channel except at x/(H-h) = 1.6. At this point, marking a position within the separation bubble behind the step, disagreeing results occur especially concerning the wall shear stress. In our opinion neither the experimental results, because of obvious difficulties in the measurement of velocities near the wall nor the numerical schemes, which produce disagreeing results even between similar numerical schemes (see Ref. [1]) can be used as a basis to test the accuracy of the present results. It can be seen in Ref. [1] that larger differences from the average value (Av) in general occur in the finite element methods, whereas the finite difference methods show results closer to each other including present results.

Tab. 4 : Values of velocities $\bar{u} = u/U_{max}$ at points x/(H-h).

x/(H-h) \ Case		1 Min.	1 Max.	2 Min.	2 Max.	3 Min.	3 Max.	4 Min.	4 Max.
1.6	Exp	-0.04	0.898	-	-	-0.07	0.972	-0.091	0.920
	Av	-0.042	0.898	-0.038	0.715	-0.067	0.963	-0.99	0.902
	Cal	-0.03	0.880	-0.040	0.715	-0.073	0.964	-0.106	0.892
4	Exp	0.00	0.772	-	-	-0.046	0.928	-0.016	0.711
	Av	0.00	0.780	0.00	0.523	-0.064	0.903	-0.039	0.709
	Cal	0.00	0.763	0.00	0.521	-0.045	0.880	0.00	0.698
8	Exp	-	-	-	-	0.00	0.819	0.00	0.556
	Av	0.00	0.698	0.00	0.500	0.00	0.809	0.00	0.554
	Cal	0.00	0.681	0.00	0.502	0.00	0.795	0.00	0.554
16	Exp	-	-	-	-	-	-	-	-
	Av	0.00	0.665	0.0	0.497	0.00	0.716	0.00	0.502
	Cal	0.00	0.660	0.00	0.500	0.00	0.690	0.00	0.500

Reynolds number and geometry parameters in case (5) allow a comparison to the experimental investigation by Armaly et. al. [3] and the recent numerical study by Osswald et. al. [2] . Our computation was carried out on a grid with 32 * 400 grid points and a multigrid technique [8] was used to reduce computational time. The flow development for the characteristic times T = 2, T = 20 and T = 50, shown in Figs. 6-9, confirms the existance of a highly oscillatory flow with several larger and smaller separation zones along the bottom and the top wall of the channel. The existence of several small recirculation zones within the main separation zones behind can be seen from Figs. 6-9.

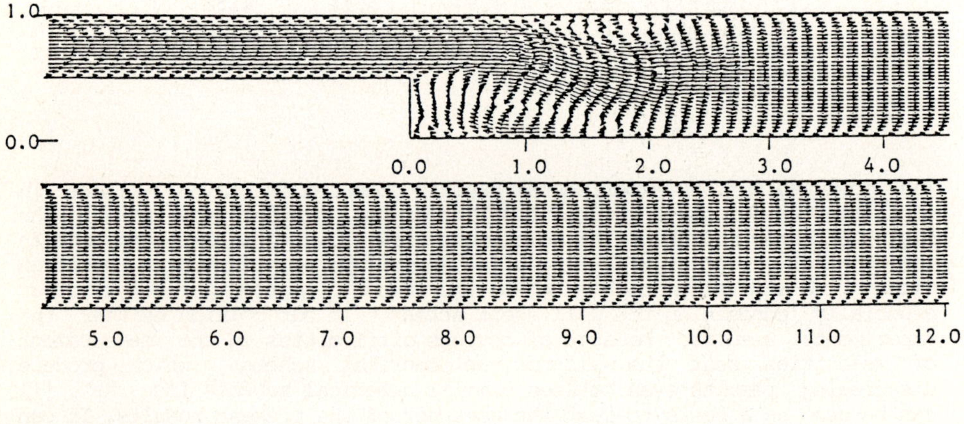

Fig. 6 : Velocity vector plot for T = 2 ; Re = 737.

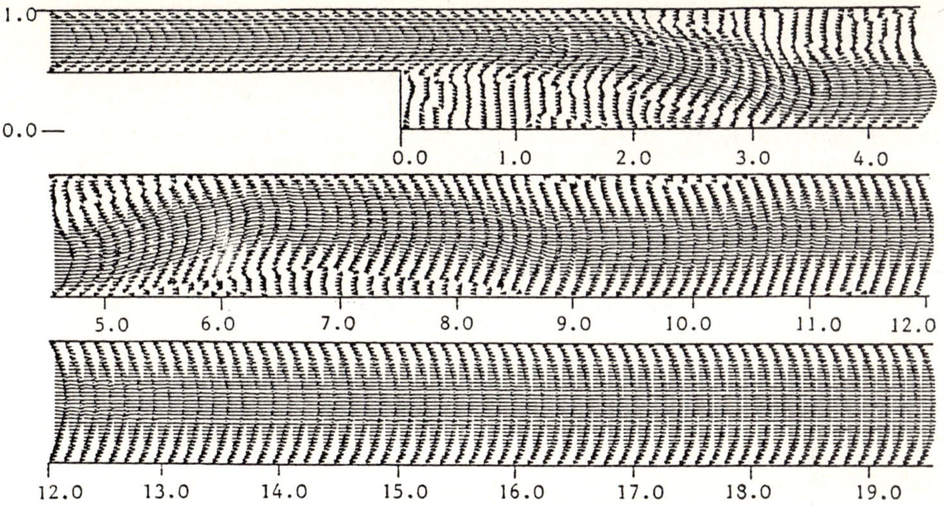

Fig. 7 : Velocity vector plot for T = 20 ; Re = 737.

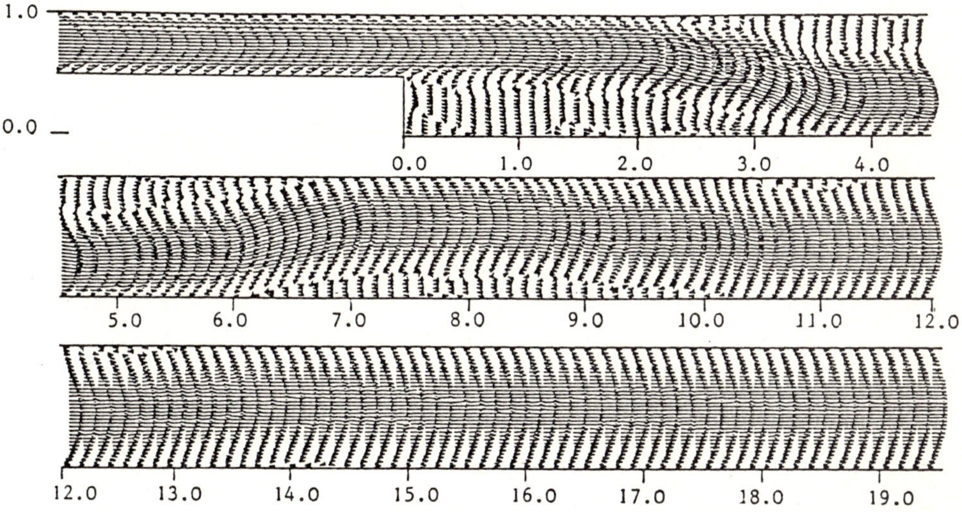

Fig. 8 : Velocity vector plot for T = 50 ; Re = 737.

During the development to the steady state, the main separation zones are convected downstream and the small recirculations within the main separations are diffused.
When steady state is reached at about T = 250 three main separations, two at the bottom wall and one at the top wall can be distinguished (see Fig. 9 for T = 300). The position of these steady state separation and reattachment points is compared in Table 5 with the results [2] and [3] .

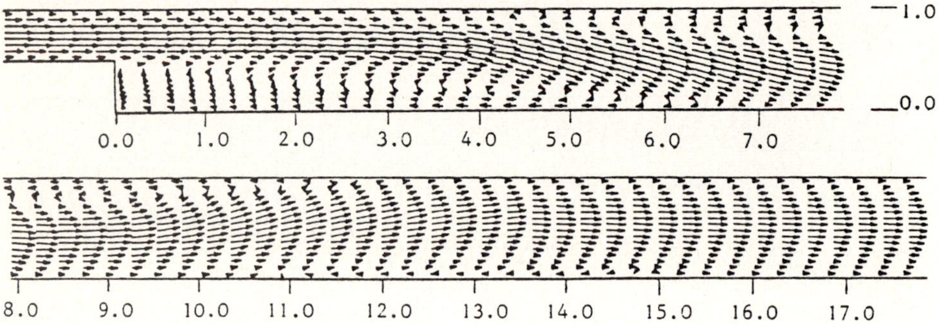

Fig. 9 : Velocity vector plot for T = 300; Re = 737.

Tab. 5 : Reattachment and separation points for the case (5).

	x_1	x_4	x_5	(x_5-x_4)	x_2	x_3	(x_3-x_2)
Osswald, Ghia [2]	6.87	5.36	14.38	9.02	13.55	17.00	3.45
Armaly Durst [3]	6.65	4.85	10.80	5.95	10.26	11.99	1.73
Present study	6.57	4.34	11.71	7.37	11.00	14.73	3.73

The length of the first separation (x_1) at the bottom wall is in good agreement for the computed and the experimental results. However our results are closer to the experiment than the computations of Osswald et. al. [2] . Our results for the length (x_5-x_4) and the position (x_4) of the upper wall separation agree well with the experimental results from Armaly et. al. [3], while Osswald et. al. [2] found remarkable differences and made threedimensional effects responsible for the disagreement.
This disagreement between the computed results from Osswald et. al. [2] on the one hand and our computations and the experimental results from Armaly et. al. [3] on the other hand for the upper wall separation can also be seen from the velocity profiles and wall shears in Figure 10.

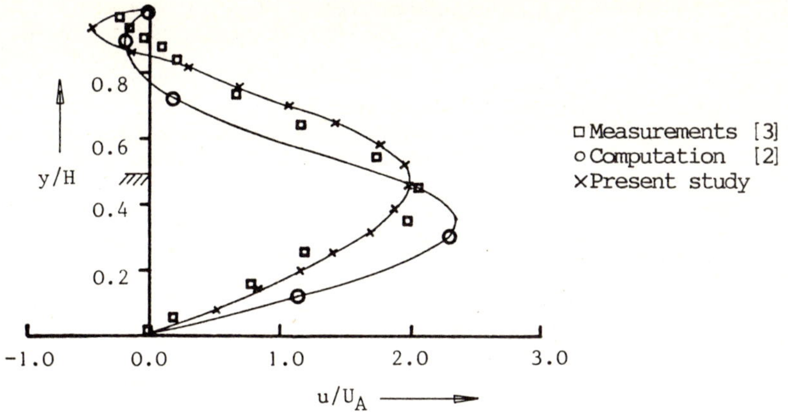

Fig. 10 : Velocity distribution at x_u = 10.11, (see Fig. 9)

We think that the threedimensionality affects strongly the flow further downstream. In the range of the upper wall separation the effect of threedimensionality is weak. While the results for second lower wall separation point (x_2) from our study agree well with the experiment [3] , disagreement appears in the length of the second lower wall separation ($x_3 - x_2$). Here threedimensional effects manifest themselves as momentum diffusion in the lateral direction and cause a shorter separation zone in the experiment.

CONCLUSION

The report of the GAMM-workshop [1] shows that the computed results of full Navier-Stokes equations for flows in a channel with a backward facing step in relatively low Reynolds number flows do not agree and depend strongly on the computational scheme. It is not possible to judge the accuracy of numerical methods since a basis of comparison is missing, except the experiment with its unknown accuracy. The present results with the SOLA-algorithm are in relative good agreement with the experiment.
For Reynolds number 1389 the SOLA-multigrid results compare well with the experiments. Computations of the same flow by Osswald et. al. [2] with the full implicit scheme in surface orientated coordinates show considerable more disagreement with the experiment than our results.
So it can be concluded that a simple semi-implicit MAC-scheme can produce results closer to the experiment than the far more involved full implicit scheme of Osswald et. al. [2]. For a fair comparison however the required CPU time should be taken into account.

REFERENCES

[1] Morgan, K ; Periaux, J; Thomasset, F;
"Analysis of Laminar Flow over a Backward Facing Step",
Notes on numerical fluid mechanics ;Vol 9, (1984).

[2] Osswald, G.A; Ghia, K.N; Ghia, U.;
"Study of Incompressible Seperated Flow using an Implicit Time-Dependent-Technique", AAIA-83-1984-CP.

[3] Armaly, B.F.; Durst, F.;
"Reattachment Length and Circulation Regions Downstream of a Twodimensional Single Backward Facing Step"; Momentum and Heat Transfer Process in Recirculating Flows, ASME HTD-VOL13, (1980), pp. 1-7.

[4] Hirt, C.W.; Nichols, B.D.; Romero, N.C.;
"SOLA - A Numerical Solution Algorithm for Transient Fluid Flows";
Los Alamos Scientific Laboratory report LA-5652,
New Mexico, (1975).

[5] Chorin, A.J.
"Numerical Solution of Navier-Stokes Equations";
Mathematics of computation, Vol 22, (1968).

[6] Peyret, R; Taylor, T.D;
"Computational methods for Fluid Flow";
Springer Series in Computational Physics,
Springer (1883).

[7] Brandt, A.; Dendy, J.E.(Jr.); Ruppel, H.;
"The Multigrid Method for Semi-Implicit Hydrodynamic Codes";
J. Comput. Phys., 34, 1980, pp. 348-370.

[8] Brockmeier, U.;Mitra, N.K.; Fiebig, M.;
"Implementation of Multigrid in SOLA Algorithm",
presented in 2nd European Conference on Multigrid Methods,
Cologne (1985).

[9] Kueny, J.L.; Binder, G.;
"Viscous Flow over Backward Facing Steps",
An Experimental Investigation". pp 32 - 62 in Ref. [1] .

CALCULATIONS AND EXPERIMENTAL INVESTIGATIONS OF THE LAMINAR UNSTEADY FLOW IN A PIPE EXPANSION

F. Durst, J.C.F. Pereira, G. Scheuerer

Lehrstuhl für Strömungsmechanik, Universität Erlangen-Nürnberg
Egerlandstr. 13, D-8520 Erlangen, Germany

SUMMARY

Numerical calculations and experimental investigations of the unsteady, laminar flow in a sudden pipe expansion are presented. The calculation procedure is based upon a numerical solution of the Navier-Stokes equations for axisymmetric flows. A finite-volume approach is used for discretization. The convective terms in the flow equations are approximated by a higher order differencing scheme based on quadratic upstream interpolation. This method reduces the effects of numerical diffusion. A fully implicit first order forward differencing scheme is used to discretize the temporal derivatives. The difference equations resulting from discretization are solved by a stronlgy implicit iterative procedure. The results obtained with the described numerical procedure compare favourably with the experimental data.

INTRODUCTION

The present paper deals with the numerical calculation of unsteady, separated flows in a sudden pipe expansion, the unsteadiness being caused by a piston moving in the tube with the larger diameter. The flows considered are assumed to be axisymmetric, laminar and incompressible, and can be described by the Navier-Stokes equations.

These equations are solved with a numerical method based on the finite-volume approach outlined by Patankar [1]. A higher order method based on quadratic upstream interpolation is used for approximating the convective and viscous terms in the momentum equations. Fully implicit first-order forward differences are used for the temporal discretization. The resulting sets of difference equations are solved successively with the strongly implicit method of Stone [2].

Because of the absence of reliable experimental data to assess the calculated results with, an experimental investigation was carried out in parallel. Profiles of the two velocity components were measured as a function of space and time with the aid of laser-Doppler anemometry. In this paper the computed results are compared with the experimentally obtained velocity distributions.

MATHEMATICAL MODEL

Geometry. Fig. 1 shows the geometry considered and the coordinate system. When the piston moves to the right, fluid is sucked through the inlet tube into the tube with radius R_2. Flow separation occurs at the sudden expansion.

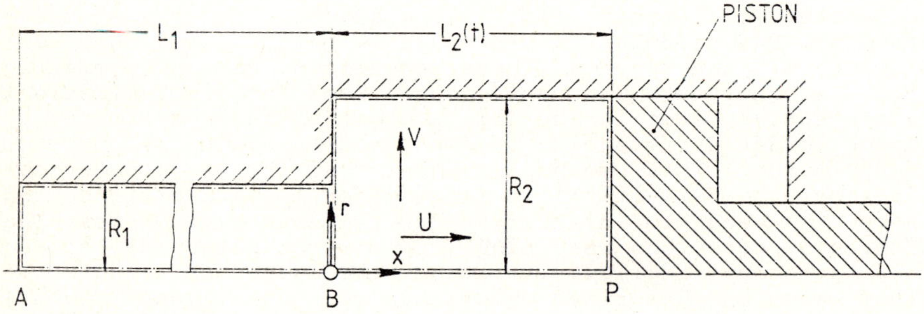

Fig. 1 Flow geometry and coordinate system

Flow Equations and Boundary Conditions. The unsteady flow in the geometry shown in Fig. 1 can be described by the Navier-Stokes equations; these are, under the assumption of axisymmetric and incompressible flow:

continuity equation:

$$\frac{\partial U}{\partial x} + \frac{1}{r}\frac{\partial (rV)}{\partial r} = 0 \quad , \tag{1}$$

x-momentum equation:

$$\frac{\partial U}{\partial t} + U\frac{\partial U}{\partial x} + V\frac{\partial U}{\partial r} = - \frac{1}{\rho}\frac{\partial P}{\partial x} + \nu \left[\frac{\partial^2 U}{\partial x^2} + \frac{1}{r}\frac{\partial}{\partial r}\left(r\frac{\partial U}{\partial r}\right)\right] , \qquad (2)$$

r-momentum equation:

$$\frac{\partial V}{\partial t} + U\frac{\partial V}{\partial x} + V\frac{\partial V}{\partial r} = - \frac{1}{\rho}\frac{\partial P}{\partial r} + \nu \left[\frac{\partial^2 V}{\partial x^2} + \frac{\partial}{\partial r}\left(\frac{1}{r}\frac{\partial (rV)}{\partial r}\right)\right] . \qquad (3)$$

In these equations U and V are the velocities in the x- and y-direction, respectively. P is the pressure; δ and ν stand for the density and kinematic viscosity of the fluid.

The actual computations were performed in two stages. Firstly, the unsteady flow in the inlet tube was calculated. The results of this computation were then used to specify boundary conditions for the second stage, namely the calculation of the flow in the tube containing the piston. Fig. 1 shows the two integration domains, which were linked at x = 0.

For the first computational domain, the boundary conditions were specified as follows: Along the tube wall, the no-slip condition was applied, i.e. U = V = 0; on the axis of symmetry the normal gradients of the x-velocity, $\partial U/\partial r$ and the normal velocity V were set to zero. The velocity profile at the inlet plane, A, (s. Fig. 1) was assumed to be independent of r and determined by the piston velocity U_p according to

$$U_A(t) = U_p(t) \quad \left(\frac{R_2}{R_1}\right)^2 . \qquad (4)$$

Eqn. (4) implies V = 0 at the inlet plane. Finally, at the outlet cross section, boundary conditions for a fully developed pipe flow were specified:

$$\frac{\partial U}{\partial x} = V = 0 . \qquad (5)$$

Note, that the application of Eqn. (5) does not allow disturbances (for instance caused by the piston movement) to spread into the inlet tube.

The U-velocity profiles calculated with the aid of Eqn. (4) were in the second stage of the computation used as inlet boundary conditions at x = 0. The same boundary conditions as prescribed above were employed along the walls and symmetry lines of the second integration domain, with the exception of the piston surface, where U was set equal to U_p.

NUMERICAL SOLUTION PROCEDURE

Coordinate Transformation. For the calculations in the tube containing the piston, a coordinate transformation (s. Gosman and Watkins [3]) was applied, as to allow the numerical grid to contract and expand with the piston movement. The x-coordinate is thereby converted in a quantity $\xi = x/L_2(t)$, where $L_2(t)$ is the distance from the expansion (x = 0) to the piston surface, s. Fig. 1; ξ is now a function of both, x and time t. Application of this coordinate transformation to the continuity and momentum Eqns. (1) to (3) yields:

continuity equation:

$$\frac{1}{L_2}\frac{\partial U}{\partial \xi} + \frac{1}{r}\frac{\partial (rV)}{\partial r} = 0 \tag{6}$$

x-momentum equation:

$$\frac{\partial U}{\partial t} + \frac{1}{L_2}\frac{\partial (\tilde{U}\,U)}{\partial \xi} + \frac{1}{r}\frac{\partial (rUV)}{\partial r} + \frac{U\,U_p}{L_2} = -\frac{1}{L_2}\frac{\partial P}{\partial \xi} + \nu\left[\frac{1}{L_2^2}\frac{\partial^2 U}{\partial \xi^2} + \frac{1}{r}\frac{\partial}{\partial r}\left(r\frac{\partial U}{\partial r}\right)\right] \tag{7}$$

r-momentum equation:

$$\frac{\partial V}{\partial t} + \frac{1}{L_2}\frac{\partial (\tilde{U}\,V)}{\partial \xi} + \frac{1}{r}\frac{\partial (rV^2)}{\partial r} + \frac{V\,U_p}{L_2} = -\frac{\partial P}{\partial r} + \nu\left[\frac{1}{L_2^2}\frac{\partial^2 V}{\partial \xi^2} + \frac{\partial}{\partial r}\left(\frac{1}{r}\frac{\partial (rV)}{\partial r}\right)\right]. \tag{8}$$

$\tilde{U}$ represents the velocity relative to the moving grid, and is given by

$$\tilde{U} = U - \xi\,U_p \tag{9}$$

where use has been made of $U_p = dL_2/dt$. Eqns. (6) to (8) were solved in the (ξ, r, t)-space and then transformed back to the physical space.

Discretization Scheme. The conservative finite-volume approach described by Patankar [1] was used to discretize the continuity and momentum equations. With this method, it is necessary to approximate the convective and diffusive fluxes of the quantity under consideration in and out of a micro-control volume surrounding a grid node. Fig. 2 shows the location of one such control volume in a non-equidistant grid.

For the discretization of the convective fluxes representative values of the transported quantity $\phi = \{U,V\}$ at the various control volume faces are required. If, for instance the (i+1/2,j)-face is considered, one needs to evaluate the integral

$$\phi_{i+1/2,j} = \frac{2}{\Delta y_j + \Delta y_{j-1}} \left[\int_{-\Delta y_{i-1}/2}^{\Delta y_i/2} \phi(s,n)\,dn \right]_{s=\Delta x_i/2} \qquad (10)$$

where use has been made of the local (s,n)-coordinate system shown in Fig. 2. The ϕ-variation must be expressed in terms of values at the grid nodes. To this end, the third order quadratic upstream weighted differencing scheme developed by Leonard [4] is employed, the advantage being higher accuracy and the reduction of numerical diffusion.

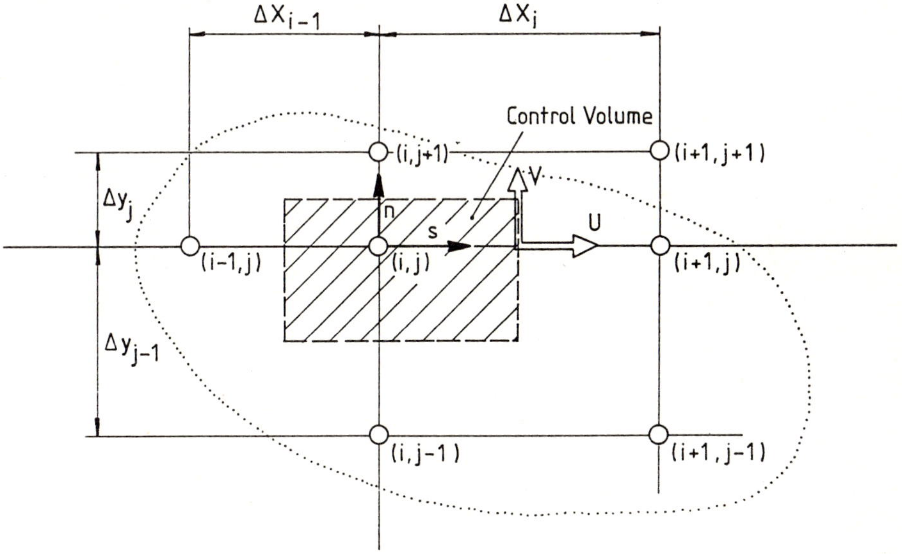

Fig. 2 Location of control volumes in the numerical grid

Leonard's method uses a quadratic interpolation surface oriented with the local velocity components in an upwind-like manner. If, for example, both velocities are positive at the (i+1/2,j)-control volume face an interpolation surface of the following form is fitted through the six points indicated in Fig. 2:

$$\phi = a_o + a_1 s + a_2 s^2 + a_3 n + a_4 n^2 + a_5 ns . \tag{11}$$

Performing the integration in Eqn. (10) with the aid of the relation above yields:

$$\begin{aligned}\phi_{i+1/2,j} &= a_o + a_1 \frac{\Delta x_i}{2} + a_2 \frac{\Delta x_i^2}{4} + \\ &\underline{\frac{1}{4}(a_3 + a_5 \frac{\Delta x_i}{2})(\delta y - 1)\Delta y_{j-1}} + \frac{a_4}{12} \frac{\delta y^3 + 1}{\delta y + 1} \Delta y_{j-1}^2\end{aligned} \tag{12}$$

where $\delta y = \Delta y_j / \Delta y_{j-1}$. The underlined term in Eqn. (12) is small for grid contraction ratios $0.8 < \delta y < 1.2$ and was neglected in this work. After substitution of the appropriate values of the coefficients a_i, one obtains for the face-value $\phi_{i+1/2,j}$:

$$\begin{aligned}\phi_{i+1/2,j} &= \underline{\frac{\phi_{i,j} + \phi_{i+1,j}}{2}} - \frac{1}{4} [\frac{\delta x^2}{1 + \delta x} \phi_{i-1,j} + \frac{\delta x}{1 + \delta x} \phi_{i+1,j} - \\ &- \delta x \phi_{i,j}] + \frac{1}{12} \frac{\delta y^3 + 1}{(\delta y + 1)^2 \delta y} [\phi_{i,j+1} + \delta y \phi_{i,j-1} - (1+\delta y)^2 \phi_{i,j}]\end{aligned} \tag{13}$$

with $\delta x = \Delta x_i / \Delta x_{i-1}$. Analogous expressions were derived for the remaining control volume faces and combinations of velocity vectors, respectively. Note, that Eqn. (13) reduces to central differencing if only the underlined term is considered.

To approximate the diffusive term the gradient $(\partial \phi / \partial x)_{i+1/2,j}$ at the control volume face is required. Application of Eqn. (11) results in

$$(\frac{\partial \phi}{\partial x})_{i+1/2,j} = \frac{\phi_{i+1,j} - \phi_{i,j}}{\Delta x_i} \tag{14}$$

which is identical to the expression obtained from central differencing.

For the temporal discretization a fully implicit first order forward differencing scheme was employed, i.e. the term $\partial\phi/\partial t$ was expressed as

$$(\frac{\partial \phi}{\partial t})_{t+\Delta t} = \frac{\phi^{t+\Delta t} - \phi^t}{\Delta t} \tag{15}$$

and all the spatial derivates and source terms were evaluated the new time level $t + \Delta t$.

Solution Algorithm. To link the pressure to the velocity field the SIMPLE-algorithm described by Patankar [1] was employed. It makes use of a pressure correction equation derived from the contiunity equation. At each time level the two momentum equations and the pressure correction euqation were solved iteratively until convergent solutions were achieved. The scheme was then advanced in time.

The sets of difference equations resulting from the discretization were solved successively with the strongly implicit method developed by Stone [2], which is an iterative procedure based upon a modified LU-triangularization of the coefficient matrices. It has been found by Durst et al. [5] to converge faster for the model problem than a previously employed line-iterative scheme.

A numerical grid of 45 x 44 non-uniformly distributed nodes was used for the inlet tube. The grid points were contracted in the vicinity of the tube wall and close to the inlet (plane A) where high velocity gradients occured. The time-increments chosen were $\Delta t = 0.001$ s, corresponding to a maximum piston displacement of 12 μm. The calculations in the inlet tube proved to be grid independent. The grid inside the cylinder had 35 x 30 nodes, also placed non-uniformly. The time-intervals used were $\Delta t = 0.005$ s, and proved to be sufficiently small to obtain solutions not influenced by the time increment. The calculations were performed on a CDC CYBER 845 computer.

EXPERIMENTAL INVESTIGATION

In order to have a basis for appraising the results obtained from the calculation procedure an experimental investigation using laser-Doppler anemometry was carried out in a geometry identical to the one shown in Fig. 1. A mixture of Dibutylphtalate and Diesel oil having the same refractive index as the Duran glass used to build the test section, was employed as working liquid. In order to eliminate wall curvature effects in the laser-Doppler measurements of the velocity components, the piston-cylinder assembly was placed in a rectangular tank filled with the working liquid.

The laser-Doppler system used for the experiments was operated in dual beam, forward scattering mode and set up to record the axial and radial velocity components. The beam of a 15 mW He-Ne-laser was split into two parts by a beam splitter prism in an integrated transmission optical unit. This unit also included two Bragg cells to produce a frequency shift of 300 kHz between the two beams, which were thereupon focussed by a biconvex lens. The scattered light was collected in a lens and focussed on the pinhole of a photomultiplier.

The Doppler signals were amplified and band-pass filtered to remove the low-frequency pedestal and high-frequency electronic noise, respectively. A frequency tracking demodulator was used to process the signals. The analog output of the frequency tracker was directly fed into a minicomputer. Information on the instantaneous piston velocity was stored simultaneously with the measurement of the velocity components and allowed the velocities to be obtained as a function of the piston location. By passing an external mechanism through an adjustable gating circuit of the piston control box, data acquisition could be obtained in prescribed intervals of the piston movement. This system allowed to measure the velocities over a full cycle either by recording the velocity continuously over one piston cycle, or by sampling at a succession of piston positions by adjusting the gating circuit. Further details on the experimental investigations are provided in Durst et al. [6].

RESULTS AND DISCUSSION

In this section the numerical results obtained for the flow impulsively started at a time t = 0 by a piston movement in the positive x-direction are presented and discussed together with the experimental data.

The test case considered corresponds to a maximum piston velocity of 11.9 mm/s, equivalent to a Reynolds number $Re = 2\,U_p R_2/\nu$ of 98. The initial clearance of the piston, i.e. the distance between the plane x = 0 and the piston surface, was 40 mm. Fig. 3 shows a comparison of the predicted and measured velocity components at a time t = 0.71 s. Note, that the scale for the U-velocity is four times the one for the V-velocity. The agreement between the experimental axial velocities and the predictions in the inlet tube ($x < 0$) is very good, s. Fig. 3a. This is due to the relatively small time-steps used in the calculations leading to a high accuracy of the implicit temporal discretization, but also, and more important, to the diminishing back-influence of the flow in the larger tube at the piston position ($L_2/R_2 = 2.07$) the comparison is made. At earlier times or smaller piston displacements L_2, the flow in the piston-tube exerted a noticeable influence on the velocity profile at the inlet section x = 0, and then the agreement between experiments and predictions was not as good as is shown in Fig. 3. In general it was found that the chosen inlet profiles influenced the predictions in the piston-tube substantially. For instance, in an earlier stage of the work, profiles corresponding to a fully developed flow and a constant velocity are employed. In both cases serious deviations from the experimental results were obtained. This sensitivity to the inlet conditions should be felt in more complex, practical applications also (e.g. the flow in reciprocating engines), and makes predictions for design purposes extremely difficult.

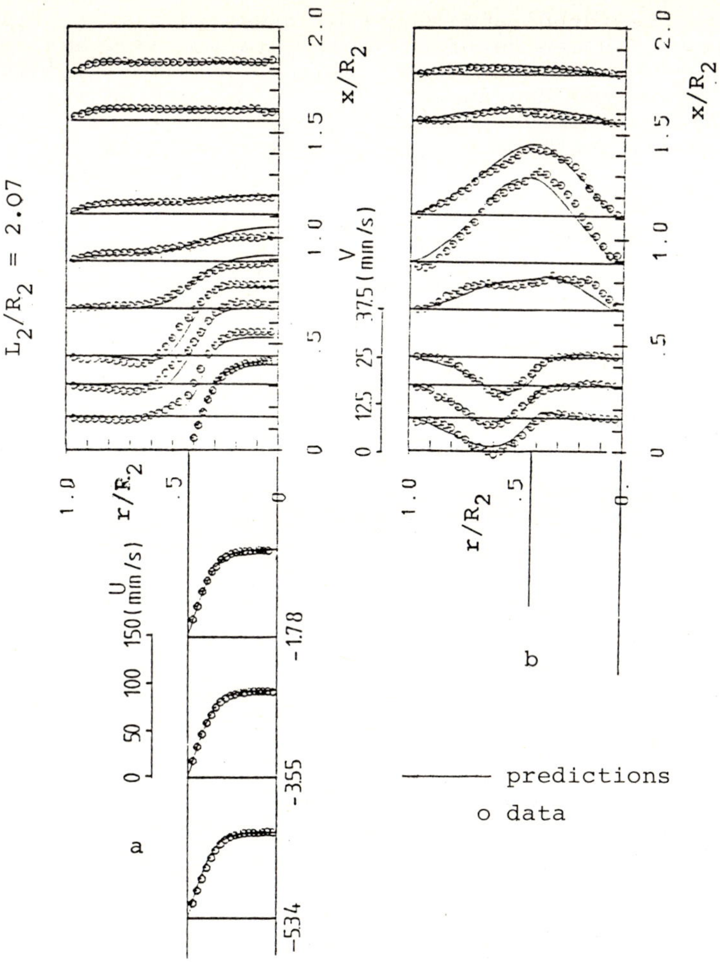

Fig. 3 Comparison of predicted and measured velocity profiles at t = 0.71 s; a) axial velocity; b) radial velocity

Inside the cylinder the velocity of the incoming jet is predicted in good agreement with the data. For $x/R_2 < 0.5$ the flow close to the centre-line is almost parallel to the x-axis; it is then retarded and turned upwards in the r-direction. Close to the piston, radial velocities are very small and the axial velocity becomes almost constant and equal to the piston velocity. Above the inlet jet and close to the inlet plane, a weak recirculation zone with negative U-velocities is formed.

Fig. 4 shows predicted and measured streamlines at a later time t = 1.44 s. The centre of the recirculation zone has now been convected in the positive x-direction and the axial length of the vortex has increased. The inlet jet still exhibits streamlines almost parallel to the x-axis. The agreement between predictions and experiments is remarkable.

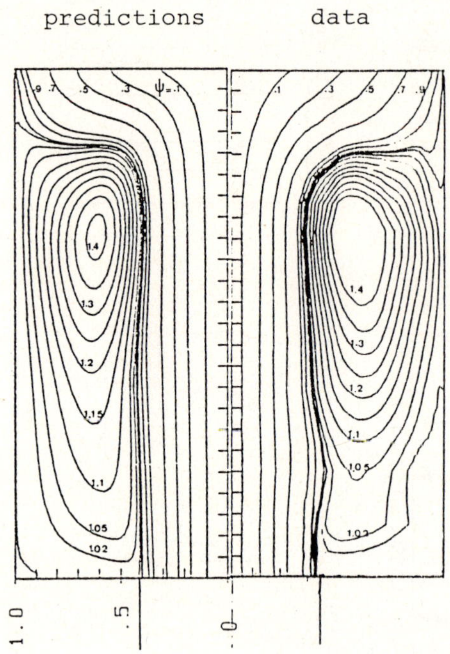

Fig. 4 Predicted and measured streamlines at t = 1.44 s

Axial velocity profiles at t = 2.92 s (L_2/R_2 = 3.1) are plotted in Figure 5. As was already mentioned, the agreement between the calculations and the data increased with increasing distance L_2. At t = 2.92 s, the piston has moved for some time with constant speed, and the effects of the initial piston acceleration have vanished. The inlet velocity corresponds closely to a fully developed pipe flow, and, as a consequence, very good agreement in the jet flow is achieved. The broadening of the jet as it impinges on the piston surface is also well simulated. Close to the piston, the main vortex displays negative U-velocities up to 30% of the piston velocity, and discrepancies arise in this region.

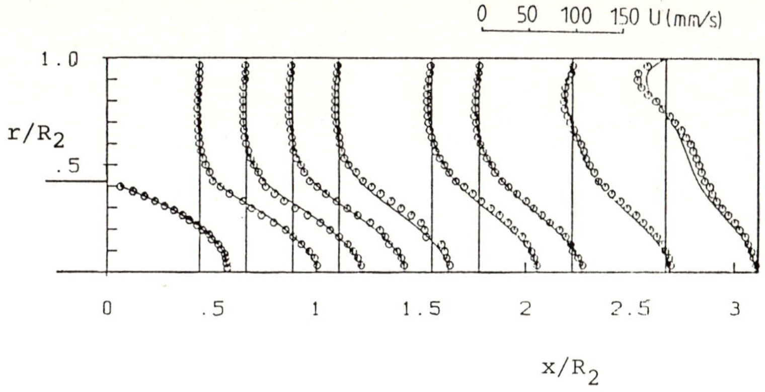

Fig. 5 Comparison of predicted and measured velocity profiles at $t = 2.92$ s;

All the predictions shown so far were obtained with the quadratic upstream weighted discretization scheme described above, the reason being the reduction of numerical diffusion. To emphasize this property of the discretization method comparative calculations were performed with the conventional hybrid-scheme, i.e. a combination of first order upwind and second order central differencing in space, s. Patankar [1]. Fig. 6 shows streamlines calculated with the two schemes and compares them to the measured ones. For the low Reynolds numbers considered here, the differences are not too dramatic. It can be seen, however, that the secondary flow on the cylinder wall is not resolved by the hybrid scheme. Also, inside the recirculation region, where very small velocities occur, qualitative differences between the two schemes arise. The only explanation for these discrepancies is the presence of false diffusion due to the use of first-order upwind differencing in the hybrid-scheme. It causes a more pronounced spreading of the inlet jet close to the piston surface. As a consequence, negative velocities inside the vortex are smaller than predicted by quadratic upstream interpolation or measured. These small differences are responsible for the absence of the recirculating flow region on the cylinder wall.

In general the quadratic upstream interpolation scheme yielded a much more reliable simulation of the flow pattern than the hybrid scheme. This was especially the case for the higher Reynolds numbers not shown here for reasons of brevity.

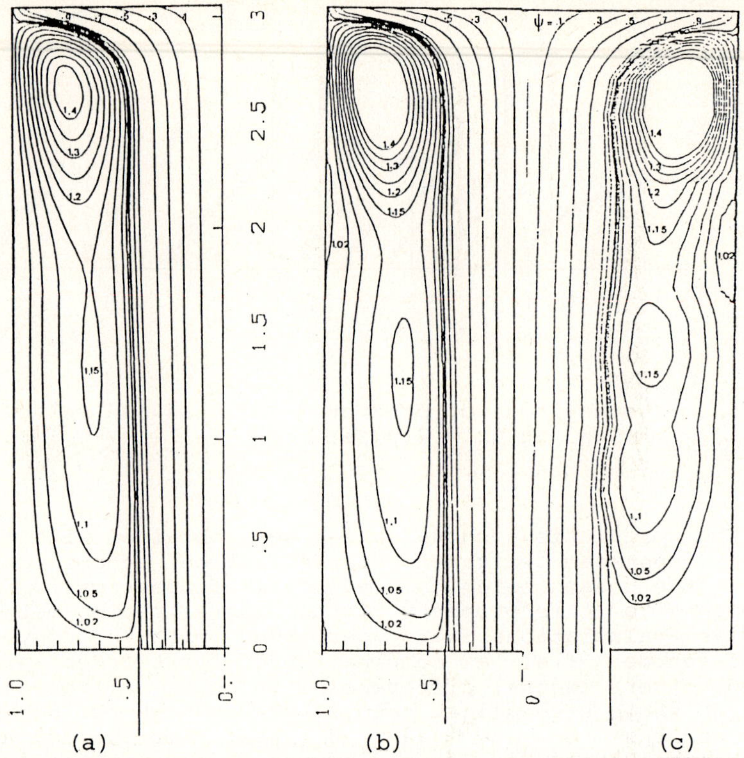

Fig. 6 Comparison of predicted and measured streamline at t = 2.92 s; a) hybrid-scheme; b) quadratic upstream interpolation scheme; c) data

CONCLUSIONS

The comparison of the predicted results with the experimental data shows that the present method allows the calculation of laminar, unsteady separated flows with good accuracy. This is mainly due to the use of the third order quadratic upstream weighted differencing scheme used for the discretization of the convective terms, which diminishes the problems of numerical diffusion.

It was found that the computed results in the tube containing the piston were noticeably dependent on the prescribed inlet profiles at the pipe expansion. The two-pass calculation procedure employed in this work, where the flow in the inlet tube and the piston-tube was calculated separately, did not prove to be completely satisfactory for all test cases considered. This was especially true for situations with small initial piston clearances and shortly after the onset of the flow. Present work is therefore directed towards calculating the inlet flow and the flow in the piston chamber simultaneously. Because of the relatively large number of grid nodes required for such calculations, more efficient solution algorithms like multigrid methods and better ways of coupling the momentum and pressure correction equations, s. Issa [7], are investigated in parallel. This is necessary to reduce the otherwise very high computing times to reasonable bounds.

ACKNOWLEDGEMENTS

The work documented in this paper was sponsored by the Deutsche Forschungsgemeinschaft. The authors gratefully acknowledge this support. The authors are also grateful to Dipl.-Phys. F. Ernst and Mr. H. Weber for their contributions in the experimental part of the work. The manuscript was typed by Miss A. Messner in her usual expert manner.

REFERENCES

[1] PATANKAR, S.V.: "Numerical Heat Transfer and Fluid Flow", Hemisph. Publ. Corp., New York, 1980.

[2] STONE, H.L.: "Iterative Solution of Implicit Approximations of Multidimensional Partial Differential Equations", SIAM J. Numer. Anal., 5 (1968), pp. 530-558.

[3] GOSMAN, A.D., WATKINS, A.P.: "A Computer Prediction Method for Turbulent Flow and Heat Transfer in Piston/Cylinder Assemblies", Proc., 1st Symp. on Turb. Shear Flows, Pennsylvania State University, (1977), p. 523.

[4] LEONARD, B.P.: "A Stable and Accurate Convective Modelling Procedure Based on Quadratic Upstream Interpolation", Comp. Meth. Appl. Mech. and Eng., 19 (1979), pp. 59-98.

[5] DURST, F., GUGGOLZ, M.G., PEREIRA, J.C.F.: "Comparison of Two Methods for the Iterative Solution of the Approximated Fluid Flow Equations", Report-No. 35/N/83, (1983), Lehrstuhl für Strömungsmechanik, Universität Erlangen-Nürnberg.

[6] DURST, F., ERNST, F., PEREIRA, J.C.F.: "LDA-Measurements of Time-Dependent, Separated, Internal Flows", presented at 2nd Intern. Symp. on Applications of Laser Anemometry to Fluid Mechanics, Lisbon, 1984.

[7] ISSA, R.: "Solution of the Implicitly Discretised Fluid Flow Equations by Operator-Splitting", Internal Report (1984), Fluids Section, Dept. Mech. Eng., Imperial College, London.

CALCULATION OF THE FLOW - FIELD CAUSED BY SHOCK WAVE AND DEFLAGRATION INTERACTION

F. Ebert & S. U. Schöffel
Arbeitsgruppe für Mechanische Verfahrenstechnik und Strömungsmechanik, Universität Kaiserslautern
Erwin-Schrödinger-Straße, D-6750 Kaiserslautern, FRG

SUMMARY

The vortex theorem of V. Bjerknes describes the creation of vorticity and circulation, apart from that resulting from viscosity, in stratified, nonhomogeneous fluids [1]. We compared our nonreactive calculations with the Markstein-experiment on shock wave and flame interactions in a stoichiometric mixture of n -butane and air. Analogue to the simulation in [5] most of the major experimental observations are reproduced. The results of our numerical simulations and the estimates of Picone's nonlinear theory are quite consistent. The shock - wave and flame interaction can be identified as a mechanism for generating and amplifying flame - turbulence. The nonlinear integration of the vorticity transport equation by our numerical scheme supplies a framework for calculating the strength and the large scales on which turbulence is generated. Lee and Moen [2] point out that the relative ease of obtaining transition from weak deflagration to supersonic detonation in closed vessels can be explained on the basis of the Taylor interface stability due to multiple shock wave - flame interactions. This instability plays a particular role for vented explosions. The continuous regeneration of the flame folds by this instability mechanism is analogous to the role played by repeated obstacles. This is a result of the study in [2]. In further calculations we want to inquire the effect of confining walls and the influence of transverse pressure waves. Therefore we will compare the results of absorbing with reflecting boundaries. If the heat release is in phase with these pressure waves a strong, possibly resonant coupling between the transverse waves and the flame results. Lee and Moen depict that the rapid amplification of the pressure waves can easily lead to transition to detonation.

ABSTRACT

By interaction of acoustic waves (shocks) with entropy waves (deflagrations,'hot spots') large - scale vorticity fields can be produced in nonhomogeneous fluids. This mechanism is of fundamental importance for the combustion of explosive gas - air - mixtures, since the release of chemical energy produces both pressure and density disturbances in the fluid.

The increase of circulation by the interaction of the disturbances can be reasoned from the vortex theorem of V. Bjerknes [1]. Here we study the motion of a weak, plane shock - wave running through an azimuthally symmetric region characterized by density and temperature distributions similar to those found in an expanding, cylindrical flame. Markstein has studied this case experimentally for a reactive stoichiometric mixture of n -butane and air. He used a long, vertical shock tube with a 30 cm combustion chamber at the bottom and a diaphragm situated 90 cm from the top [3]. A weak shock with a pressure ratio of 1.3 (shock Mach number $\approx$ 1.12) passes through a roughly spherical flame approximately 15 cm from the bottom of the chamber.

INTRODUCTION

The mathematical model used for this simulation is based on the Euler - equations of gasdynamics. The equations are solved by a two step second-order, explicit predictor - corrector Mac Cormack - scheme which is an analogue of the famous Lax - Wendroff - scheme. Due to a theorem of Godunov all symmetric schemes which are higher than of first order degree are not monotonicity - preserving. In order to prevent spurious oscillations (wiggles, ripples) we apply a flux - limiter due to Harten [4], the so - called modified flux-approach. Calculations in two space coordinates are accomplished by a method of fractional steps, the Strang - type operator splitting [4].
The code has been tested extensively by studying the reflection and difraction of shocks with wedges and forward - facing steps.
To analyze the fluiddynamic aspects of vorticity generation we neglect chemical reaction kinetics and diffusive transport processes. Since the transition time for a shock is quite short compared to the time scale of flame propagation we may neglect the flame dynamics for calculating the large - scale eddies.

RESULTS

Our simulation provides a picture of the formation of a toroidal vortex ring. The production of vorticity corresponds to a significant distortion of the heated region. There are striking similarities between the simulation and Markstein's experimental Schlieren photographs (fig.2). By the generated vorticity local reactionrates change which produce additional local pressure waves. The interaction and amplification of the gasdynamic fluctuations of pressure, entropy and vorticity progressively reduces the length scales of inhomogeneities in the flow field. Due to cascade - like mechanism a fine - grained turbulent burning-zone developes. Picone et.al.

require a nonlinear integration of the vorticity transport equation [5]:

$$\frac{d\underset{\sim}{\omega}}{dt} + \underset{\sim}{\omega}\nabla \cdot \underset{\sim}{v} = \underset{\sim}{\omega} \cdot \nabla\underset{\sim}{v} + (\nabla\rho x \nabla p)/\rho^2 \qquad (1)$$

where

$$\underset{\sim}{\omega} = \nabla x \underset{\sim}{v} \qquad (2)$$

is the vorticity, $\underset{\sim}{v}$ is the fluid velocity, ρ is the density and p is the pressure. The nonlinear analysis proposed by Picone et. al. gives a vorticity strength and a mixing time - scale which is consistent with our simulation results and with Markstein's experiment (Fig. 2). The analysis is based on the misalignment of the local pressure and density gradient, which yields a so - called baroclinic source - term of vorticity. It is proportional to the cross product of the gradients.
We generalized Picone's analytical results for the circulation to arbitrary stratified fluids. With the same assumptions we could show that the created circulation does only depend on the upper and lower boundary values of the density and not on the particular form of the density profile. By the latter only the shape and position of the vortices are influenced. Furthermore we extended Picone's analytical results from cylindrical to spherical geometry. Thus, we could determine the mixing time by means of Biot - Savart's law. In fig. 2 the pressure gradient associated with the planar shock is parallel to the axis of the shock tube while the density gradient of the spherical flame is radially directed. Fig. 1 shows a density contour diagram shortly after the simulation starts. The dimensions of the chamber correspond to those in fig. 2. The x - axis is chosen horizontal for our simulation. Fig. 3 defines terms for calculation of vorticity strength via nonlinear analysis. Fig. 7 - 14 show contour and 3-d plots of the evolution of the density, velocities u in x - direction and v in y - direction, temperature T and vorticity ω, respectively. The time steps are chosen according to the CFL - (Courant - Friedrichs - Lewy)-stability criterion. Contrary to fig. 1 we started with a shock which propagated from the right to the left. As initial conditions we took the Rankine - Hugoniot - conditions for the shock and a Bennett - profile for the density distribution of the flame. Picone et.al. took the same functional form in their simulation [5]. In the present finite - volume - scheme we used 151 x 51 mesh points including the fictitious cells. Fig. 7 shows for Kend = 160 the reflected rarefaction wave and transmitted shock. The curvature is due to the increased sound speed inside the 'hot spot'. In contrary to [5] we first studied absorbing boundary conditions at the walls. Thus we did not get a Mach stem formation like Picone, which is caused by reflection of the curved shock by the walls. After the reflection process at the end-wall (bottom) the curvature remains, but is reduced. Motivated by a question posed in a seminary lecture at the Institute of Fluid Mechanics of the Technical University in Munich one of the authors recognized that the plane shape of the transmitted shock wave is transformed into a cylinder. This can be verified by drawing circles around the center of the chamber. Thus at Kend $\simeq$ 180 we calculated the shock - shock interaction of two cylindrical waves. The incident shock coming from the unburned dense gases accelerates the flame in the

stabilizing´ direction with the velocity $\underset{\sim}{v}$ opposing the ´perturbation´. Thus, the curved flame front folds back, causing an indentation in the center and thereby reversing the original shape. After the transmitted shock reflects from the closed end of the duct and catches up with the inverted flame again, the acceleration is reversed, with the light combustion products accelerating towards the denser unburned gas. For the ´unstable´direction the surface perturbations grow further, and the indentation at the center rapidly elongates to form a "spike" as a result of the second interaction from behind (see fig. 5 and 8 with Kend = 360). Furthermore the 3d - plot in fig. 8 for Kend = 320 shows that after the formation of the inverted "spike" on the flame front an additional shock wave emerges. The counter - rotating vortices arising from the interaction can be seen in fig. 14. One can also see the vortices behind the curved shock according to Crocco´s law.
With increasing shock strength the time - lag between the interaction and the indentation of the flame decreases and gives rapidly rise to an elongated highly turbulent funnel. This was shown not only by Markstein´s spark photographs and streak records, but also by records of transients of the radiation emitted by the flame, which could be regarded as a measure of the instantaneous burning rate (fig. 4). Markstein also noticed that the more weaker the shock the more slender was the width of the flame funnel (see fig. 5 and 6).

CONCLUSIONS

The application of low - speed turbulence modeling approaches has expired for shock - wave - turbulent flame interactions. Analytical results combined with experimental data on shock - wave - turbulent boundary - layer interactions lead Zang, Hussaini & Bushnell suggest in [6] that there are at least three further physical mechanisms which are responsible for turbulence enhancement across a shock wave. Zang et. al. point out that it is effectively impossible to isolate the basic physical mechanisms of the interaction experimentally. Also for the shock - wave -flame interaction numerical computations offer a means to examine individual effects under controlled conditions. It is indicated in our case that the neglect of viscous effects does not appear too serious. However the thermal conductivity should be taken into account for weak (subsonic) deflagrations [7].
It is well - known that gasdynamics alone describes the experimentally observed transition to detonation inadequately. Therefore we extended our schemes with global, systematically reduced reaction kinetics.
At the moment we are carrying out further non - reactice calculations with less complex high resolution schemes analyzed by Sweby [8], using Davis´particular flux limiter [9]. Latter removes the requirement for upwind weighting. With Sweby´s artificial dissipation term the Lax - Wendroff - scheme can be hybridized as well to get a TVD - (Total Variation Diminishing) - scheme. The notion of TVD - schemes was introduced by Harten (see [4]). One can show that such schemes guarantee convergence by fulfilling a certain entropy inequality. This is an admissibility criterion which ensures uniqueness of a scalar Cauchy - initial - boundary - value problem. It selects the physically relevant solution from the weak solutions. The name of the criterion is derived from the fact that it is an analogue to the second law of thermodynamics for the gas dynamics equ-

ations. In the fundamental theory of systems of conservation laws associated with famous names like Lax, Majda or Bardos one of the objectives is to 'incorporate the effects of the small scale diffusion processes on the large scale quantities without resolving the small scale effects in detail' (quotation from [10], p. 8).
We have prepared our Fortran 77 - program in order to implement the spatially five - point numerical flux functions of Harten, van Leer and Chakravarthy/Osher (see [11]). Particularly we want to test the symmetric and upwind TVD - schemes of Dr. H.C. Yee at NASA Ames [12]. Her upwind TVD - scheme is based on a second - order scheme developed by Harten; her symmetric scheme is a generalization of the TVD Mac Cormack - scheme we used for the calculations presented here.

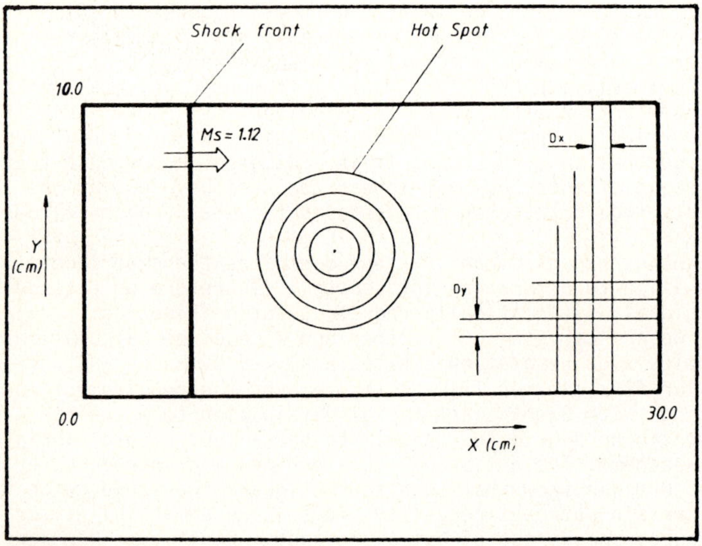

Fig.1: Density contour diagram shortly after begin of the simulation

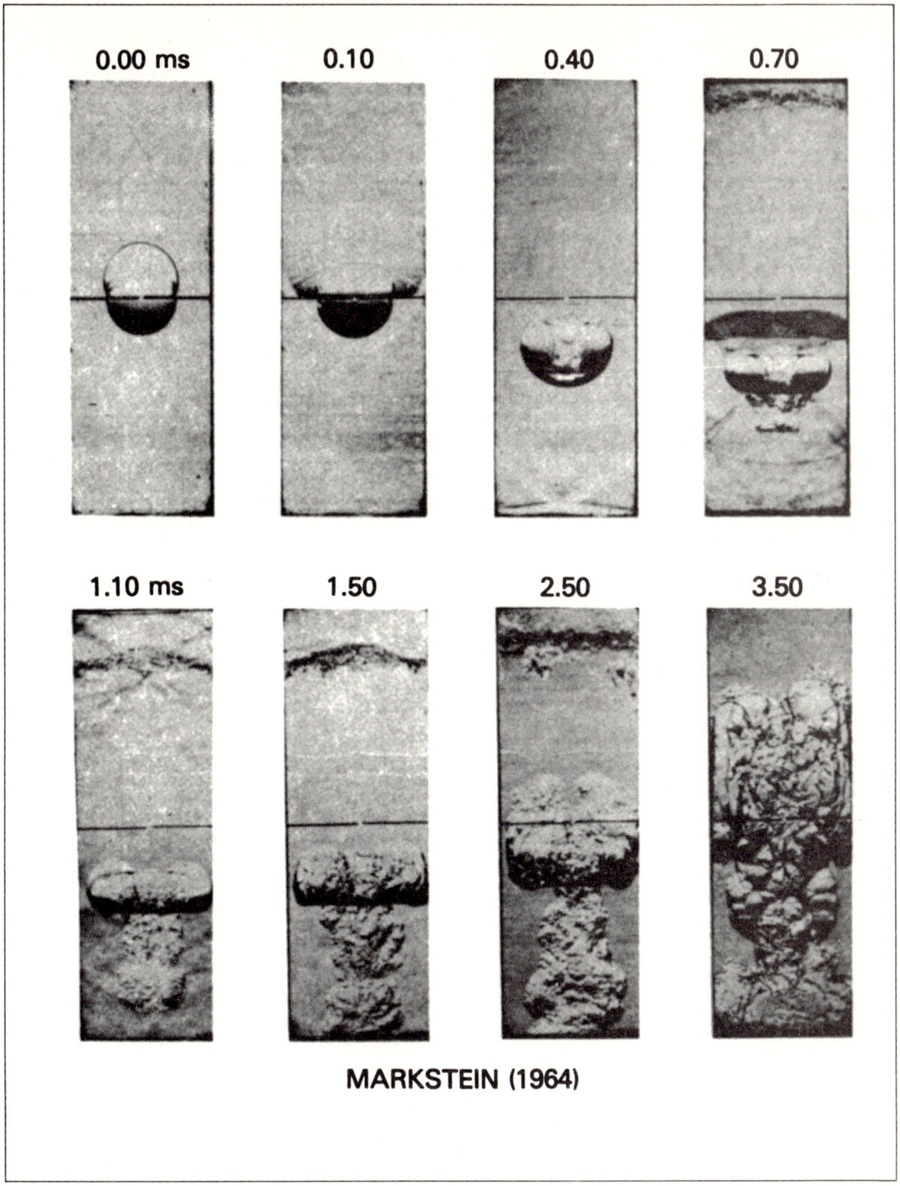

Fig.2: Markstein - experiment; iginition of stoichiometric n - butane - air mixture at center of combustion chamber 8.70 ms before origin of time scale (Rudinger, 1958 and Markstein, 1964) (Reprint in /5/ by permission of Pergamon Press, Inc.)

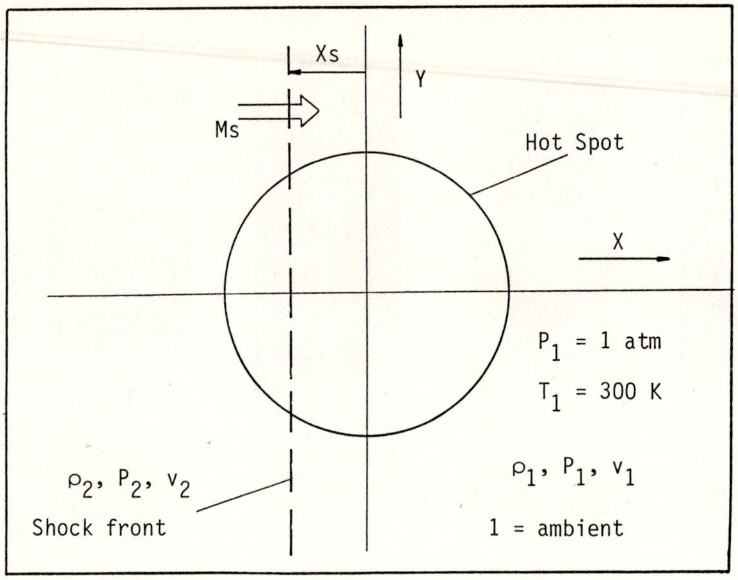

Fig.3: Diagram defining terms for calculation of circulation via nonlinear analysis

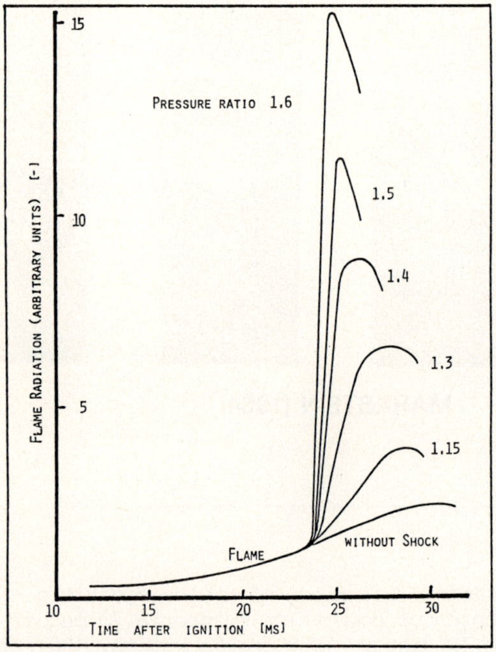

Fig.4: Flame radiation transients during interaction of stoichiometric butane - air flames with shock waves of various pressure ratios (Markstein)

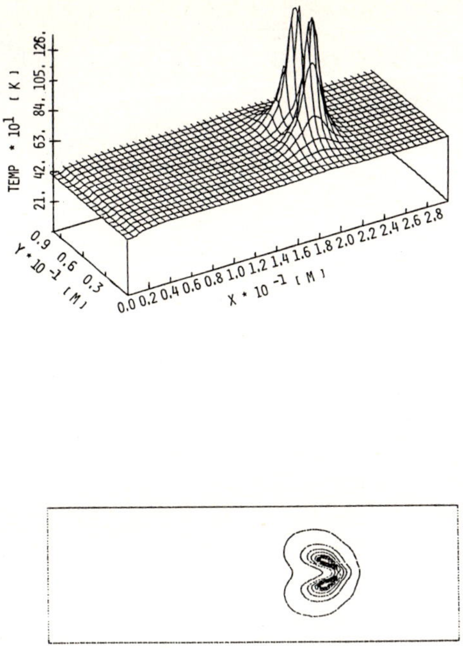

Fig.5: Temperature distribution for Kend = 360 (here the incident shock passed the flame from the left to the right!)

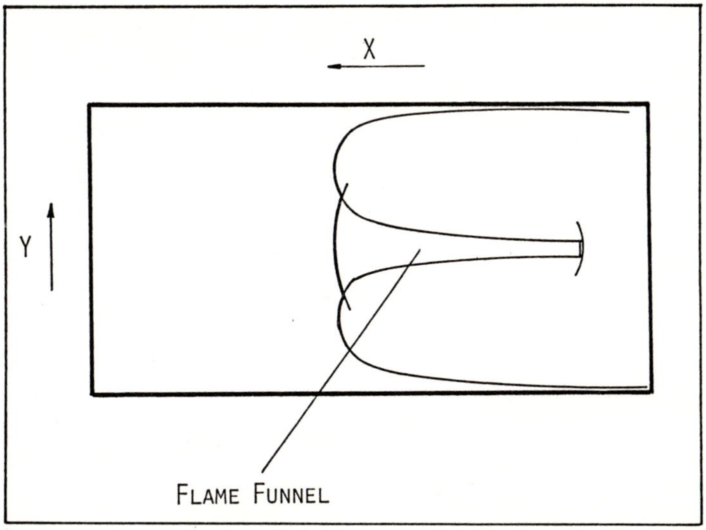

Fig.6: Flame funnel formed during interaction with a weak shock wave, of pressure ratio 1.10 2.5 msec after beginning of the interaction (Markstein 1957)

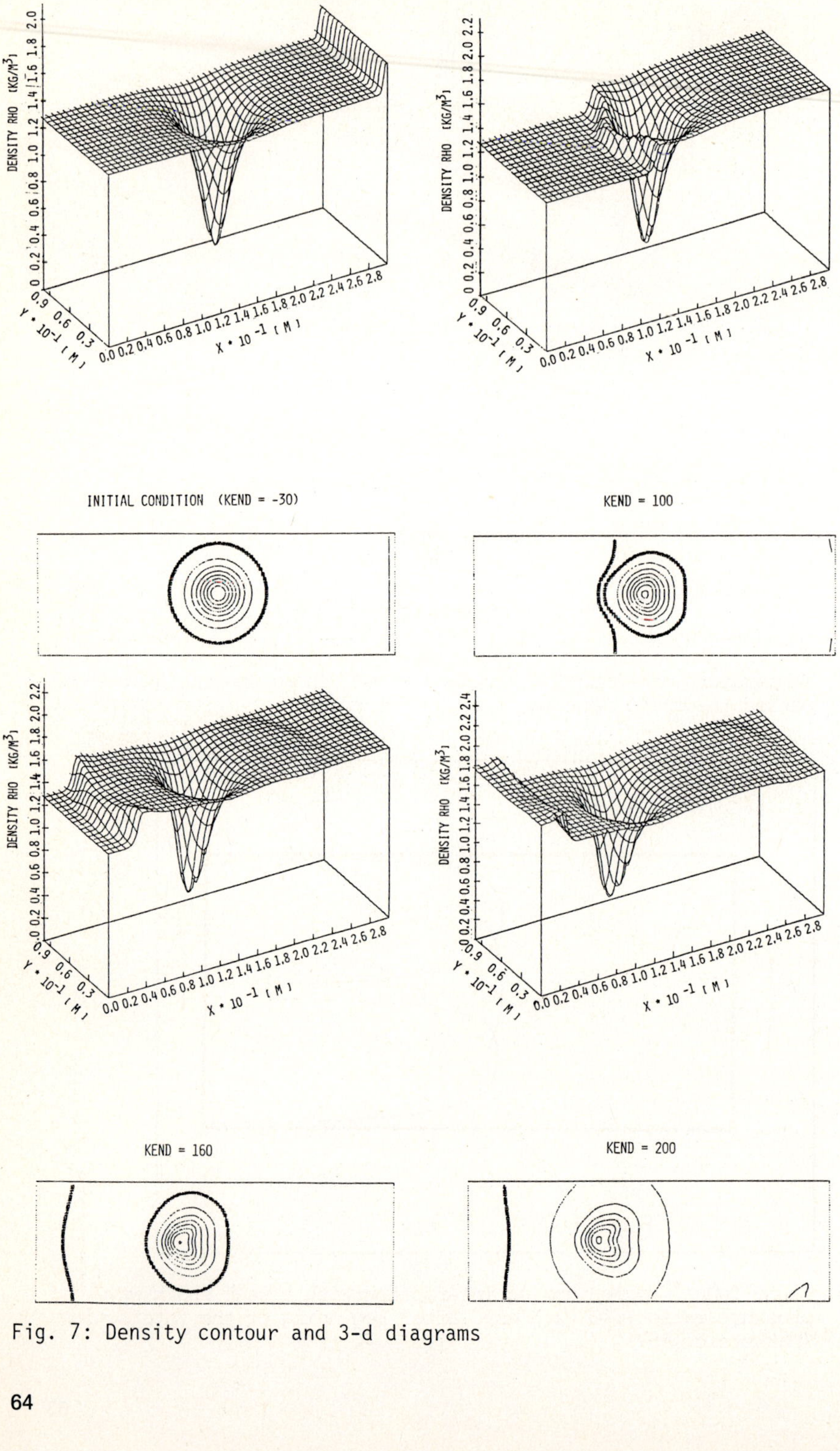

Fig. 7: Density contour and 3-d diagrams

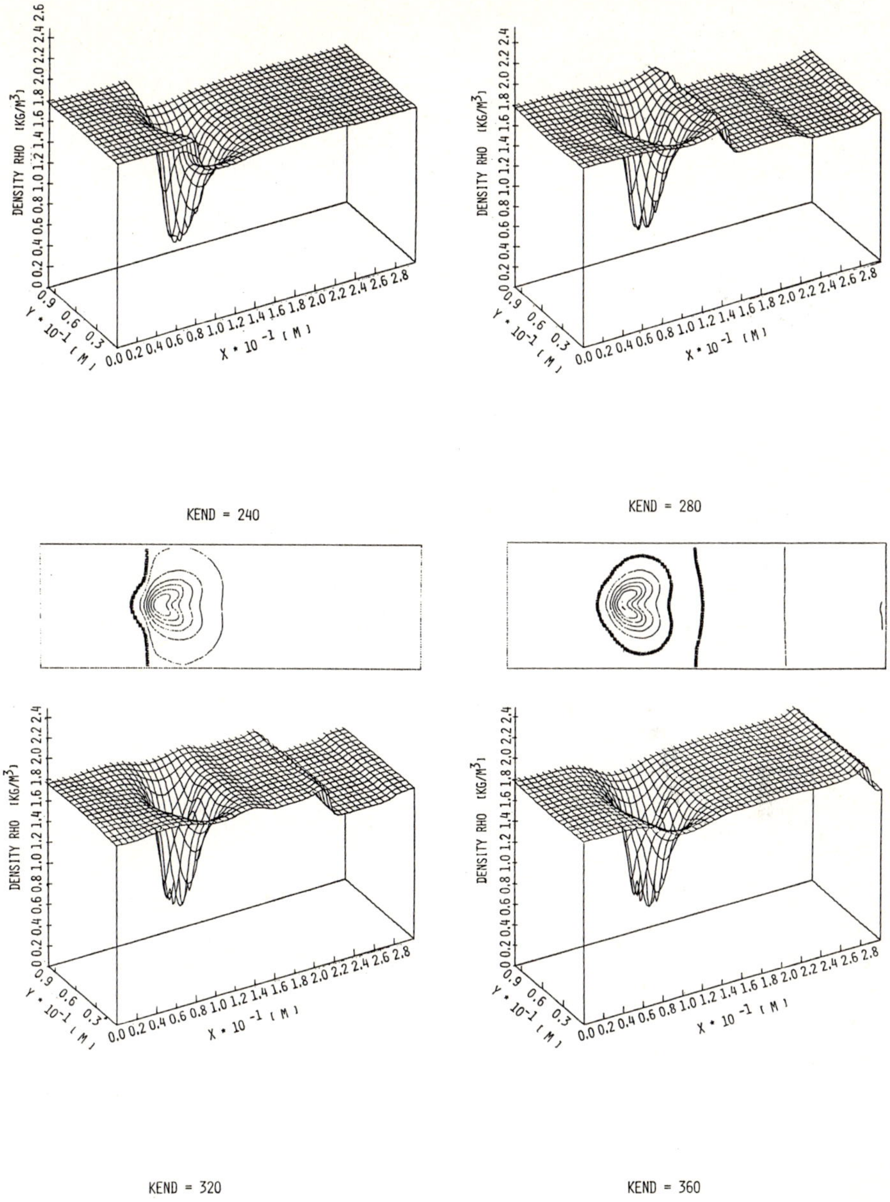

Fig.8: Density diagrams (continued)

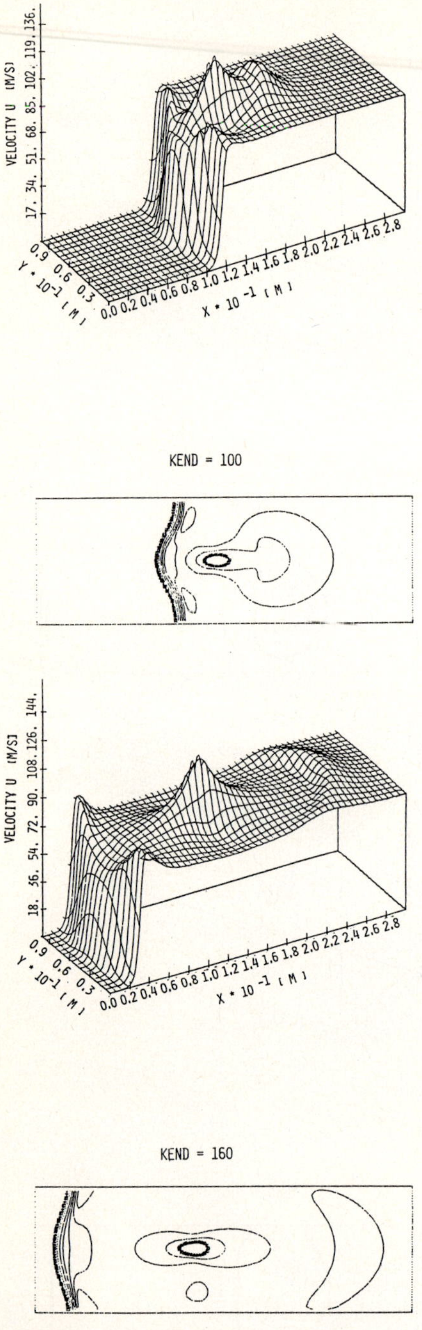

Fig.9: Velocity u in x - direction

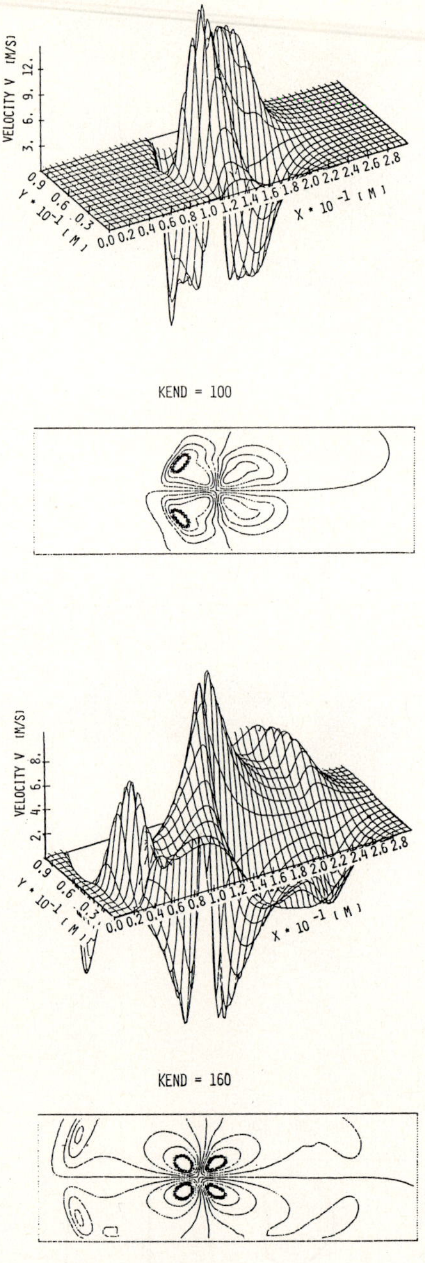

Fig.10: Velocity v in y - direction

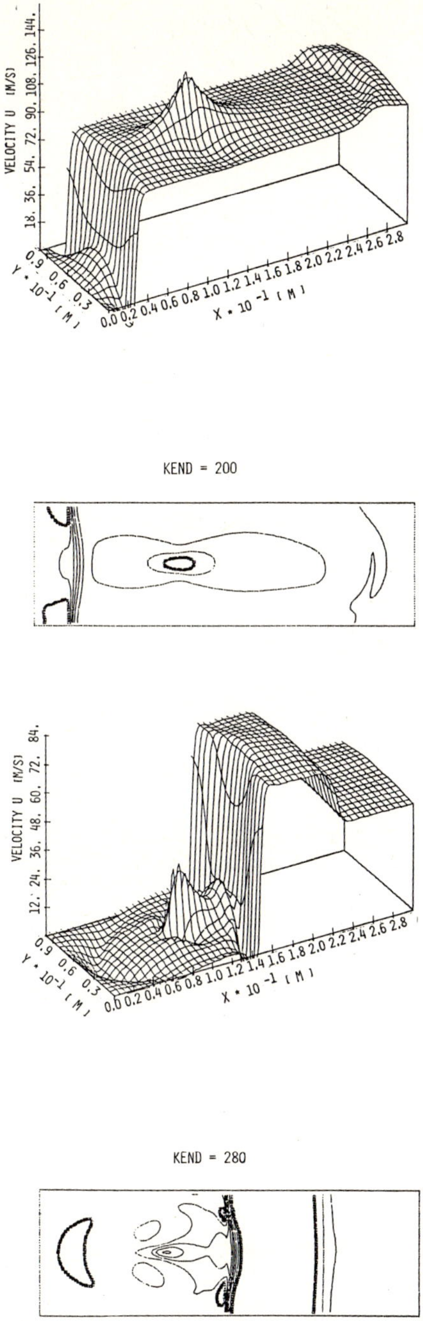

Fig.11: Velocity u (continued)

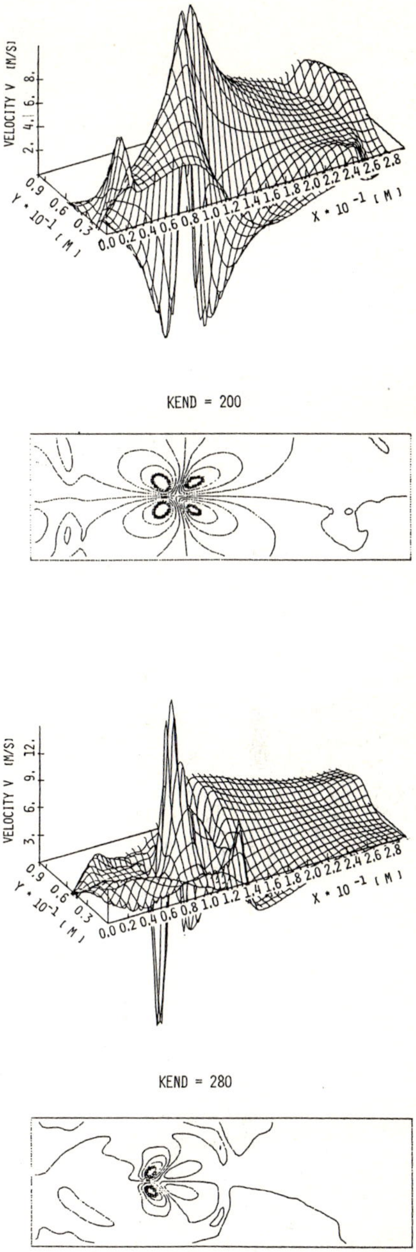

Fig.12: Velocity v (continued)

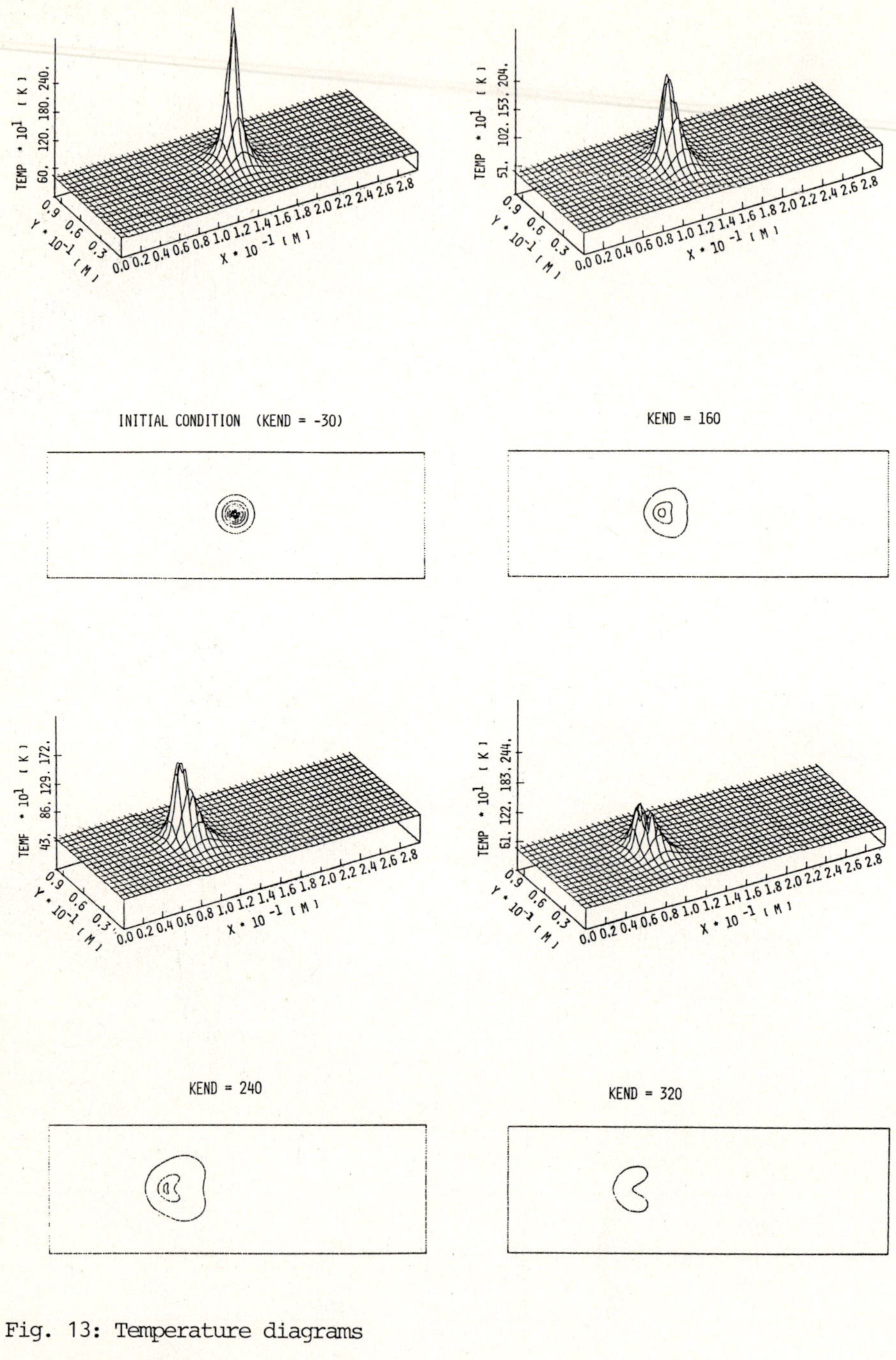

Fig. 13: Temperature diagrams

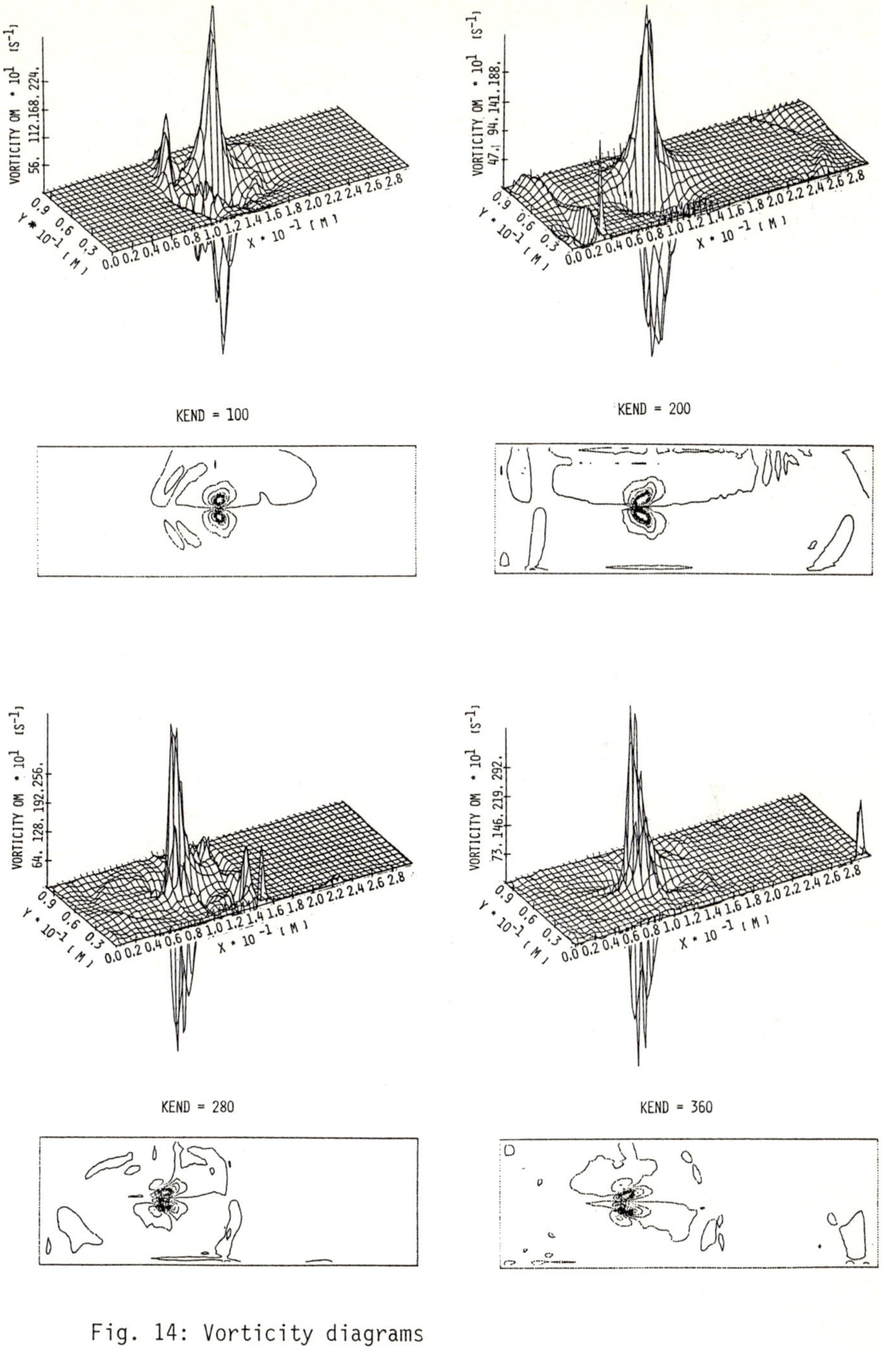

Fig. 14: Vorticity diagrams

REFERENCES

[1] YIH, Chia - Shun: Stratified Flows, Academic Press, 1980.

[2] LEE, J.H.S. & MOEN, I.O.: The Mechanism of Transition from deflagration to detonation in vapor cloud explosions, Prog. Energy Combust. Sci., Vol.6, pp. 359 - 389, 1980.

[3] MARKSTEIN, G.H.: Experimental studies of flame - front instability. Nonsteady Flame Propagation, AGARDograph N$\underline{\text{o}}$75, Pergamon Press, Oxford, pp. 75 - 100.

[4] HARTEN, Amiram: High - Resolution Schemes for Hyperbolic Conservation Laws, J. Comp. Phys., Vol. 49, 1983, pp. 357 - 393.

[5] PICONE, J.M., ORAN, E.S., BORIS, J.P. & YOUNG, T.R., Jr.: Theory of Vorticity Generation of Shock - Wave and Flame Interactions, NRL Memorandum Report 5366, Washington, D.C., July 1984.

[6] ZANG, Th.A., HUSSAINI, M.Y. & BUSHNELL, D.M.: Numerical Computations of Turbulence Amplification in Shock - Wave Interactions, AIAA - Journal, vol. 22, N$\underline{\text{o}}$1, Jan. 1984.

[7] WILLIAMS, F.A.: Combustion Theory, Addison - Wesley, Reading, MA, 1964.

[8] SWEBY, P.K.: High Resolution Schemes using Flux Limiters for Hyperbolic Conservation Laws, SIAM J. Numer. Anal., 21, 995 - 1011 (1984).

[9] DAVIS, S.F.: TVD Finite Difference Schemes and Artificial Viscosity, ICASE Report N$\underline{\text{o}}$ 84-20, June 1984.

[10] MAJDA, A.: Compressible Fluid Flow and Systems of Conservation Laws in Several Space Variables, Applied Mathematical Sciences, Vol. 53, Springer 1984.

[11] YEE, H.C.: On the Implementation of a Class of Upwind Schemes for Systems of Hyperbolic Conservation Laws, NASA Technical Memorandum 86839, Sept. 1985.

[12] YEE, H.C.: On Symmetric and Upwind TVD Schemes, NASA Technical Memorandum 86842, Sept. 1985, also: Proceedings of the 6th GAMM Conference on Numerical Methods in Fluid Mechanics, Göttingen, DFVLR, Sept. 25 - 27, 1985.

A MULTI-LEVEL DISCRETIZATION AND SOLUTION METHOD FOR POTENTIAL FLOW PROBLEMS IN THREE DIMENSIONS

W. Hackbusch
Institut für Informatik
und Praktische Mathematik
Universität Kiel
Olshausenstrasse 40
D-2300 Kiel 1, Germany

Z. P. Nowak
Institute of Applied Mechanics
and Aircraft Technology
Warsaw Technical University
Nowowiejska 24
PL-00-665 Warsaw, Poland

1. Introduction

The advantages of using the multi-level methods for the solution of the boundary integral equation problems are well known. The general convergence proofs for such methods were given in [4] and, independently, in [8]. Special cases were considered in [14], [13], [12], [7], [6]. The practical calculations, using the multi-level solution methods for the various boundary integral problems of the fluid mechanics, were presented in [14], [13], [1], [7], [6]. As follows from the theoretical studies and the numerical experimentation, several iterations of a multi-level method are sufficient to obtain the discrete solution with an error which is less than the discretization error. Moreover, the convergence rates of the multi-level methods increase together with the number of the equations to be solved. It thus seems that the minimum for the cost of the solution phase has been closely approached.

In contrast, the discretization phase for the boundary integral methods is still very costly. The calculations of the large arrays of the influence coefficients can be made faster, if not less costly, by the use of the vector or parallel computers. However, the large storage is still required.

The aim of the present work is to study the possibilities of diminishing the cost of the discretization phase by using several levels of discretization, similar to the fine and coarse grids of the multi-level solution methods. A succession of nested panel systems is constructed by a process similar to subdivision. The collocation points are located in the middle of panels of the finest system. Let h denote the characteristic panel size for the finest system. At the distance less than a fixed radius ρ from the collocation point the panels of the finest system are used for the quadrature. Further away, the quadrature is performed on the panels of the gradually coarser systems with the panel size roughly proportional to the distance from the collocation point. As a result, the approximately uniform distribution of the quadrature error is obtained at the distances larger than ρ. The additional error, introduced into the discrete solution by increasing the panel size, is shown to be O(h) times less than the discretization error of the usual single-level method, which uses only the elements of the finest system for the quadrature. As shown by the results of the calculations, a significant reduction of the computer time and storage requirements can be achieved with the aid of such a multi-level procedure, approaching 80% for two or three discretization levels.

A non-standard two-level method is used for solving the discrete systems. The present approach combines the discretization and solution phases in a consistent manner.

The problem of determining the three-dimensional flow around a wing has been chosen for the study of the simple general idea and for the numerical experimentation. One of the earliest and simplest integral formulations [9] of this problem has been used in combination with the first-order accurate basic discretization scheme. It seems that the present approach can be extended to other integral formulations, arising in aerodynamics and other applications.

2. Formulation of the problem

The problem of determining the non-lifting incompressible irrotational flow around an impermeable body can be mathematically expressed as the Neumann boundary value problem for the perturbation velocity potential:

$$\Delta\varphi = 0 \tag{1a}$$

in the region R_e outside the body, with the boundary condition

$$\frac{\partial\varphi}{\partial\nu}(p) = - V_\infty \cdot \nu(p) \tag{1b}$$

at the points p on the body surface S, where V_∞ is the velocity of the undisturbed flow far from the body and $\nu(p)$ is the outer normal at p. Additionally,

$$\varphi(p) \to 0 \quad \text{when} \quad |p| \to \infty \ , \ p \in R_e . \tag{1c}$$

The total force on S, calculated from the solution φ of the above problem, is equal to zero. To obtain lift, one must introduce additional impermeable surfaces, immersed in the flow region. These surfaces, called *vortex sheets* or *wakes*, originate at the sharp edges on S, such as the leading or the trailing edge of a wing.

The wake sheet is an idealization of a thin layer present in a real viscous flow downstream of a wing. The inclusion of these sheets prevents the derivatives of the velocity potential, i.e., the velocity components , from becoming infinite at the edge points.

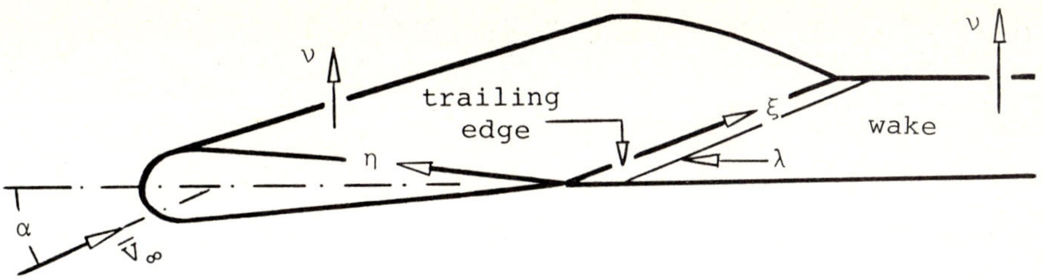

Figure 1. The wake originating at the trailing edge of a wing.

Let W denote the union of all the vortex surfaces present in the flowfield. The Neumann problem (1a-c) could be reformulated by replacing S with S∪W and R_e with $R_e \backslash W$. The shape of W could be determined from the additional nonlinear condition of the pressure equality on both sides of W. Usually, however, the shapes of the sheets are simply assumed and only the *Kutta condition* of the finite velocity at the edges is enforced.

For the boundary integral method the solution of the lifting flow problem is sought in the form of the combination of the surface potentials:

$$\varphi(p) = -\frac{1}{4\pi}\int_S \frac{\sigma(q)}{|p-q|}\, dS(q) + \frac{1}{2}\int_{S\cup W} a(q,p)\mu(q)\, dS(q) \tag{2}$$

$p \in R_e \backslash W$,

where

$$a(q,p) = \frac{1}{2\pi}\frac{\partial}{\partial \nu_q}\left(\frac{1}{|p-q|}\right).$$

Substituting (2) into the Neumann condition (1b) on S, we arrive at the integral equation

$$\sigma(p) = f(p) + (A_{11}\sigma)(p) + (A'_{12}\mu)(p), \quad p \in S, \tag{3}$$

where $f(p) = -2V_\infty \cdot \nu(p)$, and A_{11}, A'_{12} are the integral operators

$$(A_{11}\sigma)(p) = \int_S a(p,q)\sigma(q)dS(q) \tag{4}$$

$$(A'_{12})(p) = -\not\int_{S\cup W} \frac{\partial}{\partial \nu_p} a(q,p)\mu(q)dS(q)\ .$$

In the last formula $\int$ denotes the limit of the corresponding proper integral, obtained when $p \in S$ is approached from $R_e \backslash W$. For the non-lifting flow we may put $\mu=0$, which gives the Fredholm integral equation of the second kind

$$\sigma(p) = f(p)+(A\sigma)(p), \tag{5}$$

where $A := A_{11}$.

In the case of the lifting flow problem one of the functions σ or μ can be chosen arbitrarily on S. The various boundary integral methods of aerodynamics differ in the choice of these functions (for a review see e.g. [10]). Here, we shall present one of the succesful choices, which serves as a basis for a method due to Hess [9]. We shall confine our attention to the case of the flow around a wing.

Let ξ, η denote the arc-length coordinates: ξ measured along the trailing edge and η measured along the contours of the wing cross-sections, starting from the trailing edge (Fig.1). The coordinate η will be scaled by the total length of the current cross-section, so that its maximum value is always 1.

In the method of Hess, the distribution of μ on S is assumed in the form

$$\mu(\xi,\eta) = (1-2\eta)\ \hat{\mu}(\xi), \tag{6}$$

where $\hat{\mu}$ is a function to be determined. When a trailing edge point $t(\xi)$ is approached, the limits for $\mu(\xi,\eta)$ are $\hat{\mu}(\xi)$ and $-\hat{\mu}(\xi)$, on the upper and lower sides of S, respectively. Let us denote

$$\omega(\xi) = \lim_{p \to t(\xi);\, p \in W} \mu(p) \quad .$$

We must set

$$\omega(\xi) = 2\hat{\mu}(\xi). \tag{7}$$

Otherwise, the contact of the three doublet sheets distributed on the upper surface of the wing, the lower surface and the wake, with the respective local densities: $\hat{\mu}(\xi)$, $-\hat{\mu}(\xi)$ and $\omega(\xi)$ would produce a line vortex along the trailing edge, resulting in the arbitrarily large velocities forbidden by the Kutta condition.

It will be assumed that the wake surface is generated by the trailing edge bisectors or the straight lines parallel to V_∞, issuing from the trailing edge, and that μ is constant along these generators. Hence, μ on the wake surface is equal to the appropriate values $\omega(\xi)$. By (6) and (7) also the distribution of μ on S can be expressed with the aid of $\omega(\xi)$. Consequently, (3) can be rewritten in the form

$$\sigma(p) = f(p)+(A_{11}\sigma)(p)+(A_{12}\omega)(p), \quad p \in S, \tag{8}$$

where A_{12} is an integral operator transforming the function $\omega(\xi)$ into a function defined on S.

Since $\omega(\xi)$ can still be chosen arbitrarily, one can additionally require that the impermeability condition (1b) should be satisfied along a certain line λ on W, placed near the trailing edge (Fig. 1). Let $p(\xi) \in \lambda$ lie on the wake generator, originating at $t(\xi)$. Using (8) for $p(\xi)$, with $\sigma(p(\xi))=0$, we obtain the additional linear relation

$$(A_{21}\sigma)(\xi)+(A_{22}\omega)(\xi) = g(\xi), \qquad 0\leq\xi\leq\xi_{max} \quad , \tag{9}$$

where

$$(A_{21}\sigma)(\xi) = (A_{11}\sigma)(p(\xi)),$$

$$(A_{22}\omega)(\xi) = (A_{12}\omega)(p(\xi)),$$

$$g(\xi) = -f(p(\xi)),$$

and ξ_{max} is the total length of the trailing edge.

The solutions for σ and ω (which gives μ on $S\cup W$) can be used for determining the velocity distribution

$$V(p) = V_{\infty} - \frac{1}{4\pi} \int_S \nabla_p\left(\frac{1}{|p-q|}\right)\sigma(q)\, dS_q + \frac{1}{2} \int_{S\cup W} \nabla_p\, a(q,p)\mu(q)\, dS(q), \quad p\in S. \tag{10}$$

3. The discretization of the continuous problem

3.1. The construction of the approximating panel systems

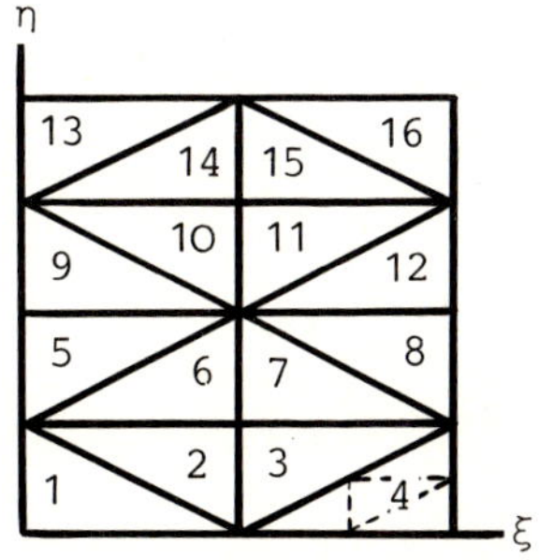

Fig. 2a The parameter plane

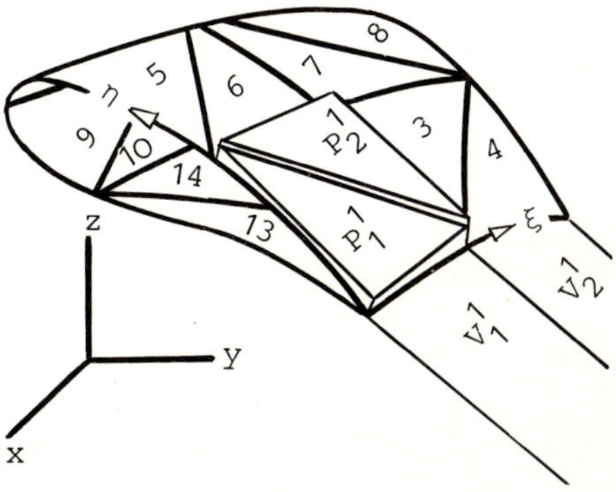

Fig. 2b The physical space

The surface S of the body is assumed to be the image of a rectangle on an auxiliary parameter plane (ξ,η), under the transformation $(\xi,\eta)\rightarrow(x,y,z)$ possessing continuous derivatives up to the third order. As mentioned previously, the wake is generated by the straight lines issuing from the trailing edge in the preassumed direction.

In the exemplary situation, illustrated in Figs. 2a and 2b, the auxiliary variables ξ and η have the same meaning as in Fig. 1.

For the discretization of continuous problems (5) or (8,9) we shall need a sequence of the panel systems $\hat{S}_1$, $\hat{S}_2$, $\hat{S}_3,\ldots$, and the corresponding systems $\hat{W}_1$, $\hat{W}_2$, $\hat{W}_3,\ldots$, of the wake strips, approximating the surfaces S and W with increasing accuracy. To obtain the system $\hat{S}_1$, a curvilinear grid is first generated on S (bold lines in Fig. 2b), as the image of a triangular grid on the parameter plane. The centroids of the curved surface elements are then determined to the second order accuracy with respect to their size. For that purpose we can simply use the images of the centroids of the cells on the parameter plane, except in the areas where the mapping from the rectangle to S is close to singular. Let p_i^1 be the approximate centroid of the i-th element of S, corresponding to the i-th cell on the parameter plane. The vertices of the i-th element are projected into the plane, tangent to S at p_i^1, to give the vertices of the panel P_i^1, approximating this element (the exemplary panels P_1^1 and P_2^1 are shown in Fig. 2b). The directions of the local normals to S will be used for the projection. It will be important to note that the point p_i^1 is a second-order accurate approximation of the centroid of P_i^1. Proceeding in this manner for all the curved elements we obtain the panel system $\hat{S}_1$.

If the mapping $(\xi,\eta)\rightarrow(x,y,z)$ is non-singular then the system $\hat{S}_2$ is constructed by splitting each cell on the parameter plane into the four identical cells (as shown by broken lines for the cell 4 in Fig. 2a), numbering the elements of the new partition in an arbitrary manner and repeating the above procedure. The further splitting will give $\hat{S}_3$, $\hat{S}_4,\ldots$ For the non-singular mapping the sizes of the panels of a system $\hat{S}_l$ are comparable, i.e.,

$$(h_{max})_l/(h_{min})_l \leq C \quad , \quad l=1,2,\ldots, \tag{11}$$

where $(h_{max})_l$, $(h_{min})_l$ denote the sizes of the largest and the smallest panels of $\hat{S}_l$, respectively, and C is a generic constant, independent of l. In such a case $(h_{max})_l$ can be called the characteristic panel size of $\hat{S}_l$. For brevity we shall write h_l instead of $(h_{max})_l$. For $m=l+1, l+2,\ldots$, we have

$$C^{-1}h_l/2^{m-l} \leq h_m \leq Ch_l/2^{m-l} \tag{12}$$

with C close to the unity. Let p_j^l denote the tangency point for a panel P_j^l belonging to the system $\hat{S}_l$. The distance ρ_j^l of p_j^l from the boundary of P_j^l satisfies the inequality

$$\rho_j^l \geq C\, h_l \tag{13}$$

for all P_j^l, $j=1,2,...,n_l$, contained in $\hat{S}_l$. The panel systems $\hat{S}_1$, $\hat{S}_2$, $\hat{S}_3$,..., have the nested structure: each panel P_j^l, $j=1, 2,..,n_l$, of the system $\hat{S}_l$ is the approximation of 4^{m-l} panels of the system $\hat{S}_m$, $m>l$. In the sequel, such groups of panels will be denoted by $G_m(P_j^l)$. Each group $G_m(P_j^l)$ has a central element P_i^m touching S at the point p_i^m, which will be denoted by $p_{l,j}^m$. Let us note that for our construction of the panel tangency points as the images of the cell centroids, we have $p_j^l=p_{l,j}^m$ for all $m>l$, i.e., the central panel of a group $G_m(P_j^l)$ touches S at the same point as the larger panel P_j^l. Let c_j^l denote the centroid of P_j^l. In general we shall require that

$$| p_j^l - p_{l,j}^m | \leq C\, h_l^2 , \tag{14}$$

$$| p_j^l - c_j^l | \leq C\, h_l^2 , \tag{15}$$

for all $l=1,2,...$; $j=1,2,..,n_l$, and $m>l$. As we have seen, both these inequalities are satisfied for our construction of the panel systems when the mapping $(\xi,\eta)\to(x,y,z)$ is non-singular.

In a neighbourhood of a singularity the equal division of the cells on the parameter plane will produce the degenerating panel systems, which do not satisfy the conditions (11), (12), (13) and (15). In such regions an irregular partition of the rectangle and unequal division of cells will be necessary. Instead of using the parameter plane, the panel constructions could be carried out directly in the physical space, using a local approximation of S. The details of such constructions will not be discussed here.

In the case of the lifting flow problem (8,9) the system $\hat{W}_l$ of the wake strips is added to $\hat{S}_l$. As shown in Fig. 2b for the system $\hat{W}_1$, the wake strips are the plane quadrilateral elements bounded by the wake generators, the chords of the trailing edge segments and the corresponding segments far from the body.

3.2. The single-level discretization method

For simplicity we shall first consider the case of the non-lifting flow problem (5) with the single unknown function σ. An outline of the discretization method for the lifting flow problem will be given in Sec. 3.4.

Let $\hat{S}_m$ be one of the approximating panel systems, composed of n_m panels. Where no ambiguity occurs we shall write n instead of n_m. The discrete solution for σ is chosen to be constant over each of the n panel areas. Such a stepwise function is substituted into (5) at the tangency points $p=p_i^m \in P_i^m$, $i=1,2,..,n$. We shall first assume that the integrals (4) over the panel areas are calculated exactly (for that purpose explicit quadrature formulas are available, see e.g. [11]). The case of the numerical quadrature with the aid of the midpoint rule will be considered later on.

The above outlined discretization method, applied to (5), gives a system of n algebraic equations

$$\sigma_n = f_n + A_{n\leftarrow n}\,\sigma_n \ , \qquad (16)$$

where σ_n , f_n are the sequences of n numbers σ_i^n , f_i^n , i=1, 2,..,n, which can be identified with the values of the discrete solution and the fuction f at the corresponding collocation points p_i^m , i=1,2,..,n, and $A_{n\leftarrow n}$ is a linear n-dimensional operator. The operator $A_{n\leftarrow n}$ is defined as follows

$$(A_{n\leftarrow n}\,\sigma_n)_i = \sum_{j=1}^{n} \alpha(p_i^m; P_j^m)\,\sigma_j^n \ , \quad i=1,2,..,n,$$

where

$$\alpha(p_i^m; P_j^m) = \int_{P_j^m} a(p_i^m, q)\, dP_j^m(q). \qquad (17)$$

The integrals (17) will be called the influence coefficients. In the remainder of this Section we shall omit the superscripts m in p_i^m and P_i^m.

Let us introduce the restriction operator R_n , transforming the functions defined on S into the sequences

$$(R_n\sigma)_i = \sigma(p_i) \quad , \ i=1,2,..,n.$$

The discretization error will be defined as

$$\varepsilon_n = \sigma_n - R_n\sigma.$$

From (5) and (16) we obtain

$$\varepsilon_n = (I_{n\leftarrow n} - A_{n\leftarrow n})^{-1}\delta_n,$$

where $I_{n\leftarrow n}$ is the discrete identity operator and δ_n is the truncation error

$$\delta_n = f_n - R_n f + (A_{n\leftarrow n} R_n - R_n A)\sigma = (A_{n\leftarrow n} R_n - R_n A)\sigma.$$

The value of the truncation error, corresponding to the collocation point p_i, can be expressed in the form

$$\delta_i^n = \sum_{j=1}^{n} \sigma(p_j)\alpha(p_i; P_j) - \int_S a(p_i, q)\sigma(q)\,dS(q) \ , \quad i=1,2,...,n. \qquad (18)$$

For a point p, close to S, let N(p) denote its projection on S in the direction of the local normal $\nu(N(p))$. It can be shown that the width of the gaps and the overlaps between the adjacent $N(P_j)$ is $O(h_m^4)$. Let us divide these gaps and overlaps in an arbitrary manner, thus surrounding each $N(P_j)$ with a border set F_j of the width $O(h_m^4)$. Now, (18) can be rewritten in the form

$$\delta_i^n = \sum_{j=1}^{n} \{ \int_{P_j} [a(p_i,q)\sigma(p_j) - a(p_i,N(q))J_j(q)\sigma(N(q))]dP_j(q) \tag{19}$$

$$- \int_{F_j} \gamma_j(q)a(p_i,q)\sigma(q)dS(q)\} \quad , \ i=1,2,\dots,n,$$

where $J_j(q) = dS(q)/dP_j(q)$ under the transformation N, and $\gamma_j(q)$ is equal to 1 or -1 depending on whether the point q belongs to a gap or an overlap. Let c_j denote the centroid of P_j, $j=1,2,\dots,n$. For $i \neq j$ the terms d_{ij} of the sum (19) can be split as follows

$$d_{ij} = \int_{P_j} [a(p_i,q) - a(p_i,N(q))]\sigma(p_j)dP_j(q)$$

$$+ \int_{P_j} [a(p_i,N(q)) - a(p_i,c_j)][\sigma(p_j) - \sigma(N(q))]dP_j(q)$$

$$+ \int_{P_j} a(p_i,N(q))[1-J_j(q)]\sigma(N(q))dP_j(q) \tag{20}$$

$$+ a(p_i,c_j) \int_{P_j} [\sigma(p_j) - \sigma(N(q))]dP_j(q)$$

$$- \int_{F_j} \gamma_j(q)a(p_i,q)\sigma(q)dS(q)$$

$$=: T_{ij}^1 + T_{ij}^2 + T_{ij}^3 + T_{ij}^4 + T_{ij}^5 \ .$$

For $i=j$ we can drop the term T_{ij}^4 and omit $a(p_i,q)$, $a(p_i,c_j)$ in the expressions for T_{ij}^1 and T_{ij}^2.

Let $B(p_i;\rho)$ denote the open sphere centred at p_i, with the radius ρ. If S is a surface of the class C^3 then the solution σ is of the class C^2 on S ([3], p. 171, Theorem 2). In such a case a straightforward analysis of the terms appearing in (20), using (13), shows that the total contribution of all the panels P_j in $B(p_i;\rho)$ to the truncation error (19) is of the order $O(h_m)$, due to the influence of the terms T_{ij}^1. The contribution of all the panels lying outside $B(p_i;\rho)$ is $O(h_m^2)$. The estimate for the term

T_{ij}^4 is derived from the equalities

$$\sigma(p_j)-\sigma(N(q)) = \sigma(p_j)-\sigma(N(c_j))$$

$$+ [\sigma(N(c_j))-\sigma(N(q))] = \sigma(p_j)-\sigma(N(c_j)) \tag{21}$$

$$+ \nabla_q \sigma(N(q))\big|_{q=c_j} (c_j-q)+r_j(q),$$

where ∇_q denotes the two-dimensional gradient operator, acting on functions defined on P_j, and the remainder $r_j(q)$ is $O(h_m^2)$ when $\sigma\in C^2$ on S. By (15) we have $|(p_j-N(c_j)|=O(h_m^2)$, and hence also $|\sigma(p_j)-\sigma(N(c_j)|=O(h_m^2)$. Moreover, let us note that the linear term in (21) makes no contribution to the value of T_{ij}^4. The sum of all the terms T_{ij}^5 is $O(h_m^3)$, and hence negligible.

If the integrals over the panel areas are calculated approximately by the midpoint rule with the nodes c_j, i.e., if (17) is replaced by

$$\alpha(p_i;P_j) = a(p_i,c_j)\cdot|P_j|,$$

where $|P_j|$ is the area of P_j, then the right-hand side of (20) is augmented by the term

$$T_{ij}^6 = \sigma(p_j)\int_{P_j} [a(p_i,c_j)-a(p_i,q)]dP_j(q).$$

For the panels outside $B(p_i;\rho)$ this term can be written as

$$T_{ij}^6 = \frac{\sigma(p_j)}{2}\left[M_{\tilde{x}\tilde{x}}^j\frac{\partial^2}{\partial\tilde{x}^2}a(p_i,q)+M_{\tilde{y}\tilde{y}}^j\frac{\partial^2}{\partial\tilde{y}^2}a(p_i,q)\right]_{q=c_j} \tag{22}$$

$$+ O(h_m^5),$$

where $M_{\tilde{x}\tilde{x}}^j$ and $M_{\tilde{y}\tilde{y}}^j$ are the geometric moments of inertia of the panel P_j with respect to the axis $\tilde{x}$ and $\tilde{y}$ of a certain orthogonal coordinate system with centre at c_j. Both moments are $O(h_m^4)$.

We thus see that, in general, the contribution of the panels in the far field (i.e., outside $B(p_i;\rho)$) to the truncation error at p_i is $O(h_m)$ times less than that due to the panels in the near field (inside $B(p_i;\rho)$). Increasing the accuracy in the near field requires a more complicated second order accurate quadrature method (see e.g. [11]). Another choice would be to use smaller panels in the near field. In order to reduce the cost, the panel sizes in the far field could be increased. For the appropriate choice of the panel sizes, the density of the truncation error in the far field will be maintained at the approximately constant level $O(h_m^2)$.

3.3. The multi-level dicretization method

Let $\hat{S}_m$ and $\hat{S}_l$, $l \leq m$, be the panel systems containing n_m and n_l panels, respectively. We shall introduce the discrete restriction operator $R_{n_l \leftarrow n_m}$, transforming the sequences $\sigma_i^{n_m}$, $i=1,2,\ldots,n_m$, into the sequences of n_l elements

$$(R_{n_l \leftarrow n_m} \sigma_{n_m})_j = \sigma^{n_m}_{i(l,m;j)} \quad , \; j=1,2,\ldots,n_l \quad , \tag{23}$$

where $i(l,m;j)$ is such that the panel $P^m_{i(l,m;j)}$ is central in the group $G_m(P_j{}^l)$. If $l=m$ then $R_{n_m \leftarrow n_m}$ is identical with $I_{n_m \leftarrow n_m}$. Loosely speaking, the restriction operator $R_{n_l \leftarrow n_m}$ approximates the stepwise functions, defined on the finer systems $\hat{S}_m$, with such functions on the coarser systems $\hat{S}_l$ by assigning to each $P_j{}^l \in \hat{S}_l$ the central value of the restricted function in the group of panels, approximated by $P_j{}^l$.

Let $\hat{S}_k$, $\hat{S}_{k+1}, \ldots, \hat{S}_m$ be a subsequence of the sequence $\hat{S}_1$, $\hat{S}_2$, $\hat{S}_3$ ..., and let $p_i{}^m$ denote the collocation point on $P_i{}^m \in \hat{S}_m$. In addition to the sphere $B(p_i{}^m;\rho)$ we shall construct a sequence of the spherical shells

$$B_j(p_i{}^m;\rho) = \{q: 2^{j-1}\rho \leq |q-p_i^m| < 2^j\rho\}, \; j=1,2,\ldots,m-k+1.$$

We shall now consider the following approximation of the operator A:

$$(A_{n_m \leftarrow n_m} \sigma_{n_m})_i = \sum_{j:c_j{}^m \in B(p_i{}^m;\rho)} \alpha(p_i{}^m; P_j^m)\, \sigma_j^{n_m}$$

$$+ \sum_{l=k}^{m} \; \sum_{j:c_j{}^l \in B_{m-l+1}(p_i{}^m;\rho)} \alpha(p_i{}^m; P_j{}^l)\,(R_{n_l \leftarrow n_m} \sigma_{n_m})_j \tag{24}$$

$$=: (D_{n_m \leftarrow n_m} \sigma_{n_m})_i + (C_{n_m \leftarrow n_m} \sigma_{n_m})_i$$

for $i=1,2,\ldots,n_m$, where $c_j{}^l$ is the centroid of $P_j{}^l$. The two sums in (24) are the approximations of the near and the far field parts of the integral operator. For the far field quadrature the variable panel size is used, depending on the distance from the collocation point.

The contribution of a typical term in the second sum to the truncation error follows from (20), (21) and (22), where we put $P_j{}^l$, p_i^m , $c_j{}^l$ and $p_{l,j}^m$ (which has the same meaning as in (14)), instead of P_j, p_i, c_j and p_j, respectively. Let $h(P_j{}^l)$ and $|P_j{}^l|$ denote the size and the area of $P_j{}^l$. The contribution (20) of this panel to the truncation error, augmented by (22), can be expressed as

$$d^l_{ij} = \int_{P_j{}^l} [e_1(q)+e_2(q)+e_3(q)+O(h(P_j{}^l)^3)]dP_j{}^l(q),$$

where

$$e_1(q) = a(p^m_i, c^l_j)[(1-J^l_j(q))\sigma(N(q))+\sigma(p^m_{l,j})$$

$$-\sigma(N(c^l_j))+r^l_j(q)],$$

$$e_2(q) = \nabla_q a(p_i{}^m, q)|_{q=c^l_j} [(q-N(q))\sigma(p^m_{l,j})$$

$$+(N(q)-c_j{}^l)(\sigma(p^m_{l,j})-\sigma(N(q)))],$$

$$e_3(q) = \frac{\sigma(p^m_{l,j})}{2|P_j{}^l|}[M^{j,l}_{\tilde{x}\tilde{x}} \frac{\partial^2}{\partial\tilde{x}^2}a(p_i{}^m, q)$$

$$+M^{j,l}_{\tilde{y}\tilde{y}} \frac{\partial^2}{\partial\tilde{y}^2}a(p_i{}^m, q)]_{q=c_j{}^l} \quad .$$

Using (14) and (15) we obtain

$$|e_1(q)| \leq |a(p_i{}^m, c_j{}^l)| \; C \; h(P_j{}^l)^2 \qquad (25)$$

$$\leq C \cdot h(P^l_j)^2/|p_i{}^m-c_j{}^l|^2.$$

If the panels with the centroids in the spherical shells B_{m-l+1} $(p_i{}^m;\rho)$ have the size $2^{m-l}h_m$, as in the second sum in (24), then the upper bound for $|e_1(q)|$ is approximately constant throughout the far field and close to its value $Ch_m{}^2/\rho^2$ near the boundary of the near field $B(p_i{}^m;\rho)$. The other components of the error density can be estimated as follows

$$|e_2(q)| \leq C \; h(P_j{}^l)^2/|p_i{}^m-c_j{}^l|^3,$$

$$|e_3(q)| \leq C \; h(P_j{}^l)^2/|p_i{}^m-c_j{}^l|^4.$$

The estimate for $|e_3(q)|$ was obtained using the inequality $|P_j{}^l| \geq C \; h_l{}^2$, which is a consequence of (13). For our choice of the panels sizes the upper bounds for $|e_2(q)|$ and $|e_3(q)|$ decrease outside $B(p_i{}^m;\rho)$.

If the solution σ has large second derivatives then the constant C in (25) may be large, due to the contribution of the term $r_j{}^l(q)$.

In order to reduce this part of the error density we can use in (24), instead of the pointwise (or rather "panelwise") restriction operator $R_{n_1 \leftarrow n_m}$, an averaging restriction operator, defined as the arithmetic mean over the groups of panels:

$$(\tilde{R}_{n_m \leftarrow n_m} \sigma_{n_m})_j = \sum_{i: P_i^m \in G_m(P_j^1)} \sigma_i^{n_m} / 4^{m-1} \qquad (26)$$

for $j=1,2,\ldots,n_1$.

Then, as it can be easily shown, the contribution of the term $r_j^1(q)$ will be maintained virtually independent of the level of discretization.

In order to perform the multi-level discretization of the form (24) we first verify for the successive panels P_j^k, $j=1,2,\ldots,n_k$, of the coarsest system $\hat{S}_k$, if their centroids c_j^k belong to the exterior of the sphere $B(p_i^m; 2^{m-k}\rho)$. If $c_j^k \notin B(p_i^m; 2^{m-k}\rho)$ then the panel P_j^k is used for calculating the corresponding term of the second sum in (24). Otherwise, the panel is replaced by the group $G_{k+1}(P_j^k)$. After completing this procedure for all the panels of $\hat{S}_k$ we obtain, as a by-product, a subset of panels of $\hat{S}_{k+1}$. Again, for all the panels of this subset we verify if their centroids belong to the exterior of the sphere $B(p_i^m; 2^{m-k-1}\rho)$, and use the panels for the quadrature or replace them by groups of the smaller ones, depending on the result of the inclusion test. The approximate calculation of the integral (4) is completed when the panels of the finest level $\hat{S}_m$ with the centroids in $B(p_i^m; 2\rho)$ have been used for the quadrature.

The number $m-k+1$ of the levels of discretization is determined by ρ and the maximum size L of the body, as a consequence of the requirement that the surface should extend further than $B(p_i^m; 2^{m-k}\rho)$ for a typical collocation point p_i^m. We could choose for example

$$m-k+1 \leq \log_2(L/\rho).$$

Due to the reasons, which will be clarified in Sec. 4.1., the two sets of the influence coefficients, appearing in the two parts of the operator (24), should be stored seperately. The sequential form is the most appropriate for these sets, since the usual matrix form requires the storing of a large number of zeros for the influences of these panels of $\hat{S}_m$ which have been replaced by the larger ones in the process of the far field quadrature. If the averaging restriction operator (26) is used in (24), then instead of zeros, the matrix would contain a large number of equal terms $\alpha(p_i^m; P_j^1)/4^{m-1}$.

3.4. The discretization for the lifting flow problem

In the case of a thin body, such as the wing of an aircraft, the near field panels of the size h_m, lying on the opposite side of the surface at the distance δ from the collocation point, contribute the term $O(h_m^2/\delta)$ to the truncation error. This term becomes most significant near the trailing edges, where δ is much less than h_m. Near the cusped edges the truncation error is $O(1)$.

The multi-level discretization procedure is even more appropriate in such cases, since the inequalities such as (25) still hold in the far field with a relatively small value of C, at least outside the areas, where the solution has large first or second derivatives, or where the surface curvature is high. In such areas the computational procedure could switch to the finer levels of discretization.

In the case of the lifting flow problem the discrete solution for ω is assumed in the form of the stepwise function with the values $\omega_i^{\tilde{n}}$ on the successive strips V_i^m, $i=1,2,\ldots,\tilde{n}=\tilde{n}_m$, of the system $\hat{W}_m$. The approximation for μ on S is chosen to be the stepwise function on the panels of $\hat{S}_m$, with the values μ_i^n, $i=1,2,\ldots,n=n_m$, calculated at the corresponding points p_i^m from (6) and (7). These two functions and the stepwise function with the values σ_i^n, $i=1,2,\ldots,n=n_m$, are substituted into (8) at the body collocation points $p=p_i^m$ and into (9) at $\xi=\xi_1,\xi_2,\ldots,\xi_{\tilde{n}}$, corresponding to the centers of the trailing edge segments. The following discrete system is obtained:

$$
\begin{aligned}
&(I_{n\leftarrow n}+D_{n\leftarrow n}+C_{n\leftarrow n})\sigma_n+A_{n\leftarrow\tilde{n}}\,\omega_{\tilde{n}} = f_n ,\\
&A_{\tilde{n}\leftarrow n}\,\sigma_n+A_{\tilde{n}\leftarrow\tilde{n}}\,\omega_{\tilde{n}} = g_{\tilde{n}} ,
\end{aligned} \tag{27}
$$

where $D_{n\leftarrow n}$, $C_{n\leftarrow n}$ are such as in (24) and $A_{n\leftarrow\tilde{n}}$, $A_{\tilde{n}\leftarrow n}$ and $A_{\tilde{n}\leftarrow\tilde{n}}$ correspond to the operators A_{12}, A_{21} and A_{22} in (8) and (9). For the integrals over $\hat{S}_m$, which constitute parts of the operators $A_{n\leftarrow\tilde{n}}$, $A_{\tilde{n}\leftarrow n}$ and $A_{\tilde{n}\leftarrow\tilde{n}}$ the multi-level quadrature can be applied. For the relatively inexpensive calculations of the integrals over $\hat{W}_m$, present in the definitions of $A_{n\leftarrow\tilde{n}}$ and $A_{\tilde{n}\leftarrow\tilde{n}}$, the single level procedure is sufficient. The influence coefficients for all these operators can be stored in the form of matrices of $n\cdot\tilde{n}$ or $\tilde{n}\cdot\tilde{n}$ elements.

Additionally, the sequences of the influence coefficients must be calculated for the components of the vector integral operators in (10), to be used for the velocity calculations from the results for σ and ω. The present form of the multi-level procedure may be used also for the integrals over S, appearing in (10), since their kernels decrease at the same rate as $a(p,q)$ or faster.

4. The two-level solution method

4.1. The general presentation

Let $\hat{S}_1$ and $\hat{S}_m$, $m>1$, be the panel systems with $k=n_1$ and $n=n_m$ elements. We shall introduce the discrete prolongation operator $P_{n\leftarrow k}$, transforming the sequences σ_k into the sequences with the elements

$$(P_{n\leftarrow k}\sigma_k)_i = \sigma^k{}_{j(m,1;i)}\ , \ i=1,2,\dots,n,$$

where j(m,l;i) is such that $P_i{}^m \in G_m(P^1_{j(m,1;i)})$.

The discrete problem (27) can be solved by the following iteration using the two levels $\hat{S}_m$ and $\hat{S}_k$:

$$(I_{n\leftarrow n}+D_{n\leftarrow n}+P_{n\leftarrow k}R_{k\leftarrow n}C_{n\leftarrow n})\sigma_n{}^{new} \tag{28a}$$

$$+A_{n\leftarrow\tilde{n}}\omega_{\tilde{n}}{}^{new} = f_n+(P_{n\leftarrow k}R_{k\leftarrow n}-I_{n\leftarrow n})C_{n\leftarrow n}\sigma_n{}^{old},$$

$$A_{\tilde{n}\leftarrow n}\sigma_n{}^{new} + A_{\tilde{n}\leftarrow\tilde{n}}\ \omega_{\tilde{n}}^{new} = g_{\tilde{n}}\ , \tag{28b}$$

where $R_{k\leftarrow n}$ is the pointwise restriction operator (23). Let σ_n , $\omega_{\tilde{n}}$ be the solutions of (27). An easy calculation gives

$$\sigma_n{}^{new}-\sigma_n = Q_{n\leftarrow n}(\sigma_n{}^{old}-\sigma_n)\ ,$$

$$\omega_{\tilde{n}}{}^{new}-\omega_{\tilde{n}} = -A^{-1}_{\tilde{n}\leftarrow\tilde{n}}A_{\tilde{n}\leftarrow n}(\sigma_n{}^{new}-\sigma_n),$$

where

$$Q_{n\leftarrow n} = (I_{n\leftarrow n}+D_{n\leftarrow n}+P_{n\leftarrow k}R_{k\leftarrow n}C_{n\leftarrow n} \tag{29}$$

$$-A_{n\leftarrow\tilde{n}}A^{-1}_{\tilde{n}\leftarrow\tilde{n}}A_{\tilde{n}\leftarrow n})^{-1}(P_{n\leftarrow k}R_{k\leftarrow n}-I_{n\leftarrow n})\ C_{n\leftarrow n}\ .$$

Let us note that $C_{n\leftarrow n}$ is an approximation of the compact integral operator

$$(C\sigma)(p) = \int_{S\setminus B(p;\rho)} a(p,q)\sigma(q)\,dS(q)\ .$$

Using the methods of the theory of the collectively compact operators (cf. [2]) we can show that the norm of the product

$$(P_{n\leftarrow k}R_{k\leftarrow n}-I_{n\leftarrow n})C_{n\leftarrow n}\ , \tag{30}$$

appearing in (29), tends to zero when both n and k<n tend to infinity. Hence, if the remaining factor in (29) has a uniformly bounded norm then the acceleration of convergence, characteristic of a multi-level solution method (i.e., increasing when the panel size tends to 0) will be observed, at least for a sufficient-

ly large number of panels. It can be shown that the norm of (30) is $O(h_1)$ if, instead of $B(p_i{}^m;\rho)$, the region

$$\bigcup_{\hat{i}:P_{\hat{i}}^m\in G_m(P^1_{j(m,1;i)})} B(p_{\hat{i}}^m;\rho)$$

is used for the definition of the near field. This, however, would complicate the multi-level descretization procedure.

4.2. The solution procedure

At the beginning of each iteration (28) the right-hand side of (28a) must be calculated, involving the multiplication $C_{n\leftarrow n}\sigma_n{}^{old}$. The methods of the type (28) give the practically converged solution in a few iterations, and the first iteration will not require this costly multiplication if we choose $\sigma_n{}^{old}=0$ as the initial approximation. Let us write (28) in the form

$$A_{\bar{n}\leftarrow\bar{n}}\ \psi_{\bar{n}} = v_{\bar{n}}\quad , \tag{31}$$

where $\bar{n}=n+\tilde{n}$, $\psi_{\bar{n}}=(\sigma_n{}^{new},\ \omega_{\tilde{n}}{}^{new})$, and $v_{\bar{n}}$ represents the right-hand sides of (28). The solution of (31) can be obtained by another iterative process of the form

$$A'_{\bar{n}\leftarrow\bar{n}}\ \psi_{\bar{n}}^{new} = v_{\bar{n}} - A''_{\bar{n}\leftarrow\bar{n}}\ \psi_{\bar{n}}^{old}\ , \tag{32}$$

where $A'_{\bar{n}\leftarrow\bar{n}}$ is a sparse and easily invertible part of $A_{\bar{n}\leftarrow\bar{n}}$ and $A''_{\bar{n}\leftarrow\bar{n}}$ is a small remainder. The product $P_{n\leftarrow\kappa}R_{\kappa\leftarrow n}C_{n\leftarrow n}$ will be included in $A''_{\bar{n}\leftarrow n}$.

By the definition (23) of $R_{\kappa\leftarrow n}$, the calculations of the first n elements in $A''_{\bar{n}\leftarrow\bar{n}}\ \psi_{\bar{n}}{}^{old}$ are relatively inexpensive, since they require only these parts of the sequence for $C_{n\leftarrow n}$ which correspond to the central collocation points in the groups $G_m(P_j{}^1)$, $j=1,2,\ldots,\ \kappa=n/4^{m-1}$. The results are extended to the other collocation points by the prolongation operator $P_{n\leftarrow\kappa}$.

For the fast convergence of (32) it was enough to include into $A'_{\bar{n}\leftarrow\bar{n}}$ the influence coefficients for the pairs of the mutually most sensitive panels (such panels usually lie on the opposite sides of a thin body) and for the triads of the surface elements adjacent to the trailing edge (such as panels P_1^1 , P_{13}^1 and the wake strip $V_1{}^1$ in Fig. 2b). The corresponding triads of equations in (32) are solved independently and then the remaining pairs are dealt with (a similar method for the two-dimensional problem was presented in [13], [5, p. 326]). Several iterations (32) give the practically converged solution.

5. Results of the calculations

The methods closely related to those presented here were used for the case of the flow around a sphere and an elliptic wing.

We shall consider two kinds of errors in our results. The absolute error of a quantity (such as the potential φ, the pressure, etc.) will be defined as the maximum norm (i.e., the maximum of the absolute value) of the difference between the exact distribution of this quantity at the collocation points and the approximate distribution obtained with the aid of the single-level discretization. The relative error will be defined as the maximum norm of the difference between the distributions obtained with the aid of the single and the multi-level discretization at the same collocation points.

In the case of the problem (5) for a sphere the exact distributions of all the interesting quantities are given by analytic formulae. When the single-level discretization with 128 triangular panels was applied to this problem, the absolute error for the potential φ, defined in (2), was found to be equal to 0.0072. For the pressure $c_p=1-|V|^2$, where V is calculated from (10), the absolute error was 0.1388. For a two-level discretization method using 32 panels of the same system in the near field (i.e., 1/4 of the total number) and 24 panels of the coarser system in the far field, the relative errors for φ and c_p were equal to 0.0011 and 0.0045, respectively. These relative errors are much less than the absolute ones, and represent negligible parts of the ranges $-1/2\leqq\varphi\leqq 1/2$ and $-5/4\leqq c_p\leqq 1$ of φ and c_p on the sphere. When 16 panels of the fine system (1/8 of the total number) were used in the near field and 28 panels of the coarse system in the far field, the relative errors for φ turned out to be equal to 0.0069. For c_p the relative error was 0.021 and hence still much less than the absolute one. For the three-level method using 32 panels of the previous fine system $\hat{S}_3$ in the near field and 8+4 panels of the respective coarser systems $\hat{S}_2$ and $\hat{S}_1$ in the two parts of the far field, the relative errors for φ and c_p were found to be equal to 0.006 and 0.0132. In this case the very crude approximation of the hemispheres with 4 panels of the system $\hat{S}_1$ did not introduce any significant errors into the discrete solutions. For each collocation point the number of the influence coefficients was equal to $44\approx 128/3$.

As we have seen in Sec. 3, the multi-level scheme affects only the $O(h_m^2)$ part of the discretization error in the far field, resulting in the relative error $O(h_m^2)$ for σ, and hence also for φ, c_p, etc. Since the absolute error for σ (φ, c_p, etc.) is $O(h_m)$ the ratio of the relative and absolute errors will, in general, be less for the finer approximation of the surface in both fields.

The discrete distribution for φ and c_p, used for the above comparisons, were derived from the single-level forms of (2) and (10), respectively, in order to isolate the effect of the multi-level discretization in (16). The solution of the discrete system for a sphere was obtained by the classical multi-level method [5, pp. 309].

The calculations for the same number of 128 panels in the finest system were performed for the non-lifting flow around the elliptic wing with the axis ratio 1/5 and the symmetric NACA 0012 cross-section, at the angle of attack $\alpha=0^o$ (see Fig. 1). The absolute errors cannot be determined in this case, since the exact solution of the flow problem is not known. For the three-level discretization method, using 16 panels of $\hat{S}_3$, 12 panels of $\hat{S}_2$ and 4 panels of $\hat{S}_1$, the relative solution error for φ and c_p was found to be 0.0015 and 0.0043, respectively. For comparison, the approximate ranges of these quantities on the surface are $-0.56 \leq \varphi \leq 0.56$ and $-0.4 \leq c_p \leq 1$. The maximum value of the normal velocity component on the surface, obtained using the single-level form of (10), was found to be equal to 0.0023, which corresponds to about 0.1° error of the velocity direction. The number of the influence of coefficients for each collocation point was equal to 32=128/4.

For a finer sytem, containing 384 panels, the calculation for the lifting flow at $\alpha=5°$ was performed by the three-level procedure, using 32 panels of this system in the near field and 24 +16 panels of the two coarser systems in the far field. The system of the wake strips, containing 16 elements, was added. For each collocation point the number of panels used for the quadrature was equal to $72 \approx 384/5$. In practice, the collocation points were confined to a quarter of the wing surface, the symmetries of which have been taken into account. The error of the velocity direction on the surface was found to be of the order 1°. The results for the pressure distribution on a wing cross-section are presented in [6].

The discrete systems for the flows around the wing were solved by using certain variants of the method presented in Sec. 4. The details of the practical implementation and the comparisons of the convergence rates are given in [6]. As a rule, the two-level methods of this type give the result accurate to four decimal places after two or three iterations, the cost of which is comparable to the cost of 6 multiplications of a vector of n components by a matrix of $n \cdot n$ elements, where n is the number of the collocation points.

References

[1] A. Ålund, Iterative methods to compute the singularity distribution in three dimensional panel methods, SAAB-SCANIA Report L-O-1 R97, Linköping, 1984.

[2] P.M. Anselone, Collectively compact operator approximation theory and applications to integral equations, Prentice Hall, 1971.

[3] N.M. Günter, Die Potentialtheorie und ihre Anwendung auf Grundaufgaben der mathematischen Physik, Leipzig, 1957.

[4] W. Hackbusch, Die schnelle Auflösung der Fredholmschen Integralgleichung zweiter Art, Beiträge Numer. Math. 9, pp. 47-62, 1981.

[5] W. Hackbusch, Multigrid methods and applications, Springer, Berlin, 1985.

[6] W. Hackbusch, Z. Nowak, Multigrid methods for calculating

the lifting potential incompressible flows around three-dimensional bodies, to appear in: Multigrid Methods, Proceedings, Cologne 1985, W. Hackbusch and V. Trottenberg, eds., Lecture Notes in Mathematics.

[7] F.K. Hebeker, A boundary integral approach to compute the 3-D Oseen's flow past a moving body, Numerical Methods in Fluid Dynamics, M. Pandolfi, R. Piva, eds., pp. 124-130, Braunschweig 1984.

[8] P.W. Hemker, H. Schippers, Multiple grid methods for the solution of Fredholm integral equations of the second kind, Math. Comp., pp. 215-232, 1981.

[9] J.L. Hess, The problem of threee-dimensional lifting potential flow and its solution by means of surface singularity distribution, Computer Methods in Apllied Mechanics and Engineering 4, pp. 283-319, North Holland Publishing Company, 1974.

[10] B. Hunt, The mathematical basis and numerical principles of the boundary integral method for incompressible potential flow over 3-D aerodynamic configurations, Numerical Methods in Fluid Dynamics, pp. 49-105, B. Hunt, ed., Academic Press, 1980.

[11] F.T. Johnson, A general panel method for the analysis and design of arbitrary configurations in incompressible flows, NASA, CR 3079, 1980.

[12] Z. Nowak, Use of the multigrid method for Laplacian problems in three dimensions, in: Multigrid Methods, Proceedings, Cologne 1981, W. Hackbusch and V. Trottenberg, eds., Lecture Notes in Mathematics 960, pp. 576-598, Springer-Verlag, Berlin 1982.

[13] H. Schippers, Multigrid methods for equations of the 2nd kind with applications in fluid mechananics, Thesis, Amsterdam 1982.

[14] H. Wolff, Multiple grid methods for the calculation of potential flow around 3-D bodies, MC Report, Math. Centrum, Amsterdam 1981.

SOLUTIONS OF THE CONSERVATION EQUATIONS WITH THE APPROXIMATE FACTORIZATION METHOD

D. Hänel*, H. Henke*, A. Merten**

*Aerodynamisches Institut, RWTH Aachen, Germany

**URANIT GmbH, Jülich

SUMMARY

The approximate factorization method is a widely used method of solution for the conservation equations in fluid mechanics. This paper reviews applications of this method in the solution of the Navier-Stokes equations [1,2] of the transonic potential equation [3,4], and of the Euler equations for inviscid rotational flow [5,6].

INTRODUCTION

The principles of the factorization method and its application to flow problems was demonstrated in the past decade in a series of paper e.g. [7,8,9,10,11]. The use of this method in solutions of flow problems for different forms of the governing conservation equations, for example potential or Navier-Stokes equations, requires adaption of the method, and a lot of numerical experimentation to construct effective algorithms.

The purpose of this paper is to review applications and experiences gained with this method. In the first sections the governing equations, their discretization, and the method of solution will be outlined briefly. Thereafter numerical aspects, as, for example, the implementation of the boundary conditions and the formulation of the numerical damping will be discussed.

Finally results will be presented for transonic potential flow in inlets and over airfoils, for compressible, non-reactive, inviscid flow in piston engines by solutions of the Euler equations, and for viscous flow in rotating cylinders.

GOVERNING EQUATIONS

Plane or axisymmetric flow of a compressible fluid is described by the conservation equations for mass, momentum, and energy, which can be written in the following form:

$$U_t + F_x + G_y + S = 0. \tag{1}$$

Herein, U is the vector of conservative variables, F and G are the components of the flux vector, and S is a source term resulting from curvature terms in axisymmetric flow. For that case the coordinate y designates the radial coordinate.

The most general case considered here, is a compressible, viscous rotating flow. The vectors then have the following definition:

$$U = \begin{pmatrix} \varrho \\ \varrho u \\ \varrho v \\ \varrho w \\ E \end{pmatrix} \; ; \; F = \begin{pmatrix} \varrho u \\ \varrho u^2 + p - {}^1/_{Re}\,\tau_{xx} \\ \varrho uv - {}^1/_{Re}\,\tau_{xy} \\ \varrho uw - {}^1/_{Re}\,\tau_{x\theta} \\ u(E+p) - {}^1/_{Re}\,[u\tau_{xx} + \\ v\tau_{xy} + w\tau_{x\theta} + {}^1/_{Pr}\,q_x] \end{pmatrix}$$

$$G = \begin{pmatrix} \varrho v \\ \varrho uv - {}^1/_{Re}\,\tau_{xy} \\ \varrho v^2 + p - {}^1/_{Re}\,\tau_{yy} \\ \varrho vw - {}^1/_{Re}\,\tau_{y\theta} \\ v(E+p) - {}^1/_{Re}\,[u\tau_{xy} + \\ v\tau_{yy} + w\tau_{y\theta} + {}^1/_{Pr}\,q_y] \end{pmatrix} \; ; \; S = \frac{1}{y} G + \frac{1}{y} \begin{pmatrix} 0 \\ 0 \\ -p - \varrho w^2 + {}^1/_{Re}\,\tau_{\theta\theta} \\ \varrho vw - {}^1/_{Re}\,\tau_{\theta y} \\ 0 \end{pmatrix} . \qquad (2)$$

The definitions of the components of the stress tensor τ and of the heat flux vector $\vec{q}$ can be found in standard text books, e.g. in [12].

In the above equations all variables are made dimensionless with reference values for the density, ϱ_r, the speed of sound, a_r, and of a length L, so that only the Reynolds number Re and the Prandtl number Pr appear in the equations as:

$$Re = \frac{\varrho_r a_r L}{\mu_r} \quad , \quad Pr = \frac{c_{p_r} \mu_r}{\lambda_r} . \qquad (3)$$

All the other approximate forms of the conservation equations can be deduced from the system of equations (2). For plane flow the source term S and the circumferential velocity component w have to be set equal to zero.

The equations for inviscid flow (Euler equations) are given by the limit $Re \to \infty$.

The steady potential equation is given by the continuity equation in which the velocity is substituted by the gradient of a potential Φ, satisfying the irrotationality condition. The density can then be expressed by the isentropic relation

$$\varrho = [1 - \frac{\gamma - 1}{2} (\vec{v}^2 - u_\infty^2)]^{1/\gamma - 1} . \qquad (4)$$

The steady potential equation is solved by a time-like iteration procedure corresponding to an artificial time derivate $\partial / \partial t \; f(\Phi)$, so that the general form of the conservation equations given by Eq.(1), is also satisfied. The potential equation reads then

$$\frac{\partial f(\Phi)}{\partial t} + \frac{\partial}{\partial x} (\varrho \frac{\partial \Phi}{\partial x}) + \frac{\partial}{\partial y} (\varrho \frac{\partial \Phi}{\partial y}) = 0 \quad , \qquad (5)$$

with $f(\Phi) = \Phi$ for a symmetric factorization method (AF-1, [11]) and with $f(\Phi) = \vec{v}$ for upwind orientated factorization method AF-2, [11].

For numerical calculations of the flow along arbitrary shaped bodies surface-orientated curvilinear coordinate systems are used. It is advantageous to use such a system for a proper formulation of the boundary conditions.

Two ways are possible in order to achieve a correct, conservative discretization in the curvilinear space: Either one transformes the differential equations (1) into a new computational space, and discretices in the transformed plane, or one writes Eq.(1) in integral form by using Gauß' integral relation and discretizes the integrales for a finite volume in the Cartesian plane. In the literature the first method is very often designated as finite-difference form, and the second as the finite-volume form, but both methods are equivalent and result in the same discretization for a given computational grid. Both methods were used in previous investigations of the authors, but only the first will be described in the following, for the sake of brevity.

The transformation of Eq. (1) from the Cartesian plane (x,y,t) into a curvilinear plane $(\xi, \eta, \hat{t})$ uses the Jacobian J of the transformation

$$J = \frac{\partial(x,y,t)}{\partial(\xi,\eta,\hat{t})} \tag{6a}$$

and expressions for the derivatives, like

$$\frac{\partial f}{\partial x} = \frac{1}{J}\frac{\partial(f,y,t)}{\partial(\xi,\eta,\hat{t})} \tag{6b}$$

so that Eq. (1) can also be formulated in conservative form for curvilinear coordinates:

$$\hat{U}_{\hat{t}} + \hat{F}_{\xi} + \hat{G}_{\eta} + (S \cdot J) = 0 \,. \tag{7}$$

The fluxes $\hat{F}$ and $\hat{G}$ are now the contravariant components of the flux vector,

$$\hat{F} = F y_{\eta} - G x_{\eta} \quad , \quad \hat{G} = -F y_{\xi} + G x_{\xi} \tag{8}$$

and $\hat{U} = U \cdot J$ is the new dependent variable.

The discretization is carried out on a computational mesh, where the coordinates and the variables are defined on a nodal point (ξ_i, η_j) of the cells, the surfaces of which are located in the middle between the nodal points. The discretized equations (7) can then be written as:

$$\delta_t \hat{U}_{i,j} + \delta_{\xi} \hat{F}_{i,j} + \delta_{\eta} \hat{G}_{i,j} + (S \cdot J)_{i,j} = 0 \,. \tag{9}$$

The time derivative $\delta_t U$ will be defined later in the method of solution. The space derivatives δ_{ξ} and δ_{η} are defined as

$$\delta_{\xi} \hat{F}_{i,j} = \frac{1}{\Delta\xi}(\hat{F}_{i+1/2,j} - \hat{F}_{i-1/2,j}) \tag{10}$$

with $\hat{F}_{i+1/2,j} = \hat{F}(\hat{U}_{i+1/2,j})$

and $\hat{U}_{i+1/2,j} = \frac{1}{2}(\hat{U}_{i+1,j} + \hat{U}_{i,j})$.

This formulation leads to second-order accurate central differences for equidistant step sizes.

The metric coefficients e.g. x_ξ or x_η are discretized in similar fashion by finite differences in such away, that for reasons of consistency the metric differences satisfy:

$$\delta_\xi(\delta_\eta x)_{i,j} = \delta_\eta(\delta_\xi x)_{i,j} \; . \tag{11}$$

NUMERICAL METHOD

The numerical method of solution used in the studies is closed on the factorization method as proposed by Beam and Warming [7]. For the discretized conservation equation (9) the method can be written as a three step method:

$$(I + \Delta t\, \delta_\xi \hat{A}^n) \cdot \Delta U^x = -\Delta t [\delta_\xi \hat{F}^n + \delta_\eta \hat{G}^n + (S \cdot J)^n] \tag{12a}$$

$$(I + \Delta t\, \delta_\eta \hat{B} + \hat{C})\, \Delta \hat{U}^n = \Delta U^x \tag{12b}$$

$$(U \cdot J)^{n+1} = (U \cdot J)^n + \Delta \hat{U}^n \; , \tag{12c}$$

where $\Delta \hat{U}^n = (U \cdot J)^{n+1} - (U \cdot J)^n$

and $\hat{A} = \dfrac{\partial \hat{F}}{\partial \hat{U}} \; , \; \hat{B} = \dfrac{\partial \hat{G}}{\partial \hat{U}} \; , \; \hat{C} = \dfrac{\partial (S \cdot J)}{\partial \hat{U}} \; .$

The variables ΔU^* is an intermediate variable, defined by the factorization. The factorization error $\Delta t^2\, \delta_\xi\, \hat{A}^n(\delta_\eta\, \hat{B}+\hat{C})_2 \Delta\hat{U}^n$, is of order $O(\Delta t^2)$; it does therefore not influence the formal accuracy $O(\Delta t, \Delta x^2, \Delta y^2)$ of the scheme.

The linear stability analysis shows unconditional stability for this scheme, but the amplification factor approaches unity for large Courant numbers, as a result of the factorization error. The consequence is a decreasing rate of convergence with increasing time step, and in practice the Courant number is restricted to $O(10)$.

Higher-order approximations in time, $O(\Delta t^2)$, e.g. with 3-point backward or with Crank-Nicholson discretization [7], did not improve the time resolution significantly or did lead to instabilities. For that reason the first-order Euler backward discretization in time was used throughout all calculations.

The solution of the system of Eq. (12) leads to block-tridiagonal solution matrices, which are solved by the Gauß-elimination method. Since for large time steps the matrices are not diagonally dominant, pivoting of the matrix elements became necessary.

For the solution of the Euler equations, where $\hat{A}$ and $\hat{B}$ have real eigenvalues the time consuming block inversions can be avoided by an approximate diagonalization of the implicit operator of Eq. (12).

Defining $\hat{A} = T_\xi^{-1} \Lambda_\xi T_\xi$ and $\hat{B} = T_\eta^{-1} \Lambda_\eta T_\eta$

with T as the diagonal eigenvalue matrices, and assuming that the transformation matrices T_ξ and T_η are locally constant the algorithm (12) for Euler equations becomes

$$T_\xi^{-1} (I + \Delta t \, \delta_\xi \Lambda_\xi) T_\xi \, T_\eta^{-1} (I + \Delta t \, \delta_\eta \Lambda_\eta) \, T_\eta \Delta \hat{U}^n = RHS \, . \tag{13}$$

This diagonalized version, as proposed by Pulliam [9] , saves about 50 percent of the CPU time per time step in comparison to scheme (12), as test calculations have shown [13] . The disadvantage is that this version is not conservative in time and that a source term and viscous terms cannot be diagonalized in the same way. Explicit treatment of these terms is possible, but naturally leads to stability restrictions.

BOUNDARY CONDITIONS

In numerical flow calculations one has to distinguish between physical boundary conditions and boundary conditions for the variables defined for the numerical scheme.

The physical boundary conditions can be divided into three classes for most of the flow problems.

Periodic boundary conditions occur e.g. in cascade flow or through mesh cuts and are treated numerically by overlapping cells.

The boundary condition of a rigid wall is defined by the vanishing normal velocity $v_n = 0$. For viscous, no-slip flow all velocity components are set equal to zero. For inviscid flow the tangential velocity is unknown, but can be obtained by extrapolation from the interior values, as was done in the solution of the Euler equations [5,6] . It can also be obtained by solving the conservation equations in a half-volume at the wall, as it was done in potential flow calculations [3,4] . Both, inviscid and viscous flow computations require information about the pressure on the wall. It can be obtained from the normal momentum equation, e.g. for a boundary $\eta = \text{const.}$

$$p_n = J \sqrt{x_\xi^2 + y_\xi^2} \, [p_\eta (x_\xi^2 + y_\xi^2) - p_\xi (x_\xi x_\eta + y_\xi y_\eta)] \, , \tag{14}$$

where p_ξ and p_η has to be expressed by the corresponding non-conservative momentum equations. Eq. (14) is discretized by central differences in ξ and one-sided, two-point differences in η-direction.

For freestream, inflow or outflow boundaries, which in general are chosen far away from the body, local one-dimensional characteristics are used to update the boundary conditions. For subsonic flow, it follows that at the inflow boundary all quantities except one, and at the outflow boundary only one have to be prescribed. The remaining quantities are calculated from the data in the computational domain by zero-order extrapolation. For the prescribed variables, either the local Riemann invariants or combinations of these, as pressure or total enthalpy, can be used. Their choice depends on the flow problem considered.

All physical boundary conditions, as described before, are implemented in the explicit part of Eq. (12) or Eq. (13).

In addition to conditions for the physical variables U, boundary conditions become necessary for the numerical variables ΔU and ΔU^*, as defined by the implicit part of scheme (12) or (13).

For steady-state calculations, where these variables approach zero, the boundary conditions can in principle be chosen arbitrary without influencing the steady-state solution. A frequent choice of such artificial boundary conditions is to set $\Delta U = \Delta U^* = 0$, resulting in a simplified and flexible implicit algorithm. The error made by this assumption is of $O(\Delta t)$ and does not impair the formal accuracy of the whole scheme, Eq. (12) or Eq. (13). Therefore this assumption can be used also in real time calculations if the time step is chosen small enough [6]. Other choices of boundary conditions for ΔU and ΔU^* are possible e.g. [4], but some of them can decrease the rate of convergence significantly or can lead to instabilities, as was found in a study [4]. The same study [4] also showed that boundary conditions for ΔU and ΔU^*, derived in accord with the physical boundary conditions, yield the best rate of convergence. Such conditions can be derived by implementing the physical boundary conditions in the basic unfactored implicit scheme and by carrying out the time-linearization and the factorization in the same way as for inner points afterwords. Then the boundary conditions are included in the implicit part through the Jacobian $\hat{A}$ and $\hat{B}$ of the scheme (12), and are, therefore, consistent with the physical explicit conditions. They then result in a fully implicit algorithm for inner as well as for boundary points with a higher accuracy in time and improved rate of convergence. This type of boundary conditions was used in time dependent Navier-Stokes calculations [1,2], and in transonic potential flow calculation [3,4].

An example of the influence of the boundary conditions for the intermediate variable ΔU^* on the rate of convergence is given in Fig. 1, taken from [4]. The rms-value of the residual is plotted as function of the number of time steps for the solution of the linear diffusion equation with scheme Eq. (12), and for different boundary conditions for the intermediate variable ΔU^*. The fully implicit boundary conditions, BC3, as described before, shows the best rate of convergence, whereas artificial conditions BC1, BC2, and BC4, described in detail in [4], lead to partially significant reductions of the rate of convergence.

NUMERICAL DAMPING

The centrally discretized factorization scheme (12,13) is not or only weak dissipative in the numerical sense, if physical diffusion is absent or only weak in comparison to the inertial terms. As a consequence, the high-frequency error components, caused e. g. by the nonlinear convection terms, are not sufficiently damped out and lead to wiggled difference solutions, especially in regions with large gradients. Therefore artificial damping terms have to be added, preferably in such a way that high frequency components are sufficiently filtered, without impairing the accuracy of the solution. Such terms can be constructed with even-order derivatives, as proposed by Richtmyer and Morton [14]. For the factorization scheme such terms, as combinations of second and fourth terms are used in the explicit as in the implicit part of the scheme.

The influence of the different damping terms on the convergence and on the accuracy of the solution was studied within this framework in [13]. For smooth regions, best filtering is achieved with fourth order terms of the form

$$D^{(4)} = \varepsilon^{(4)} \Delta x^4 \frac{\partial^4 u}{\partial x^4} \tag{15}$$

added to the explicit and implicit part of the scheme (12) or (13). But for scheme (12), with block matrices, the compuational work in the implicit part is increased significantly by these terms. Therefore in the implicit part only second order-terms of the form

$$D^{(2)} = \varepsilon^{(2)} \Delta x^2 \frac{\partial^2 u}{\partial x^2} \tag{16}$$

are implemented. They do not change the tridiagonal matrix structure [7]. Unconditional stability is achieved if $\varepsilon^{(2)} \geq 2 \varepsilon^{(4)}$.

These damping formulations lead to sufficiently smooth numerical solutions in regions, where the gradients of the flow variables are not too large. Near shock discontinuities the noncharacteristic and large 5-point extension of the fourth-order terms generate oscillations of the solution. Jameson [15] therefore employed a nonlinear combination of fourth and second order terms, whereby the fourth-order term is switched off near the shock. This nonlinear damping applied in the factorization method, also leads to an improvement in the solution, as shown in Fig. 2a and 2b. In this figure, taken from [13], the pressure distribution along the walls is plotted for a transonic flow through a plane channel. The difference between the use of linear fourth-order damping, Fig. 2a, and of nonlinear damping, Fig. 2b, is clearly indicated.

The damping formulations, as discussed before, reveal the necessity to control the damping by the solution itself without any empirical parameters. A useful criterion for the derivation of such damping formulations is given by the development of difference schemes with TVD properties. Studies with TVD terms applied to the factorization method were published recently by Yee [16].

Another way to avoid the empirical damping formulations consists in substituting the central differences, Eq. (10), of the Euler terms by upwind differencing and flux vector splitting [17]. This results in a robust and second-order accurate, implicit algorithm for Euler and Navier-Stokes equation as first results [18] obtained by the authors, have shown.

APPLICATIONS

The factorization technique as briefly described before was applied in several investigations to various flow problems. The applications will be described in the following, and some of the results will be presented.

a) Solutions of the potential equations [3,4]

For solutions of the steady conservative potential equation the time-like relaxation equation (5) was used with two variants for the factorization method AF-1 and AF-2 [11]. For transonic flow upwindig of the difference approximation, Eq. (10), in supersonic regions was achieved with the artificial density concept [11]. Variable time steps were used in a sequence to accelerate the rate of convergence. This acceleration technique is described in [11].

The algorithm for the potential equation was formulated in curvilinear coordinates and applied to external and internal flows as well. As an example, Fig. 4a and 4b shows the mesh and the pressure distribution over a NACA 00 12 airfoil in transonic flow at $Ma_{\infty} = 0.8$. The results are compared with those described in [19].

The transonic flow through an inlet of a safety valve was studied in [3]. Fig. 3 shows the isomach lines in a plane inlet for transonic flow. The potential flow calculations were used together with boundary-layer calculations, in order to find out separation-free inlet curvatures.

b) Solution of the Euler Equations [5,6]

The solution algorithm for the Euler equations was formulated in curvilinear coordinates for plane and for axisymmetric flow with swirl. As method of solution the block (12) and the diagonalized (13) scheme were used in [5].

The method of solution was applied in a study of the vortical flow in a combustion chamber of a piston engine [6]. Fig. 5 shows a result of this investigation, namely the velocity vectors of the flow during the intake stroke. The inlet valves are simulated as circular rings and the outward moving piston has a curved bottom.

c) Solution of the Navier-Stokes Equation [1,2]

For the calculation of the compressible, viscous and axisymmetric flow in strongly rotating cylinders the complete Navier-Stokes equations were solved in a cylindrical coordinate system. Sufficient resolution of the boundary layers in the sidewalls and endcaps of the cylinder requires large mesh streching in axial and radial direction for this problem. Because of the strong radial density ratios which are of the order of magnitude of 10^6, the dependent variables were chosen as deviations from the state of solid body rotation. Despite of the resulting nonconservative form of the Navier-Stokes equations the factorized algorithm (12) could be applied in the same way as described before.

As an example of a time-dependent solution Fig. 6 shows the flow field in the transient phase of the compressible spin-up problem [2]. The calculation was started from solid body rotation which was disturbed by a sudden increase of the angular velocity, by which two counter rotating vortices were formed. The spin-up problem is a well suited test problem for the calculation of the more complex flow in a gascentrifuge for uranium enrichment. This flow is characterized by very high angular velocities, high density ratios and high Reynolds numbers. The secondary motion in the centrifuge is generated by differences in the angular velocity or temperature at the walls or by external injection. As an example of test flow in a centrifuge Fig. 7 shows the streamline plattern of the secondary flow motion [2].

CONCLUSIONS

The approximate factorization method was used to solve the conservation equation for a variety of flows. Applications range from potential flow to unsteady internal viscous flow. The advantage of the method is the flexibility and the simple matrix structure due to factorization. On the other hand the method has some severe disadvantages, as Courant number restrictions caused by the factorization error or the necessity of adding empirical damping terms. Future work is aimed at removing some of these disadvantages. Flux vector splitting or TVD schemes will be used and it can be expected that the efficiency of the implicit part of the method can be improved.

REFERENCES

[1] MERTEN, A., HÄNEL, D.: "Navier-Stokes solution for compressible flow in a rotating cylinder", in: H. Viviand (Ed.), Notes on Num. Fluid Mech., 5 (1982) pp. 197-206.

[2] MERTEN, A., HÄNEL, D.: "Implicit solution of the unsteady Navier-Stokes equations for the flow in a gascentrifuge", Proc. of the Sixth Workshop on Gases in Strong Rotation, Tokyo 1985.

[3] GIESE, U.: "Computation of inviscid transonic flow", Lecture Notes in Physics, 170 (1982) pp. 217-223.

[4] HÄNEL, D., GIESE, U.: "The influence of boundary conditions on the stability of approximate factorization methods", Notes on Num. Fluid Mech., 7 (1984) pp. 108-115.

[5] HENKE, H., HÄNEL, D.: "Numerical simulation of gas motion in piston engines", Lecture Notes on Physics, 218 (1985) pp. 267-271.

[6] HENKE, H., HÄNEL, D.: "Numerical simulation of vortex flow in a piston engine", Proc. of Int. Symp. on Diagnostics and Modeling of Combustion in Reciprocating Engines, Tokyo 1985.

[7] BEAM, R., WARMING, R. F.: "An implicit finite-difference algorithm for hyperbolic systems in conservation-law-form", Journal of Comp. Physics, 22 (1976) pp. 87-110.

[8] PULLIAM, T. H., STEGER, J. L.: "Implicit finite-difference simulations of three-dimensional compressible flow", AIAA Journal, 18 (1978) pp. 159-167.

[9] PULLIAM, T. H., CHAUSSEE, D. S.: "A diagonal form of an implicit approximate-factorization algorithm", Journal of Comp. Physics, 39 (1981) pp. 347-363.

[10] BALLHAUS, W. F., JAMESON, A., ALBERT, J.: "Implicit approximate-factorization schemes for steady transonic flow problems", AIAA Journal, 16 (1978) pp. 573-579.

[11] HOLST, T. L.: "Fast conservative algorithm for solving the transonic full potential equation", AIAA Journal, 18 (1980) pp. 1431-1439.

[12] BIRD, R. D., STEWART, W. E., LIGHTFOOT, E. N.: " Transport phenomena", New York 1960.

[13] HENKE, H., HÄNEL, D.: "Artificial damping in approximate factorization methods", to be published in: Notes on Num. Fluid Mech., (1986).

[14] RICHTMYER, R. D., MORTEN, K. W.: "Difference methods for initial value problems", New York 1967.

[15] JAMESON, A., SCHMIDT, W., TURKEL, E.: "Numerical solutions of the Euler equations by finite volume methods using Runge-Kutta Time-Stepping schemes", AIAA Paper, 81-1259 (1981).

[16] YEE, H. C.: "On symmetric and upwind TVD schemes", NASA TM 86842 (1985).

[17] VAN LEER, B.: "Flux-vector splitting for Euler equations", Lecture Notes in Physics, 170 (1982) pp. 507-512.

[18] SCHRÖDER, W., HÄNEL, D.: "A comparison of several multigrid methods for the solution of the Navier-Stokes equations", Proc. of 2nd Europ. Multigrid Conf., Köln 1985 (to be published in Lecture Notes on Mathematics, Springer Verlag, Berlin).

[19] RIZZI, A., VIVIAND, H.: "Numerical methods for the computation of inviscid transonic flows with shock waves", GAMM Workshop, Notes on Num. Fluid Mech., 3 (1981).

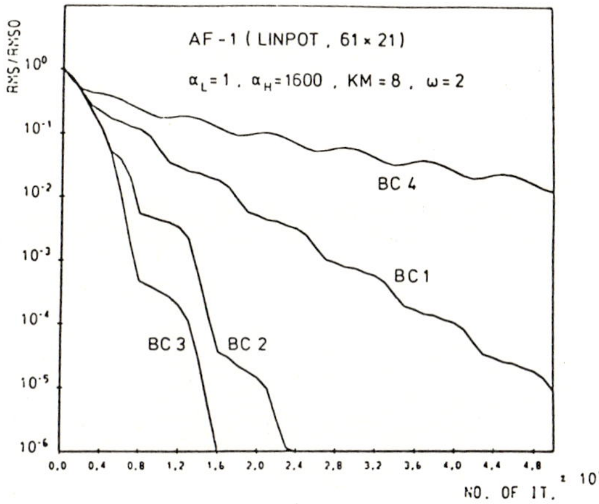

Fig. 1
Solution of the linear diffusion equation with the AF-1 method. Residual (rms) as function of the number of time steps for different types of boundary conditions for the intermediate variables in factorization methods (taken from [4]).

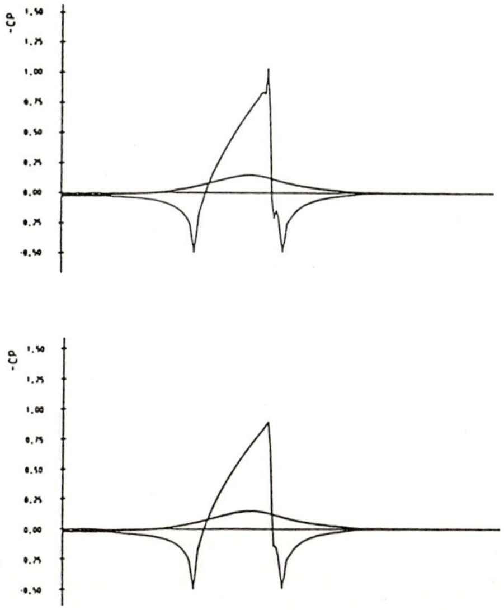

Fig. 2
Solution of the Euler equations with the diagonalized factorization method. Pressure distribution along the wall of a plane channel (as defined in [19])for transonic flow with $Ma_\infty = .837$
Influence of different numerical damping formulation (taken from [13])
a) linear fourth order damping
b) nonlinear second/fourth order damping

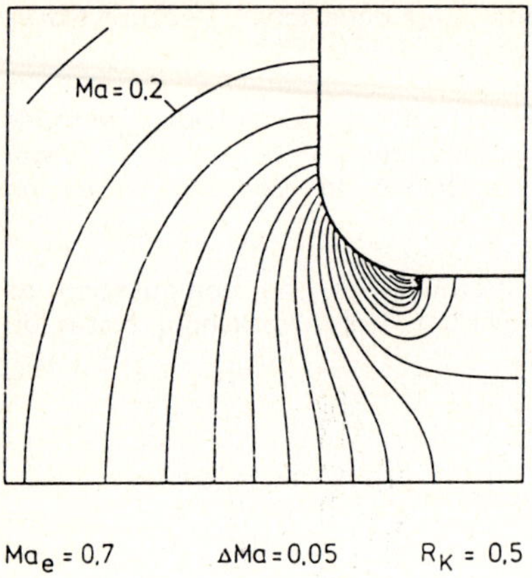

Fig. 3
Solution of the conservative potential equation
Isomach lines of the transonic flow through a plane inlet (taken from [3])

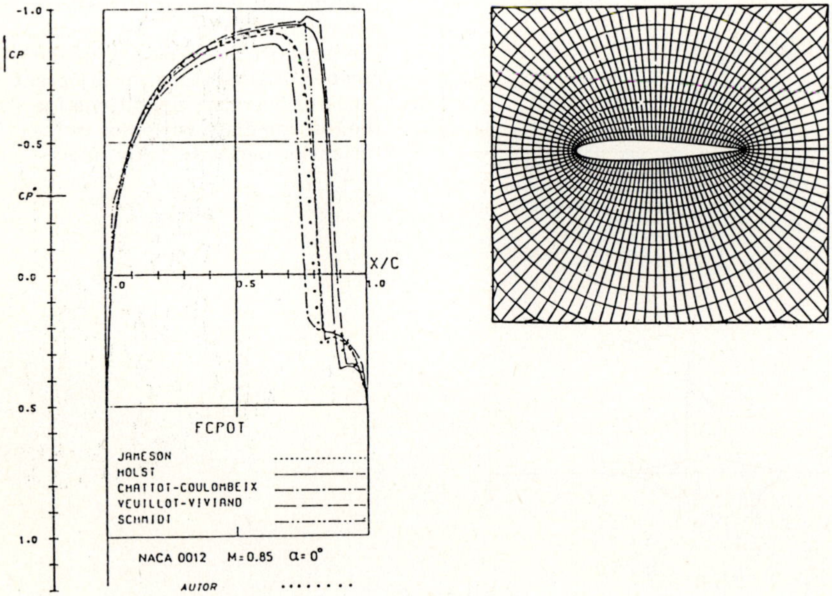

Fig. 4 Solution of the conservative potential equation
Mesh arrangement and pressure distribution for a NACA 0012 airflow ($Ma_{\infty} = 0.85$, $\alpha = 0^{\circ}$).

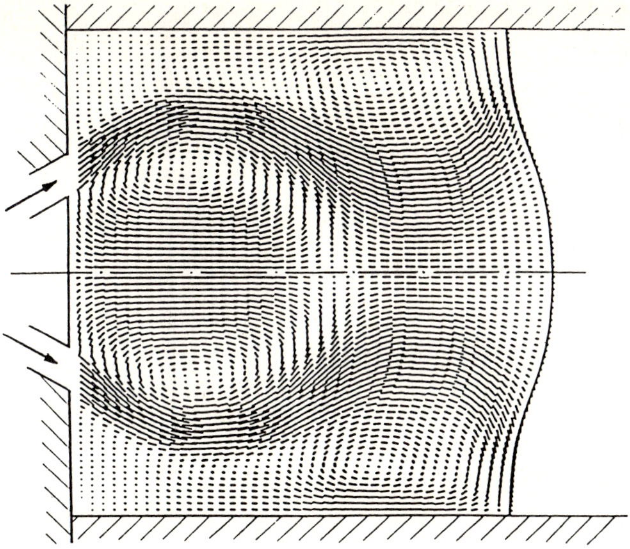

Fig. 5 Solution of the Euler equation
Velocity pattern of the vortical flow during the injection stroke of a piston engine.

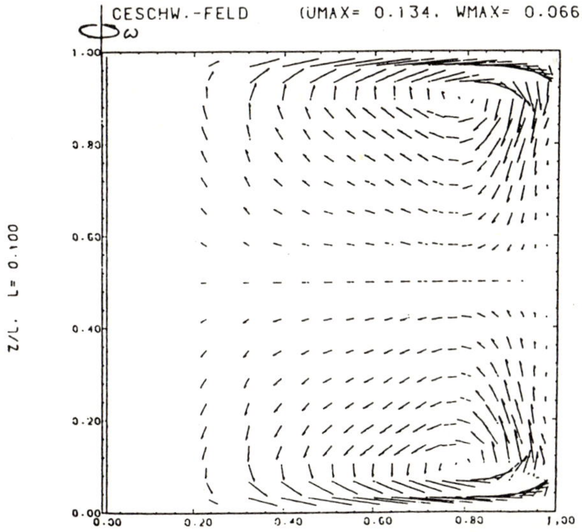

Fig. 6 Solution of the complete Navier-Stokes equations
Flow pattern of the secondary motion in a rotating cylinder for a sudden spin-up of the cylinder (taken from [1]).

$\Omega_o = 500\ sec^{-1}$ $\quad$ $Re = 5 \cdot 10^2$

$\Omega_1 = 525\ sec^{-1}$ $\quad$ $t = 0.025\ sec$

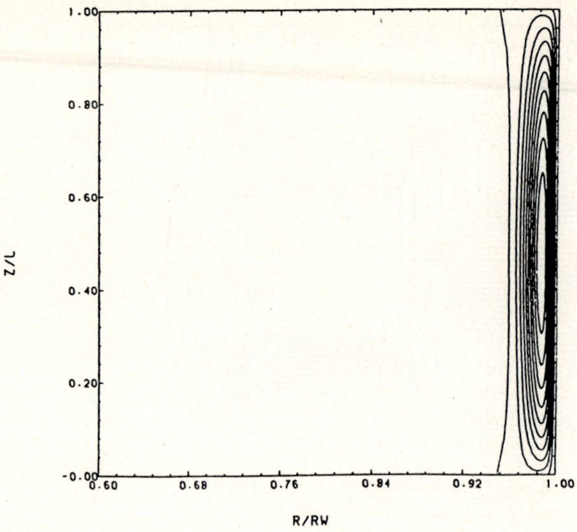

Fig. 7 Solution of the complete Navier-Stokes equations
Streamlines of the steady secondary motion in a gascentrifuge with thermal generation of the flow (taken from [2]).

$\Omega \cdot R = 600$ m/s
$Re = 5.47 \cdot 10^6$

INVISCID AND VISCOUS FLOW THROUGH ROTATING MERIDIONAL CONTOURS

T. Heiter, E. Steck, K.O. Felsch
Institut für Strömungslehre und Strömungsmaschinen
an der Universität Karlsruhe
Kaiserstrasse 12, D-75oo Karlsruhe 1, Germany

SUMMARY

A numerical procedure is presented for the calculation of the inviscid and viscous flow through rotating axially symmetrical casings with arbitrarily shaped meridional contours.
A rotating vaneless annular gap (the so-called meridional contour) serves as model for the investigations and can be regarded as a first step to determine the flow field in impellers of turbomachines.
The problem is solved by means of a finite difference method. By applying a nonorthogonal mesh which is based on a numerically generated coordinate transformation a wide range of possible applications is given.

PHYSICAL MODEL AND ASSUMPTIONS

On the left side Fig. 1 shows the meridional plane of a mixed-flow impeller of a pump. For our calculations we make the abstraction that there are no blades within the axially symmetrical calculation domain, which consists of an axially directed inlet region, a rotating mixed-flow casing and a radial diffuser as outlet region.
The flow field is axially symmetrical and stationary. The fluid is incompressible. We consider both an ideal fluid for the inviscid flow and a Newtonian fluid for the laminar flow.
The investigations are carried through in a transformed but nonrotating absolute coordinate system.

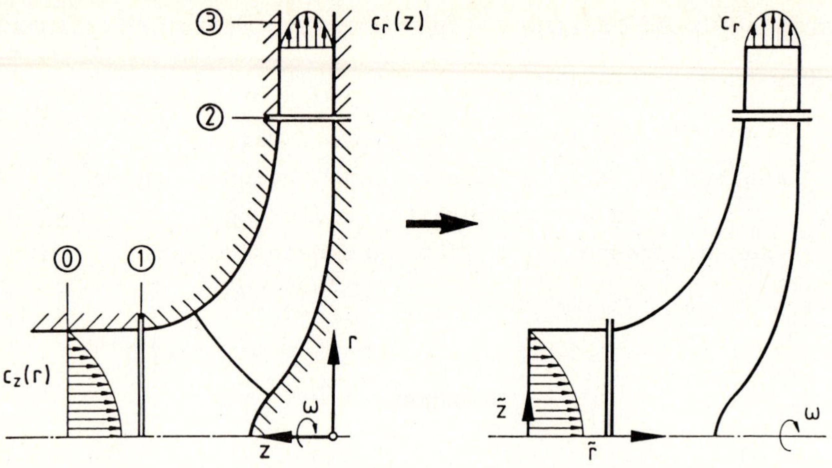

Fig. 1: Geometry of the Model

TRANSFORMATION OF THE COORDINATES

With the method of Thompson [1] it is possible to transform the axially symmetrical boundaries of the calculation domain to plaines of constant body-fitted coordinates $\tilde{R}$ and $\tilde{Z}$ respectively. (see fig. 2) Doing so one has to determine the transformation relations

$$\tilde{R} = \tilde{R}(R,Z) \qquad\qquad R = R(\tilde{R},\tilde{Z})$$
$$\text{respectively} \tag{1}$$
$$\tilde{Z} = \tilde{Z}(R,Z) \qquad\qquad Z = Z(\tilde{R},\tilde{Z})$$

as solution of two elliptic boundary value problems.

The Laplacian of the body-fitted coordinate system is

$$\underset{\tilde{R}\tilde{Z}}{\Delta} = \frac{1}{D^2}\left[(R_{\tilde{Z}}^2 + Z_{\tilde{Z}}^2)\frac{\partial^2}{\partial\tilde{R}^2} + (R_{\tilde{R}}^2 + Z_{\tilde{R}}^2)\frac{\partial^2}{\partial\tilde{Z}^2} - 2(R_{\tilde{R}}R_{\tilde{Z}} + Z_{\tilde{R}}Z_{\tilde{Z}})\frac{\partial^2}{\partial\tilde{R}\partial\tilde{Z}}\right] + F\frac{\partial}{\partial\tilde{R}} + H\frac{\partial}{\partial\tilde{Z}}, \tag{2}$$

wherein D denominates the determinant of the transformation coefficients

$$D = \frac{\partial(R,Z)}{\partial(\tilde{R},\tilde{Z})} = \frac{\partial R}{\partial \tilde{R}}\frac{\partial Z}{\partial \tilde{Z}} - \frac{\partial R}{\partial \tilde{Z}}\frac{\partial Z}{\partial \tilde{R}} = R_{\tilde{R}}\, Z_{\tilde{Z}} - R_{\tilde{Z}}\, Z_{\tilde{R}} \tag{3}$$

(For coordinates the subscripts denote partial derivatives).

Thompson called F and H forcing functions and showed that they can be chosen at random in a wide range, thus taking influence on the distribution of the lines of constant coordinates within the computational field.

The cylindrical and the surface-oriented coordinate system are linked together by means of the invariance of the Laplacian and formulate the two following elliptic differential equations:

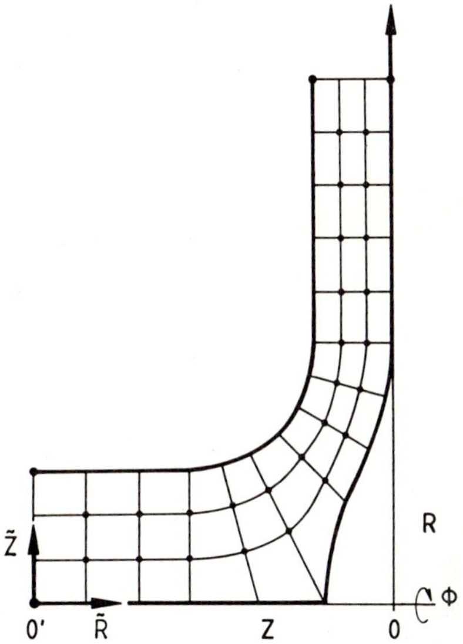

Fig. 2: Coordinate System

$$\underset{RZ}{\Delta} R = \underset{\tilde{R}\tilde{Z}}{\Delta} R = \frac{1}{R}$$

$$\underset{RZ}{\Delta} Z = \underset{\tilde{R}\tilde{Z}}{\Delta} Z = 0 . \tag{4}$$

The divergence of a vector $\vec{c}$ in surface-oriented coordinates is given by:

$$\underset{\tilde{R}\tilde{Z}}{\operatorname{div}} \vec{C} = \frac{1}{RD} \cdot \frac{\partial}{\partial \tilde{R}} \left(\frac{RD}{\sqrt{R_{\tilde{R}}^2 + Z_{\tilde{R}}^2}} \cdot C_{\tilde{R}}\right) + \frac{1}{R} \cdot \frac{\partial}{\partial \tilde{\phi}} C_{\tilde{\phi}}$$

$$+ \frac{1}{RD} \cdot \frac{\partial}{\partial \tilde{Z}} \left(\frac{RD}{\sqrt{R_{\tilde{Z}}^2 + Z_{\tilde{Z}}^2}} \cdot C_{\tilde{Z}}\right). \tag{5}$$

The equations (4) are solved with a finite difference scheme of second order consistency. The forcing functions are chosen to be zero.

The boundary conditions are determined through the distributions of the grid points on the boundaries of the computational domain.

INVISCID FLOW

As there are no blades within the rotationally symmetrical flow field ($\partial/\partial\varphi = 0$) the circumferential velocity c_φ is given by the law $r \cdot c_\varphi = \text{const}$ and can be superimposed to any solution of the through-flow problem.

The equation of continuity $\operatorname{div} \vec{c} = 0$ is identically satisfied through a stream function ψ with the definition

$$r \cdot c_z = \frac{\partial\psi}{\partial r}, \quad r \cdot c_r = -\frac{\partial\psi}{\partial z}. \tag{6}$$

Thus the condition for potential flow $\operatorname{rot} \vec{c} = 0$ yields the scalar Stokes equation

$$\Delta\psi - \frac{2}{r} \cdot \frac{\partial\psi}{\partial r} = 0 \tag{7}$$

or transformed to surface-oriented coordinates

$$\underset{\tilde{r}\tilde{z}}{\Delta} \psi - \frac{2}{r} \cdot \frac{1}{d} \cdot \left(z_{\tilde{z}} \frac{\partial\psi}{\partial\tilde{r}} - z_{\tilde{r}} \frac{\partial\psi}{\partial\tilde{z}}\right) = 0. \tag{8}$$

As boundary conditions constant values of the streamfunction ψ are used on the axis of rotation and on the hub and shroud con-

tours. On the inflow and outflow cross-sections constant values of the velocity are assumed.
The problem is solved by means of an implicit finite difference method of second order consistency using SOR.
Steck [2] has carried through a comparison between an exact solution of the frictionless rotationally symmetrical flow against a flat wall and the numerical solution. The agreement was very good.

LAMINAR FLOW

BASIC EQUATIONS

The basic equations for the laminar flow are the Navier-Stokes equation

$$\frac{d\vec{c}}{dt} = - \frac{1}{\rho} \cdot \nabla p + \nu \nabla^2 \vec{c} \tag{9}$$

and the continuity equation

$$\nabla \cdot \vec{c} = 0 \tag{1o}$$

which again is satisfied by a stream function ψ.
Due to the rotational symmetry the curl of the velocity vector field has the components

$$\nabla \times \vec{c} = \begin{bmatrix} -\frac{\partial c_\varphi}{\partial z} \\ \frac{\partial c_r}{\partial z} - \frac{\partial c_z}{\partial r} \\ \frac{1}{r}\frac{\partial}{\partial r} r c_\varphi \end{bmatrix} . \tag{11}$$

The azimuthal component is called vorticity function η

$$\eta = \frac{\partial c_r}{\partial z} - \frac{\partial c_z}{\partial r} . \tag{12}$$

The pressure derivatives in the Navier Stokes equations are eliminated by cross-differentiation. The azimuthal equation remains unchanged. The application of the definitions of ψ and η yields the

vorticity transport equation

$$\frac{\partial}{\partial r}(c_r \eta) + \frac{\partial}{\partial z}(c_z \eta) - \frac{1}{r}\frac{\partial c_\varphi^2}{\partial z} = \nu \left[\frac{\partial}{\partial r}\left(\frac{1}{r}\frac{\partial r\eta}{\partial r}\right) + \frac{\partial^2 \eta}{\partial z^2} \right] \tag{13}$$

azimuthal velocity equation

$$\frac{\partial}{\partial r}(c_r c_\varphi) + \frac{\partial}{\partial z}(c_z c_\varphi) + \frac{2}{r} c_r c_\varphi = \nu \left[\frac{\partial}{\partial r}\left(\frac{1}{r}\frac{\partial r c_\varphi}{\partial r}\right) + \frac{\partial^2 c_\varphi}{\partial z^2} \right] \tag{14}$$

Poisson equation

$$- r\eta = r \frac{\partial}{\partial r}\left(\frac{1}{r}\frac{\partial \psi}{\partial r}\right) + \frac{\partial^2 \psi}{\partial z^2} . \tag{15}$$

The system of equations is made dimensionless with the outlet radius of the rotating casing r_2 and the mean radial velocity c_{r2m} at this radius. The resulting dimensionless variables and characteristic numbers are:

$$\begin{aligned}
R &= \frac{r}{r_2} & Z &= \frac{z}{r_2} & \tilde{R} &= \frac{\tilde{r}}{r_2} & \tilde{Z} &= \frac{\tilde{z}}{r_2} \\
C_R &= \frac{c_r}{c_{r2m}} & C_Z &= \frac{c_z}{c_{r2m}} & C_{\tilde{R}} &= \frac{c_{\tilde{r}}}{c_{r2m}} & C_{\tilde{Z}} &= \frac{c_{\tilde{z}}}{c_{r2m}} \\
\Psi &= \frac{\psi}{r_2^2 c_{r2m}} & & & C_\phi &= \frac{c_\varphi}{c_{r2m}} & C_{\tilde{\phi}} &= \frac{c_{\tilde{\varphi}}}{c_{r2m}} \\
Re &= \frac{r_2 c_{r2m}}{\nu} & & & U &= \frac{\omega_o r_2}{c_{r2m}} & H &= \frac{\eta r_2}{c_{r2m}} .
\end{aligned} \tag{16}$$

Wherein Re is the Reynolds number and U is a characteristic number for the ratio of Coriolis and inertia forces.

The transformed basic equations write as follows

vorticity transport equation

$$\operatorname*{div}_{\tilde{R}\tilde{Z}}(\vec{C}H) = \frac{1}{Re}\left[\underset{\tilde{R}\tilde{Z}}{\Delta} H - \frac{1}{R^2}H\right] + \frac{H}{R}\left[\frac{R_{\tilde{R}}}{\sqrt{R_{\tilde{R}}^2+Z_{\tilde{R}}^2}} C_{\tilde{R}} + \frac{R_{\tilde{Z}}}{\sqrt{R_{\tilde{Z}}^2+Z_{\tilde{Z}}^2}} C_{\tilde{Z}}\right]$$

$$+ \frac{2C_{\tilde{\phi}}}{RD}\left[-R_{\tilde{Z}}\frac{\partial C_{\tilde{\phi}}}{\partial \tilde{R}} + R_{\tilde{R}}\frac{\partial C_{\tilde{\phi}}}{\partial \tilde{Z}}\right] \qquad (17)$$

azimuthal velocity equation

$$\operatorname*{div}_{RZ}(\vec{C}C_{\tilde{\phi}}) = \frac{1}{Re}\left[\underset{\tilde{R}\tilde{Z}}{\Delta} C_{\tilde{\phi}} - \frac{1}{R^2} C_{\tilde{\phi}}\right] \qquad (18)$$

$$- \frac{C_{\tilde{\phi}}}{R}\left[\frac{R_{\tilde{R}}}{\sqrt{R_{\tilde{R}}^2+Z_{\tilde{R}}^2}} C_{\tilde{R}} + \frac{R_{\tilde{Z}}}{\sqrt{R_{\tilde{Z}}^2+Z_{\tilde{Z}}^2}} C_{\tilde{Z}}\right]$$

Poisson equation

$$H = -\frac{1}{R}\left[\underset{\tilde{R}\tilde{Z}}{\Delta}\psi - \frac{2}{R}\frac{1}{D}(Z_{\tilde{Z}}\frac{\partial \psi}{\partial \tilde{R}} - Z_{\tilde{R}}\frac{\partial \psi}{\partial \tilde{Z}})\right]. \qquad (19)$$

BOUNDARY CONDITIONS

To solve the elliptic partial differential equations the boundary conditions have to be established on all boundaries of the computational domain.

Inflow Cross-Section

At the inflow cross-section an arbitrary velocity profile $C_{\tilde{R}}(\tilde{Z})$ is prescribed, which can be superimposed by any circumferential velocity $C_{\tilde{\phi}}(\tilde{Z})$. Let the normal component of the velocity $C_{\tilde{Z}}$ be zero then the stream function ψ and the vorticity η can immediately be calculated

$$C_{\tilde{R}} = C_{\tilde{R}}(\tilde{Z}) \qquad C_{\tilde{Z}} = 0$$

$$C_{\tilde{\phi}} = C_{\tilde{\phi}}(\tilde{Z}) \qquad \qquad (20)$$

$$\psi = \psi(\tilde{Z}) \qquad H = H(\tilde{Z}).$$

Shroud Contour

On the shroud contour the no-slip condition is used. The stream function ψ is zero and the vorticity η is calculated from the Poisson equation written for impermeable walls.

a) nonrotating parts:

$$C_{\tilde{R}} = C_{\tilde{Z}} = C_{\tilde{\phi}} = 0 \qquad \psi = 0 \qquad H = H_w \qquad (21)$$

b) rotating part:

$$C_{\tilde{\phi}} = U \cdot R. \qquad (22)$$

Axis of Rotation

On the axis of rotation both the normal velocity $C_{\tilde{Z}}$ and the circumferential velocity $C_{\tilde{\phi}}$ are zero. Due to the symmetry the vorticity is also zero. The velocity $C_{\tilde{R}}$ is determined by a limitation process according to l'Hospital's rule.

$$C_{\tilde{\phi}} = C_{\tilde{Z}} = 0 \qquad \psi = \psi_o \qquad H = 0$$

$$C_{\tilde{R}} = - \frac{\sqrt{R_{\tilde{R}}^2 + Z_{\tilde{R}}^2}}{R_{\tilde{Z}} D} \cdot \frac{\partial^2 \psi}{\partial \tilde{Z}^2}. \qquad (23)$$

Hub Contour

On the hub contour the no-slip condition is applied analogously. The difference between the stream function value on the shroud and the hub is a direct measure for the flow rate $\dot{V}$.

a) nonrotating part:

$$C_{\tilde{R}} = C_{\tilde{\phi}} = C_{\tilde{Z}} = 0 \qquad \psi = \psi_o \qquad H = H_W \qquad (24)$$

b) rotating part:

$$C_{\tilde{\phi}} = U \cdot R \ . \qquad (25)$$

Outflow Cross-Section

The outflow cross-section is located at a radial distance where the normal velocity component $C_{\tilde{Z}}$ can be assumed to be nearly zero. According to this condition $C_{\tilde{R}}$ is only a function of $\tilde{Z}$. For the last two grid lines $\tilde{R}$ = const conservation of the momentum is assumed to make the calculation of the azimuthal velocity possible. The vorticity function η is determined from the velocity profile on the outflow cross-section.

$$\begin{aligned} C_{\tilde{Z}} &= 0 \quad \rightarrow \quad \frac{\partial \psi}{\partial \tilde{R}} = 0 \\ C_{\tilde{R}} &= C_{\tilde{R}}(\tilde{Z}) \qquad C_{\tilde{\phi}} \cdot R = \text{const} \\ H &= H(\tilde{Z}) \ . \end{aligned} \qquad (26)$$

SOLUTION

The system of equations is solved by a finite difference scheme of approximately second order consistency, as the convective terms are treated with an upwind technique of the second kind described by Roache [3]. The Poisson equation and the vorticity transport equation are solved simultaneously on lines of constant coordinates $\tilde{R}$, using SOR. The calculations are based on a mesh consisting of 71 x 31 nodes.

CALCULATION OF THE PRESSURE DISTRIBUTION

After the solution of the stream function-vorticity equation system the pressure distribution is calculated from the modified Navier-Stokes equations. Introducing the total pressure of the secondary flow

$$p_t = p + \frac{\rho}{2}\left(c_r^2 + c_z^2\right) \tag{27}$$

the Navier-Stokes equations for the axial and the radial direction write as follows:

$$\begin{aligned} \frac{1}{\rho}\frac{\partial p_t}{\partial r} &= \frac{c_\varphi^2}{r} - c_z\eta + \nu\frac{\partial\eta}{\partial z} \\ \frac{1}{\rho}\frac{\partial p_t}{\partial z} &= c_r\eta - \nu\left[\frac{\partial\eta}{\partial r} + \frac{\eta}{r}\right]. \end{aligned} \tag{28}$$

With a dimensionless total pressure

$$P_t = \frac{p_t}{\rho\cdot c_{r2m}^2} \tag{29}$$

the coordinate transformation yields the two corresponding equations

$$\begin{aligned} \frac{\partial P_t}{\partial\tilde{R}} = {} & R_{\tilde{R}}\frac{C_{\tilde{\varphi}}^2}{R} - Z_{\tilde{R}}\frac{1}{Re}\frac{H}{R} - \psi_{\tilde{R}}\frac{H}{R} \\ & - \frac{1}{Re}\frac{1}{D}\left[\left(R_{\tilde{R}}R_{\tilde{Z}} + Z_{\tilde{R}}Z_{\tilde{Z}}\right)H_{\tilde{R}} - \left(R_{\tilde{R}}^2 + Z_{\tilde{R}}^2\right)H_{\tilde{Z}}\right] \end{aligned} \tag{3o}$$

$$\frac{\partial P_t}{\partial \tilde{R}} = R_{\tilde{R}} \frac{C_{\tilde{\Phi}}^2}{R} - Z_{\tilde{Z}} \frac{1}{Re} \frac{H}{R} - \psi_{\tilde{Z}} \frac{H}{R} - \frac{1}{Re} \frac{1}{D} \left[(R_{\tilde{Z}}^2 + Z_{\tilde{Z}}^2) H_{\tilde{R}} - (R_{\tilde{R}} R_{\tilde{Z}} + Z_{\tilde{R}} Z_{\tilde{Z}}) H_{\tilde{Z}} \right] . \qquad (31)$$

The integration is carried through along lines $\tilde{Z}$ = const in the middle of the flow field and then following lines $\tilde{R}$ = const in the direction of the impermeable boundaries. At the inflow cross-section a constant static pressure p is assumed. The boundary conditions for the stream function, the vorticity and the circumferential velocity are equal to those of the flow calculation.

RESULTS

The computational model allows to investigate the influence of the Reynolds number, the rotation and the other boundary conditions for a wide range of arbitrarily shaped meridional contours. Some of these variations are pointed out. Figure 3 shows streamlines for different Reynolds numbers. To distinguish plainly the effects of this characteristic number the contour is not rotating. An axial velocity profil belonging to the fully developed pipe flow is assumed in the entrance cross-section of the computational domain.
At the shroud contour for Re = 1oo a little region of reverse flow can be seen which is growing with increasing Reynolds number while the seperation point moves upstream. This effect is in accordance with the experience.
The variation of the axial velocity profile (see figure 4) results in a considerable decrease of reverse flow caused by the fact, that there is more flow energy in the region near the shroud contour now. The velocity profile with the distinct maximum near the wall is simulating the effects of a nozzle in the inflow cross-section.
For the next picture (figure 5) the hub geometry is changed. As the hub body is removed step by step the deceleration of the flow increases leading to flow separation at the end.

The influence of the number U characterizing the rotation is shown in figure 6 for a constant Reynolds number of Re = 3oo. Starting with U = 1, i.e. the circumferential velocity at the outlet of the rotating contour is equal to the mean through flow velocity, the rotation increases and finally for U = 4 the region of reverse flow has vanished.
To investigate the validity and accuracy of the model a comparison with an analytical solution is carried through. There is a solution of Homann [4] for the viscous flow against a flat plate. In figure 7 the analytical solution is drawn in with dashed lines. The Reynolds number is Re = 3oo, the fluid enters the control section on the left hand side. The streamlines of the numerical solution are plotted in full lines. For a mesh having 41 x 31 node points the agreement is good. Increasing refinement of the mesh yields to an exact overlapping of the solutions.

CONCLUDING REMARKS

Based on the comparison with the analytical solution reliable results for flows without separation can be expected by the model. Considering reverse flow, the tendencies are in good agreement with the experience.
The next step will be the calculation of the three dimensional inviscid flow before the laminar flow in three dimensions is going to be studied.

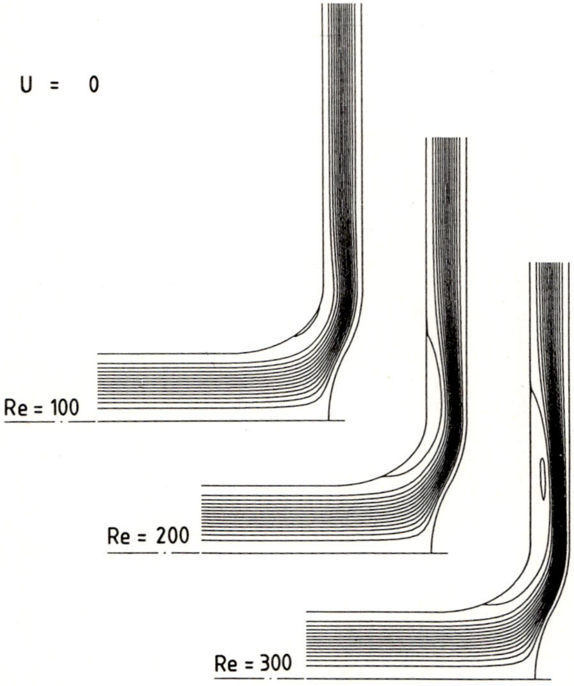

Fig. 3: Variation of the Reynolds number

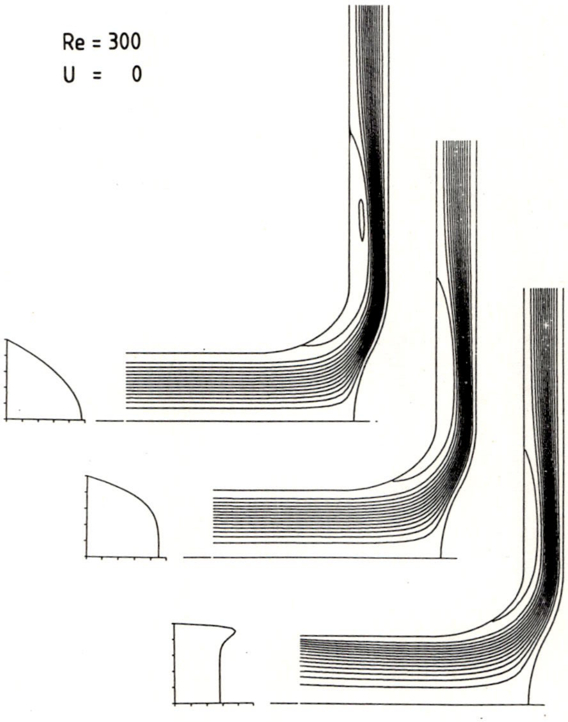

Fig. 4: Variation of the Velocity Profiles

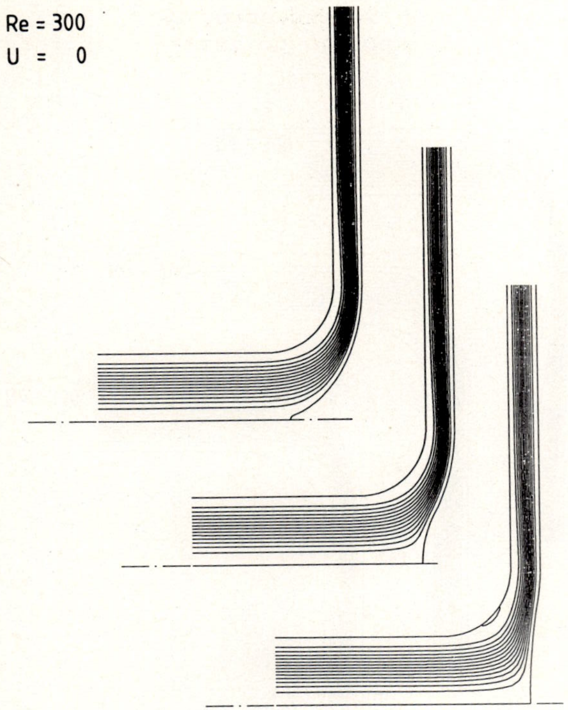

Fig. 5: Variation of the Hub Contour

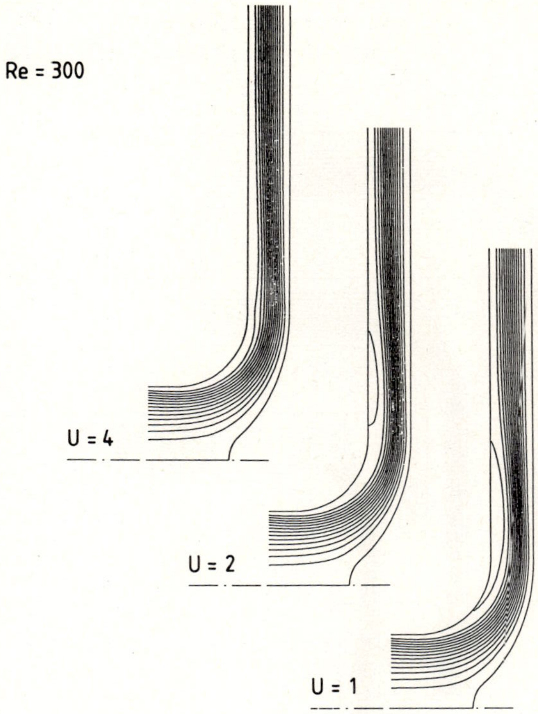

Fig. 6: Variation of the Rotation

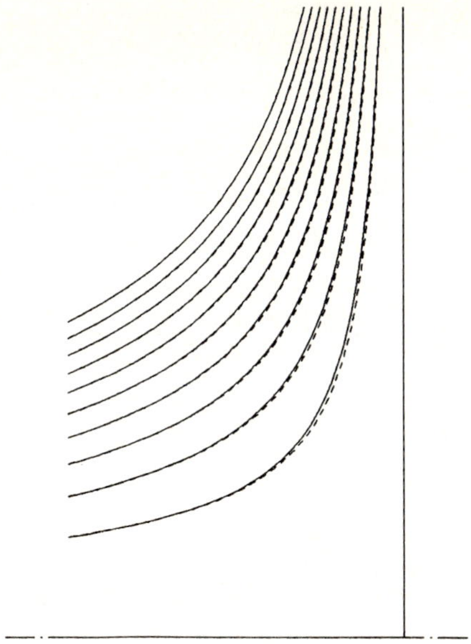

Fig. 7: Comparison to Analytical Solution

REFERENCES

[1] THOMPSON, J.F.: "Automatical Numerical Generation of Body-Fitted Curvilinear Coordinate System for Field Containing any Number of Arbitrary Two Dimensional Bodies", J. Comp. Phys. 15 (1974).

[2] STECK, E.: "Berechnung der reibungsfreien Strömung in rotationssymmetrischen Räumen mit krummflächigen Berandungen", ZAMM 64 (1984), Heft 4/5.

[3] ROACHE, P.J.: "Computational Fluid Dynamics", Albuquerque: Hermosa Publishers (1972).

[4] HOMANN, F.: "Der Einfluß großer Zähigkeit bei der Strömung um den Zylinder und um die Kugel", ZAMM 16 (Juni 1936), Heft 3.

ZONAL SOLUTIONS FOR VISCOUS FLOW PROBLEMS

E.H. Hirschel and M.A. Schmatz
Messerschmitt-Bölkow-Blohm GmbH
Postfach 801160, 8000 München 80, FRG

SUMMARY

The concept of zonal solutions for viscous flow problems, where local solutions of the Navier-Stokes equations are coupled with solutions of the Euler equations, and the boundary-layer equations, is outlined. The basic idea is to solve the Navier-Stokes equations only in those parts of the flowfield where strong interactions appear. The goal is to reduce the computation costs while improving at the same time the description of boundary-layer and separation phenomena. The governing equations, the computational grids, and the coupling procedures are presented for general steady three-dimensional viscous flows. Results are given from two-dimensional applications. Future developments are discussed.

1. INTRODUCTION

With present day computers it is not possible to compute the whole flowfield past realistic configurations in sufficient detail if viscous effects are to be included. Even with computers with much higher performance than today the costs would be prohibitively high in an industrial environment. It appears sensible therefore to search for ways to reduce the costs especially of solutions of the Navier-Stokes equations (see also [1]).

A possible Ansatz is to solve these equations only in those parts of the flowfield where strong interactions arise. In the other parts, where weak interaction prevails, appropriately only the Euler and the boundary-layer equations are solved, as is shown in Fig. 1.

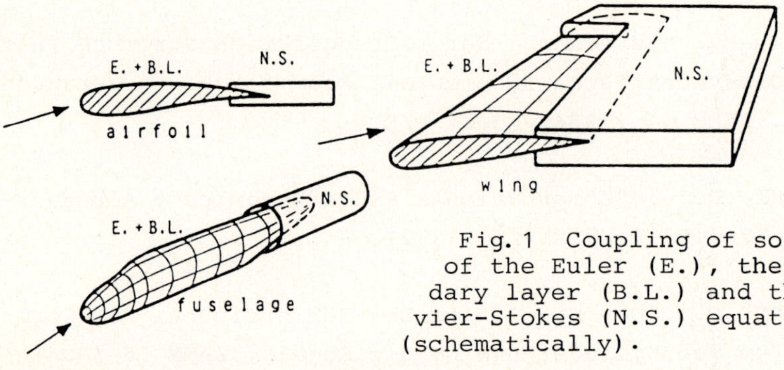

Fig. 1 Coupling of solutions of the Euler (E.), the boundary layer (B.L.) and the Navier-Stokes (N.S.) equations (schematically).

The expected benefits of such a procedure are the following:

- <u>Reduction of storage and of computation time.</u> In a Navier-Stokes solution today the boundary-layer region in direction normal to the surface (i.e. the boundary-layer stem) is resolved at most with approximately twenty cells. This is roughly one third to one fifth of the number used in (non-transformed) boundary-layer solutions. Each of these cells is active at every time step. An also time-stepping Euler solution would need only two to three cells in the boundary-layer stem, while a coupled space-marching boundary-layer solution is performed only after, say, every one hundred time steps (if explicit Euler and Navier-Stokes codes are used). Therefore the work performed by the Navier-Stokes solution per stem is six to seven times larger than that of the Euler and boundary-layer solution together. Thus, if the Navier-Stokes equations are solved only where necessary, i.e. where strong interactions arise, much can be gained with zonal solutions. Of course, the smaller the Navier-Stokes regimes are on a configuration, Fig. 1, compared to the Euler and boundary-layer regimes, the higher are the gains. The goal of the present work is to develop zonal solutions with only ten to twenty per cent more effort than that for a pure Euler solution.

- <u>Improvement of the solution for the boundary-layer region.</u> In those parts of the surface of the configuration, where the boundary-layer assumptions hold, the proper solution, the boundary-layer solution can be applied with an adequate number of cells per stem. With this solution it is possible to compute the development of laminar or turbulent boundary layers with sufficient accuracy to get wall-shear stress, wall-heat flux, and displacement properties. Considering the developments in boundary-layer control (passive and active laminarisation, turbulence management, etc.) it is evident that only boundary-layer solutions can give results accurate enough to apply stability and transition criteria, and the like. A side effect of zonal solutions is that also the Euler solution as well as the Navier-Stokes solution can be improved by using more cells if appropriate. In Fig. 2 it is shown how for the severe case of

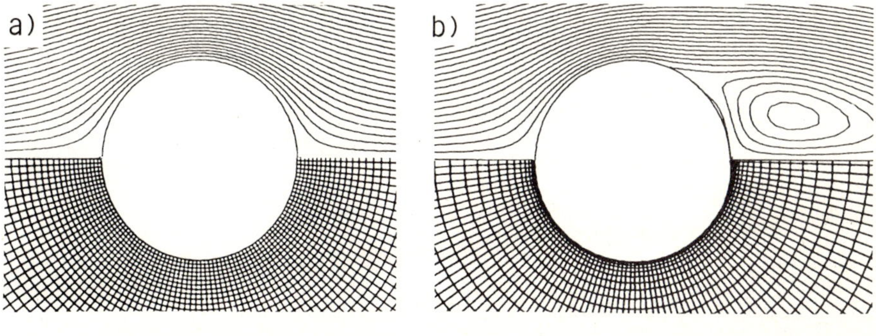

Fig. 2 Euler solution for flow past a circular cylinder [3], a) sufficiently fine and quadratic grid, b) grid too coarse: recirculation area appears (upper parts: streamlines, lower parts: grids).

an Euler solution (here method [2]) for the flow past a circular cylinder the results are influenced by the discretization [3].

- Avoidance of hydrodynamic instability phenomena. A solution of the Navier-Stokes equations for a laminar boundary layer with a discretization fine enough will allow the description of instability, transition, and turbulence at proper Reynolds numbers. Such a solution would belong to the class IV problems in the classification of [4], to the direct simulation of turbulence. Present day computers do not allow such computations on large scale. However, even in an unsufficiently resolved Navier-Stokes solution for the boundary-layer regime, unwanted instability phenomena might show up, which are avoided if the boundary-layer equations are solved instead.

Of course not every configuration is suited for the application of zonal solutions. If strong interaction regions are located too closely to each other, one should use a uniform approach. This will hold in general also for unsteady viscous flow problems. As is observed in [1], of course large eddy or direct simulation of turbulence has also to be tackled with a uniform approach.

The concept of zonal solutions for viscous-flow problems probably was used first in [5]. It appears that it has been applied thereafter only to particular cases (see [1] for references), notably to supersonic flow situations. Zonal solutions of other type, especially for block grids or moving grids with one type of equation, e.g. the Euler equations, or the Navier-Stokes equations are described for example in [6]. Zonal solutions of Navier-Stokes and Euler equations are used in many codes by simply switching off the viscous terms in the parts of the flowfield which are inviscid.

In the following a general concept of coupling boundary-layer, Euler and Navier-Stokes equations is presented. Initial applications to two-dimensional cases are reported.

2. GENERAL DESCRIPTION OF THE APPROACH

Methods for the computation of inviscid flows (Euler solutions), and for boundary layers have reached a high degree of applicability even in design aerodynamics (references relevant for the present work are [2,7,8]). Navier-Stokes codes are available,too, for instance [9,10], so that a combination of such codes in a zonal solution can be attempted.

Consider now for simplicity the two-dimensional flow past an airfoil which has been treated by the second author [11] with the present zonal solution, Fig. 3.

In most parts of the flowfield weak interaction occurs between the boundary layer and the inviscid flow. Only at the shock, perhaps behind the shock, and at the trailing edge regions of strong interaction exist. In Fig. 4 the zones of app-

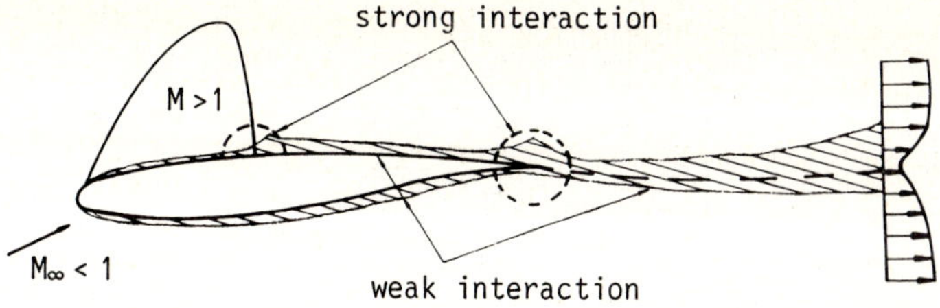

Fig. 3 Regions of weak and strong interaction on an airfoil (schematically) [11].

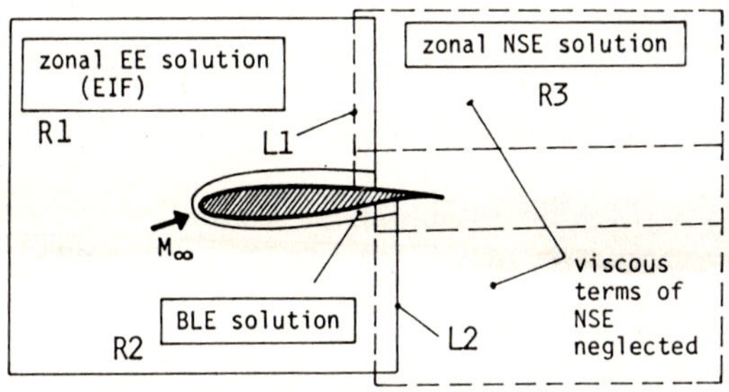

Fig. 4 Schematic of solution zones (BLE = boundary-layer equations, EE = Euler equations, NSE = Navier-Stokes equations, EIF = equivalent inviscid flow) [11].

lication of the different solutions are given schematically.

The appropriate grids are shown in Fig. 5. Grid a) for the Euler solution is a typical C-type grid with approximately 3 cells in the boundary-layer region at the rear part of the airfoil. Grid b) is similar to grid a), but has a much finer spacing (approx. 20 cells) near the wall and, of course, also in the wake region. The two grids overlap in five cells each. Ideally the grid should capture the wake in such a sense that the wake centerline is a gridline, and the other family of gridlines lies orthogonal to it. Then a boundary-layer turbulence model can be used directly in the wake. The situation is similar on any other configuration, especially on finite span lifting wings. However, the role of turbulence with regard to the lateral wake shear, which carries the induced drag [12], is not clear. Probably, as any inviscid lifting wing theory shows, it does not play a large role.

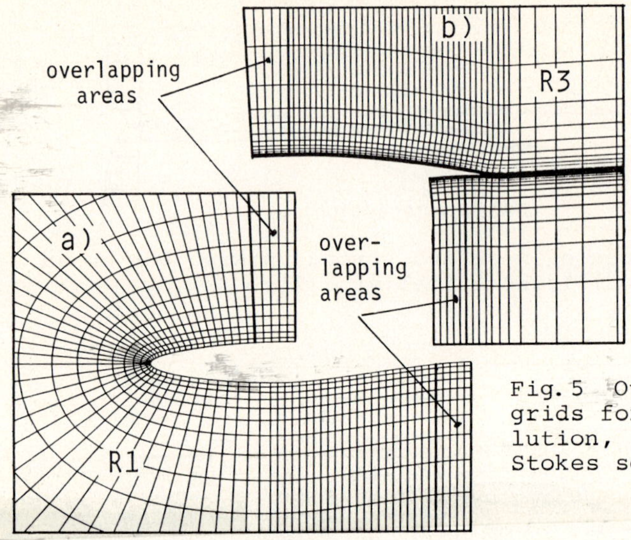

Fig. 5 Overlapping grids for a) Euler solution, b) Navier-Stokes solution.

The solution now proceeds in the following way (see also Fig. 4):

1. Computation of the Euler flow (EE) on the whole configuration, areas R1, R2 and R3, with N1 time steps.
2. Computation of the boundary layers (BLE) in R2, computation of the equivalent inviscid source distribution [13].
3. Computation of the Euler flow (EE) in R1 and R2 with the equivalent inviscid source distribution (weak interaction) in order to get the equivalent inviscid flow (EIF) [14], N2 time steps.
4. Computation of BLE in R2, equivalent inviscid source distribution.
5. Definition of the boundary values on L1 (see chapter 4).
6. Computation of the Navier-Stokes flow (NSE) in R3 with N3 time steps. Note that outside the region with viscous effects the viscous terms are neglected in NSE in [9].
7. Definition of the boundary values on L2.
8. Repetition of the preceeding steps beginning with step 3, until convergence is reached.

With the methods used in [11]: Euler solution [2], Navier-Stokes solution [9], boundary-layer solution [15], the following number of time steps are typical for a lifting transonic airfoil case without separation: $N1 \approx 400$, $N2 \approx 350$, $N3 \approx 150$, with fifteen coupling cycles, until the asymptotic steady state is reached.

3. EQUATIONS AND SOLUTION METHODS

The general governing equations are the conservative Reynolds-averaged Navier-Stokes equations for three-dimensional compressible flow in general non-orthogonal curvilinear coordinates $x^i = x^i(x^{j'})$, $i,j = 1,2,3$:

$$\frac{\partial}{\partial t}[q\sqrt{g}] + [(\underline{q}\underline{v}+\underset{\sim}{b})\sqrt{g}\,\underline{g}^i]_{,i} = 0 \;, \qquad (1)$$

where $\underline{q} = (\rho,\rho\underline{v},e)^T$, $e = \frac{p}{\gamma-1} + \frac{\rho}{2}(\underline{v})^2$,

$$\underset{\sim}{b} = (\underline{b}_\rho, \underline{\underline{b}}_m, \underline{b}_e)^T, \quad \underline{b}_\rho = 0, \quad \underline{\underline{b}}_m = p\underline{\underline{T}} + \underline{\underline{\tau}},$$

$$\underline{b}_e = -k\,\mathrm{grad}\,T + p\underline{v} + \underline{\underline{\tau}}\underline{v}, \quad \underline{\underline{\tau}} = -\lambda\,\mathrm{div}\,\underline{v}\,\underline{\underline{I}} - \mu[(\mathrm{grad}\,\underline{v}) + (\mathrm{grad}\,\underline{v})^T] \;.$$

The Einstein summation convention is used in eq.(1), with $[\;]_{,i} = \partial[\;]/\partial x^i$. $x^{j'}$ are Cartesian reference coordinates, $\sqrt{g}$ is the inverse of the Jacobian of the transformation relations, $\sqrt{g}\,\underline{g}^i$ the contravariant surface normal of the surfaces x^i = constant. The letters $\rho, p, T, \mu, \lambda, k$ denote density, pressure, temperature, coefficients of viscosity and heat conductivity coefficient, respectively, $\underline{v}$ is the velocity vector with Cartesian components. Details of the formulations can be found in [9] and [13]. Effective transport coefficients are introduced with the Boussinesq Ansatz, the algebraic two-layer eddy-viscosity model of [16] for instance can be used for closure of the equations.

In [11] method [9] was used as two-dimensional code. Presently a three-dimensional code basing on method [2] with an implicit time step is under development.

The Euler equations are found from eq.(1) by neglecting all diffusive terms:

$$\underline{\underline{b}}_m = p\,\underline{\underline{I}}\,, \quad \underline{b}_e = p\underline{v}\,, \quad \underline{\underline{\tau}} = 0 \;. \qquad (2)$$

In [11] method [2] was used in a two-dimensional version. Presently an implicit time step is introduced into method [2], too.

The first-order boundary-layer equations in locally monoclinic non-orthogonal coordinates $x^\alpha = x^\alpha(x^{j'})$, $\alpha = 1,2$ [13] read:

$$(k_{01}\,\rho\,v^i)_{,i} = 0 \;, \qquad (3)$$

$$\rho\,[v^i v^\alpha_{,i} + k_{\alpha 1}(v^1)^2 + k_{\alpha 2}v^1 v^2 + k_{\alpha 3}(v^2)^2] =$$

$$= k_{\alpha 6}p_{,1} + k_{\alpha 7}p_{,2} + (\mu\,v^\alpha_{,3})_{,3} \;, \qquad (4)$$

$$c_p\rho\;v^i T_{,i} = Pr^{-1}_{ref}(k\,T_{,3})_{,3} + E_{ref}\{v^\alpha p_{,\alpha} +$$

$$+\mu\,[k_{41}(v^1_{,3})^2 + k_{42}v^1_{,3}\,v^2_{,3} + k_{43}(v^2_{,3})^2]\} \;. \qquad (5)$$

These are the continuity equation, the boundary-layer equations and the energy equation. v^i are the contravariant components of the velocity vector in the x^α-coordinates, k_{mn}, $k_{\alpha n}$ are metric factors (see [13]), c_p, Pr_{ref}, E_{ref} are the specific heat at constant pressure, the Prandtl number, and the Eckert number at reference conditions, respectively.

The equivalent inviscid source distribution can be computed once the boundary-layer profiles are known:

$$(k_{01}\,\rho_o v_o^3)_{inv.} = [k_{01}\,\rho_e\, v_e^\alpha\, \delta_{1x}(\alpha)]_{,\alpha} \quad , \tag{6}$$

with the "components" of the displacement thickness,

$$\delta_{1\,x^\alpha} = \int_o^\delta (1 - \frac{\rho\, v^\alpha}{\rho_e\, v_e^{(\alpha)}}\, dx^3 \quad , \tag{7}$$

where x^3 is the coordinate normal to the surface [13], and δ the boundary-layer thickness. In [11] the boundary-layer method [15] was used, presently a finite-difference code is developed for the general three-dimensional case.

The present zonal solution will finally be made in a closely coupled form. Euler and Navier-Stokes solution are integrated in one code where relation (2) is used to switch from the Navier-Stokes solution to the Euler solution. The latter of course will be made on a coarse grid in the boundary-layer region. The boundary-layer equations are applied as before. This closely coupled zonal solution simplifies the coupling procedure (chapter 4), and will improve the performance of the solution.

4. COUPLING PROCEDURES

In a zonal solution for viscous flows the following coupling procedures are necessary (see Fig. 4):

1a. Euler solution → boundary-layer solution,

1b. boundary-layer solution → Euler solution,

2a. boundary-layer solution → Navier-Stokes solution (downstream coupling),

2b. Navier-Stokes solution → Euler solution in the viscous region (upstream coupling),

3a. Euler solution → inviscid part of Navier-Stokes solution (downstream coupling),

3b. inviscid part of Navier-Stokes solution → Euler solution (upstream coupling).

Coupling 1a. is made in the natural way of boundary-layer theory. The weak interaction 1b. is achieved via eq. (6). Coupling 2a. concerns the inflow boundary L1 (Fig. 4). There the boundary conditions for the Navier-Stokes solution are simply the boundary-layer profiles. The mass flow added to the equivalent inviscid flow (EIF) is removed in this way. Coupling 2b., which yields the outflow condition on L2 in the viscous region for the equivalent inviscid flow solution is more complicated. On L2 the velocity field from outside the boundary-layer region is extrapolated towards the wall. The pressure in this region is taken directly from the Navier-Stokes solution. The density is found with the condition of locally constant entropy in direction normal to the wall. Coupling 3a. and 3b. in the inviscid region is made with characteristic boundary conditions because of the time-transient solution approach used. Details are given in [11,17].

A problem occurs in coupling 2a. In first-order boundary-layer theory the tangential flow profiles have a discontinuity in the first derivative at the boundary-layer edge, Fig. 6a).

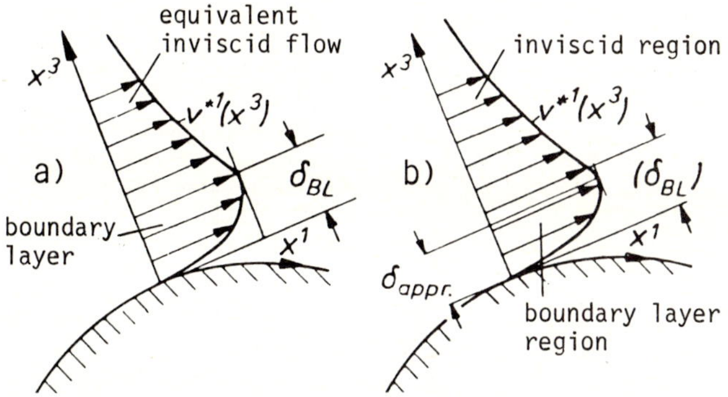

Fig. 6 Profiles of the tangential velocity component v^{*1} of a two-dimensional flow past a curved surface: a) classical boundary-layer approximation, b) real flow situation (to be predicted by Navier-Stokes solution) [13].

In reality the transition is smooth, Fig. 6b). The defect is very small and is negligible for very high Reynolds numbers. However, with a boundary-layer solution of higher order, a correct description of the outer-edge flow of the boundary layer will be possible also at lower Reynolds numbers (see also chapter 6).

In [11] the couplings 2a. - 3b. are made with an overlapping technique, Fig. 7. This technique was developed originally for

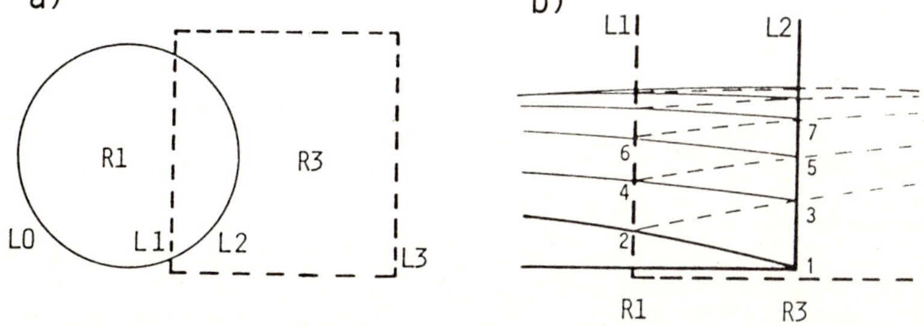

Fig. 7 Overlapping technique [18], a) computational subdomains, b) convergence history (schematically).

elliptic boundary-value problems [18]. It allows in an iterative way a fast build up of the solutions in both R1 and R3. In a closely coupled solution the overlapping technique will be used for the couplings 3a. and 3b. only.

The coupling region, of course, must always lie upstream of strong-interaction areas. Usually the first boundary-layer solution gives an indication where to place it.

RESULTS

The present zonal solution was applied to several two-dimensional test cases in order to validate the concept. In the following results of an application to the RAE2822 airfoil are given. The transonic case with $M_\infty = 0.73$, $Re = 6.5 \cdot 10^6$ at an angle of attack $\alpha = 3.19°$ was considered (see also [11]). Transition was inforced on both the suction and the pressure side at $x/c = 0.03$. This flow was investigated experimentally in [19]. It has served as test case for many authors. It must be noted, however, that the computations are made with a corrected angle of attack $\alpha_C = 2.79°$, taking into account the finite channel width. It has been confirmed in the present study by changing the position of the upper and lower bounds of the computational area that this correction is justified.

In Fig. 8 the grid in the neighbourhood of the airfoil is shown (see also Fig. 5). In R1 77 x 24 cells were used, and in R3 98 x 32. The overlapping was made with 5 cells in stream-

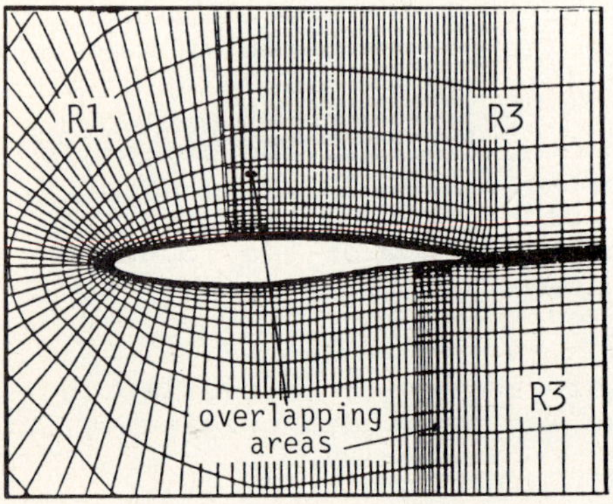

Fig. 8 Part of overlapping grids for RAE2822 [11].

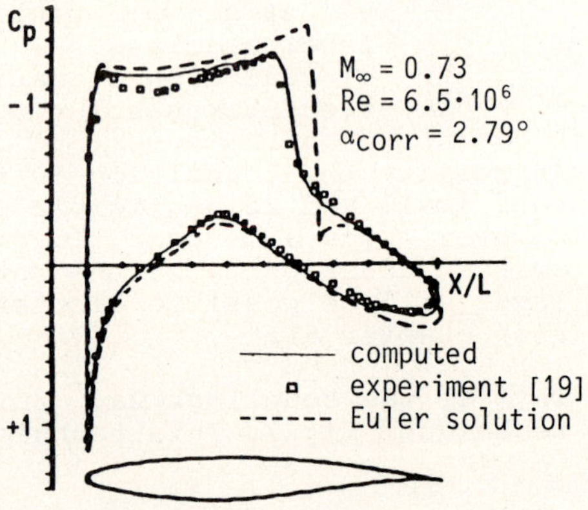

Fig. 9 Computed and measured pressure distribution, RAE2822 [11].

wise direction. The computed pressure distribution, Fig. 9, compares well with the measured except for the small suction peak of the experiment. This is probably due to an insufficient resolution near the nose of the airfoil, and is observed also in other solutions (see e.g. [20]). Also the inforced transition in the solution as well as in the experiment may have an influence.

In Figs. 10 and 11 the iso-Mach lines and the isobars are given. It is seen that away from the surface discrepancies appear in the overlapping region. In a closely coupled solution these discrepancies will disappear. The lines in the Navier-

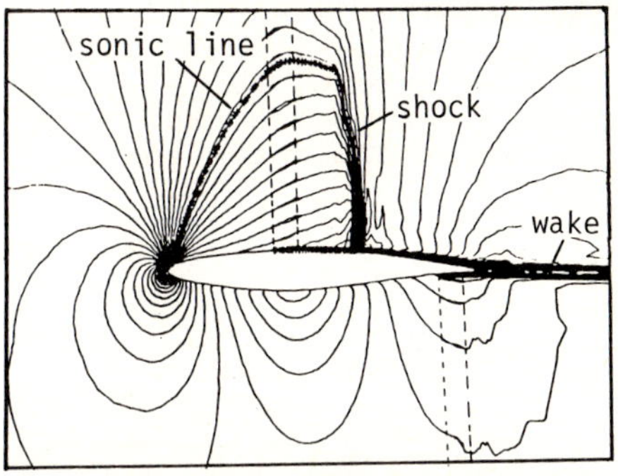

Fig. 10 Computed iso-Mach lines, RAE2822.

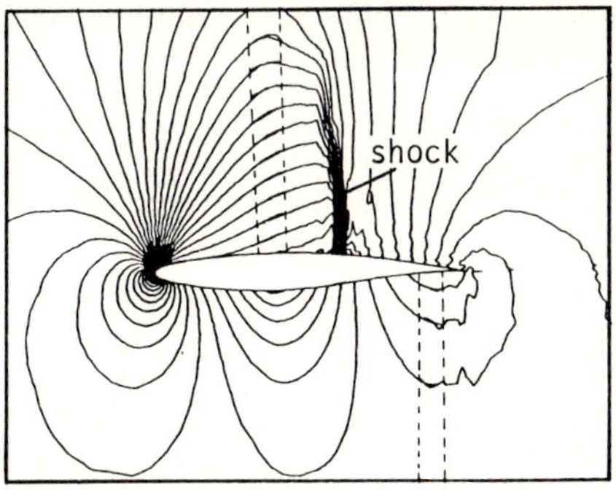

Fig. 11 Computed isobars, RAE2822.

Stokes part (R3) exhibit the wiggles typical for centered finite-volume schemes.

The vector plot of the velocity field near the trailing edge of the airfoil, Fig. 12, contains part of the Navier-Stokes

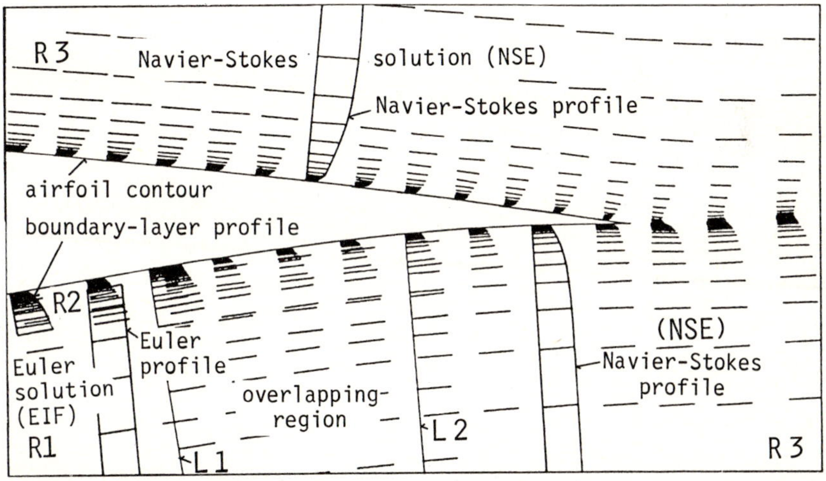

Fig. 12 Vector plot of velocity field near the trailing edge of RAE2822.

solution on the suction, and on the pressure side, part of the Euler solution (EIF) on the pressure side, the boundary-layer solution in this part, and the overlapping region. Note that the grids are not totally conform in this region, and note also the few Euler cells in the boundary-layer region.

A comparison of the computed and the measured skin-friction coefficients on the suction side is shown in Fig. 13. There is no smooth blending of the computed data in the overlapping region. Here the coarse discretization for the Navier-Stokes solution in the boundary-layer region may play a role. It is also possible that the coupling procedures 2a. and 2b. give rise to the discrepancies. The small recirculation area at $x/c \approx 0.6$ is observed for this case in almost any Navier-Stokes solution. Here the turbulence model may play a role, but also other factors, like unsufficient resolution or the shock-capturing. A finer resolution of the experimental data, on the other hand, might change the picture, too.

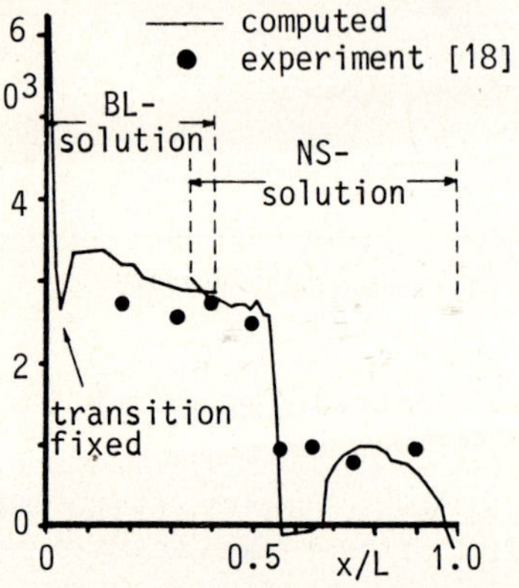

Fig. 13 Computed and measured skin-friction coefficient on suction side, RAE2822.

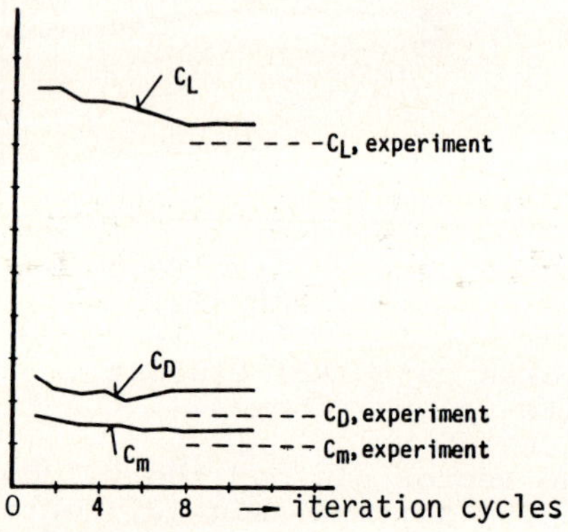

Fig. 14 Convergence history, RAE2822.

In Fig. 14 finally the convergence history is given. After approximately eight iteration cycles the force and moment coefficients don't change anymore. The experimental data are approached within the usual limits. Initially 1900 explicit time steps were made with the Euler solution [2] until the shock had reached approximately its final position. Then in each iteration cycle the Navier-Stokes solution [9] was performed with 150 implicit time steps, the Euler solution with 300. The boundary-layer solution was made once in each iteration cycle.

CONCLUSIONS

The above results show that the concept of the present zonal solution is sound. All applications sofar gave results well comparable with that of uniform Navier-Stokes solutions. Certain weakness will be overcome by a closely coupled solution with an implicit time step. The zonal solution of course has its real field of application in three-dimensional problems. The first steps are made presently to tackle such problems.

For two reasons it appears necessary, at least for three-dimensional viscous fuselage flow problems, to use higher-order boundary-layer theory in the weak-interaction region. One can be called the inverse Mangler effect, which occurs in fuselage base regions. The surface of the configuration contracts there and the boundary-layer thickness grows much faster as it would with the same adverse pressure gradient on a non-contracting surface. This can also be observed on upswept fuselage rear parts, where the still attached boundary layer might become very thick (see e.g. [21]). The other reason is that many verification experiments are made at relatively low Reynolds numbers which also leads to large boundary-layer thicknesses compared to the surface curvature radii of the configuration (fuselage, swept-wing leading edges, etc.).

A higher order boundary-layer solution like [22] takes both into account the centrifugal force effects on the pressure distribution in the boundary-layer, and the deviations from the surface metric for thick boundary layers, especially in lateral direction. A positive side effect is the smooth blending into the external inviscid flow, see Fig. 6b. Of course, curvature effects must and can be included to a certain degree in the turbulence model.

REFERENCES

[1] SHANG, J.S.: "An Assessment of Numerical Solutions of the Compressible Navier-Stokes Equations", J. of Aircraft, Vol. 22, No. 5 (1985), pp. 353-370.

[2] EBERLE, A.: "3-D Euler Calculations Using Characteristic Flux Extrapolation", AIAA-Paper 85-0119 (1985).

[3] KRUKOW, G.: "Untersuchung des diffusiven Drehungstransportes und der Auswirkungen geometrischer Singularitäten auf Lösungen der Euler'schen Bewegungsgleichungen", Diploma Thesis, Technical University München, MBB/LKE122/S/PUB/218 (1985).

[4] CHAPMAN, D.R.: "Computational Aerodynamics, Development and Outlook", AIAA Journal, Vol. 17, No. 12 (1979), pp. 1293-1313.

[5] BRILEY, W.R.: "A Numerical Study of Laminar Separation Bubbles Using the Navier-Stokes Equations", J. Fluid Mechanics, Vol. 47 (1971), pp. 713-736.

[6] RAI, M.M.: "Navier-Stokes Simulations of Rotor-Stator Interaction Using Patched and Overlaid Grids", AIAA Paper 85-1519 (1985).

[7] EBERLE, A., SCHÄFER, O.: "High-Order Characteristic Flux Averaging for the Solution of the Euler Equations", Proc. 6th GAMM-Conference on Numerical Methods in Fluid Mechanics, D. Rues, W. Kordulla (eds.), Notes on Numerical Fluid Mechanics, Vol. 13, Vieweg, Braunschweig-Wiesbaden (1986), pp. 78-85.

[8] HIRSCHEL, E.H.: "Computation of Three-Dimensional Boundary Layers on Fuselages", J. of Aircraft, Vol. 21, No. 1 (1984), pp. 23-29.

[9] KORDULLA, W., MacCORMACK, R.W.: "Transonic-Flow Computation Using an Explicit-Implicit Method", Proc. 8th Internat. Conf. on Numerical Methods in Fluid Dynamics, E. Krause (ed.), Lecture Notes in Physics, Vol. 170, Springer, Berlin-Heidelberg-New York (1982), pp. 420-426.

[10] KORDULLA, W.: "The Computation of Three-Dimensional Transonic Flows With an Explicit/Implicit Method", Proc. 5th GAMM-Conference on Numerical Methods in Fluid Mechanics, M. Pandolfi, R. Piva (eds.), Notes on Numerical Fluid Mechanics, Vol. 7, Vieweg, Braunschweig-Wiesbaden (1984), pp. 193-202.

[11] SCHMATZ, M.A.: "Calculation of Strong Interactions on Airfoils by Zonal Solutions of the Navier-Stokes Equations", Proc. 6th GAMM-Conference on Numerical Methods in Fluid Mechanics, D. Rues, W. Kordulla (eds.), Notes on Numerical Fluid Mechanics, Vol. 13, Vieweg, Braunschweig-Wiesbaden (1986), pp. 335-342.

[12] HIRSCHEL, E.H.: "On the Creation of Vorticity and Entropy in Solutions of the Euler Equations for Lifting Wings", MBB-LKE122-AERO-MT-716 (1985).

[13] HIRSCHEL, E.H., KORDULLA, W.: "Shear Flow in Surface-Oriented Coordinates", Notes on Numerical Fluid Mechanics, Vol. 4, Vieweg, Braunschweig-Wiesbaden (1981).

[14] LOCK, R.C., FIRMIN, M.C.P.: "Survey of Techniques for Estimating Viscous Effects in External Aerodynamics", in: "Numerical Methods in Aeronautical Fluid Dynamics", P.L. Roe (ed.), Academic Press, London-New York-Paris (1982), pp. 337-430.

[15] HERRING, H.J., MELLOR, G.L.: "Computer Program for Calculating Laminar and Turbulent Boundary-Layer Development in Compressible Flow", NASA CR-2068 (1972).

[16] BALDWIN, B.S., LOMAX, H.: "Thin Layer Approximation and Algebraic Model for Separated Turbulent Flows", AIAA-Paper 78-257 (1978).

[17] SCHMATZ, M.A.: "Berechnung der zähen Wechselwirkungen in Flügelströmungen mit Hilfe lokaler Lösungen der Navier-Stokesschen Gleichungen", Doctoral Thesis in preparation.

[18] SCHWARZ, H.A.: "Über einen Grenzübergang durch alternierendes Verfahren", Gesammelte Mathematische Abhandlungen, Vol. 2, Springer Verlag, Berlin (1890), pp. 133-143.

[19] COOK, P.H., McDONALD, M.A., FIRMIN, M.C.P.: "Aerofoil RAE2822 - Pressure Distributions, and Boundary Layer and Wake Measurements". Experimental Data Base for Computer Program Assessment, AGARD-AR-138 (1979), pp. A6-1 to A6-77.

[20] KORDULLA, W.: "Experiences With an Unfactored Implicit Predictor-Korrektor Method", Proc. 6th GAMM-Conference on Numerical Methods in Fluid Mechanics, D.Rues, W.Kordulla (eds.), Notes on Numerical Fluid Mechanics, Vol. 13, Vieweg, Braunschweig-Wiesbaden (1986), pp. 185-192.

[21] HIRSCHEL, E.H.: "Three-Dimensional Boundary-Layer Calculations in Design Aerodynamics", Proc. IUTAM-Symp. on Three-Dimensional Turbulent Boundary Layers, H.H. Fernholz, E. Krause (eds.), Springer, Berlin-Heidelberg-New York (1982), pp. 353-365.

[22] MONNOYER, F.: "Second-Order Three-Dimensional Boundary Layers", Proc. 6th GAMM-Conference on Numerical Methods in Fluid Mechanics, D. Rues, W. Kordulla (eds.), Notes on Numerical Fluid Mechanics, Vol. 13, Vieweg, Braunschweig-Wiesbaden (1986), pp. 263-270.

NAVIER - STOKES COMPUTATION OF TWO DIMENSIONAL LAMINAR WAKES OF A CIRCULAR CYLINDER IN CHANNEL FLOWS

P. Kiehm, N. K. Mitra and M. Fiebig
Institut für Thermo- und Fluiddynamik
Ruhr - Universität Bochum
4630 Bochum 1, Fed. Rep. Germany

ABSTRACT

The two dimensional unsteady Navier - Stokes and continuity equations with primitive variables are solved for the incompressible laminar fluid flow behind a circular cylinder confined between two solid walls. With a modified version of the Marker-and-Cell (MAC) technique the influence of blockage ratio and inlet profile on the wake flow are studied at Reynolds numbers in the range of 50 to 500.
The flow investigations show that the periodicity of the wake becomes apparent at higher Reynolds numbers with increasing blockage ratio, defined by cylinder diameter D through channel width B, e.g. at Re = 53 for D/B = 0.2, and at Re = 170 for D/B = 0.5, when a parabolic inlet profile at the channel entrance is prescribed. The channel walls damp the periodicity and a parabolic flow profile will evolve far downstream. Calculations were performed with a parabolic and a uniform channel inlet profile for channel blockage ratios of D/B = 0.2, 0.33 and 0.5. Computed wake length for D/B = 0.2 compare well with available experimental results. Computations of friction factor and pressure loss along the channel wall in front of the cylinder and in the near wake are strongly dependent on the velocity profile at the channel inlet.

INTRODUCTION

Although vortex shedding in the wake of a circular cylinder in an infinite medium has often been numerically investigated in two dimensions [1,2,3] the effects of confining walls on the wake has received relatively little attention. In crossflow heat exchangers flows around a circular cylinder inside a channel occur. Hence, such studies are useful in understanding the characteristics of plate - fin heat exchangers.
Numerical and experimental investigations have shown that the vortex shedding behind a circular cylinder in an unbounded medium starts when the Reynolds number exceeds 40, and the wake flow becomes periodic [4].
However, if the cylinder is confined in a channel, the channel walls will damp the periodic vortical flow.
The purpose of this work is to study numerically the influence of channel walls on the wake flow behind a circular cylinder and to investigate the effect of the cylinder on the friction factor and the pressure loss along the walls of the channel for different blockage ratios and Reynolds numbers.

COMPUTATIONAL DOMAIN

Figure 1 shows the computational domain. The calculation area consists of two solid walls with no-slip conditions and a circular cylinder between these walls.

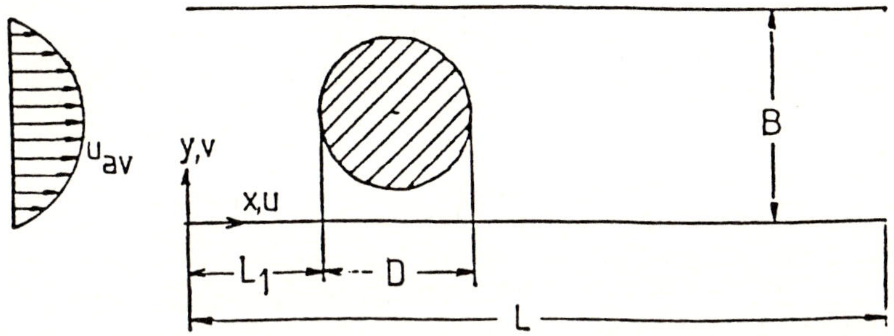

Fig. 1 : Configuration definition for 2-D channel flow

The flowfield around a circular cylinder in a channel is characterized by the following parameters (Fig. 1):

- the Reynolds number
- the blockage ratio D/B
- the position of the cylinder in the channel L_1/L and
- the velocity profile at the channel inlet.

These parameters determine the structure of the vortex street in the wake.

BASIC EQUATIONS

The basic equations to be solved for the calculation of the incompressible laminar fluid flow are the time-dependent Navier - Stokes and the continuity equations with primitive variables. In nondimensional conservative form these equations read as

$$\text{I.)} \quad \frac{\partial u}{\partial t} + \frac{\partial u^2}{\partial x} + \frac{\partial uv}{\partial y} = - \frac{\partial p}{\partial x} + \frac{1}{Re} \left(\frac{\partial^2 u}{\partial x^2} + \frac{\partial^2 u}{\partial y^2} \right),$$

$$\text{II.)} \quad \frac{\partial v}{\partial t} + \frac{\partial uv}{\partial x} + \frac{\partial v^2}{\partial y} = - \frac{\partial p}{\partial y} + \frac{1}{Re} \left(\frac{\partial^2 v}{\partial x^2} + \frac{\partial^2 v}{\partial y^2} \right), \tag{1}$$

$$\bar{D} = \frac{\partial u}{\partial x} + \frac{\partial v}{\partial y} = 0. \tag{2}$$

The Reynolds number is defined by $Re = (u_{av} \cdot D)/\nu$, where u_{av} stands for the average velocity at the channel inlet and D denotes the cylinder diameter.

METHOD OF SOLUTION

A modified version of the Marker and Cell (MAC) method is employed to obtain numerical solutions of the flow equations [5,6]. In the numerical scheme the computational domain of Fig. 1 is divided into a nonequidistant cartesian mesh, combined with a polar grid in the region of the cylinder (Fig. 2).

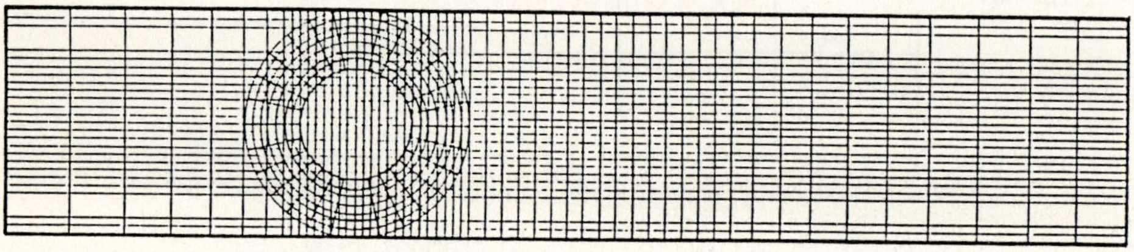

Fig. 2 : Mesh grid in the computational domain

The velocity components are defined at the cell faces to which they are normal and the pressures are determined at the cell centers (Fig. 3). The main advantage of the MAC - method is that pressure values on solid surfaces are not needed. The use of primitive variables is preferable when an extension to 3-D flow calculations is needed [7,8].

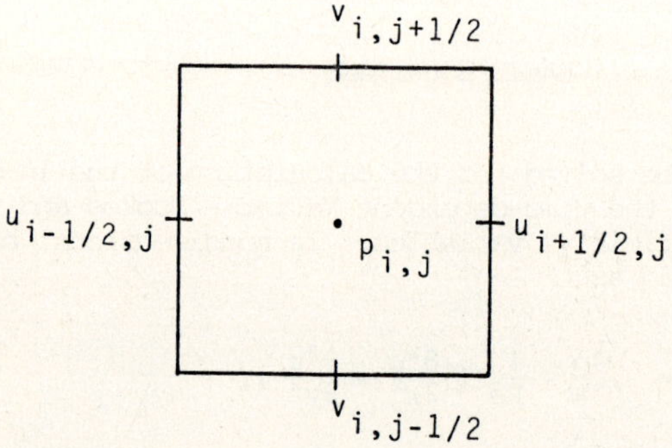

Fig. 3 : Determination of the velocity and pressure location on a staggered grid

The velocity components in r, θ and z direction on the polar grid are interpolated from the cartesian values using a interpolation method of Launder and Massey [9].
Calculations were first performed only on the cartesian grid, until a stationary or periodic solution is obtained. Second order accurate bounda-

ry conditions are used to guarantee a good description of the cylinder on the cartesian mesh.
In this way it is possible to define solid surfaces not only along the cell faces, but also to describe solid surfaces, which cut through a cell. A detailed description of this method will be reported elsewhere [10].
Then the variables u and v are transformed on the outer ring of the polar mesh, and are taken as Dirichlet - boundary conditions. Then a new calculation of the complete Navier - Stokes equations in polar coordinates is performed on the polar grid. The use of the overlapping grid has the advantage that gradients on the cylinder surface can be accurately calculated.
The MAC method is a semi - implicit time dependent scheme in which the solution is obtained in two steps. First the velocity components are all advanced by a time step δt using the previous state of flow to calculate the accelerations caused by convection, viscous stresses and pressure gradients. For this purpose eq. (1), discretized with a second order upwind scheme is used. This explicit time advancement will not necessarily lead to a velocity field with zero mass divergence in each cell. So in the second step adjustments of pressure and velocity must be made to ensure mass conservation. This is done by calculating the velocity divergence $\bar{D}$ from eq. (2) for each cell. If the magnitude of $\bar{D}$ is greater than some prescribed small value ε, the pressure in each cell is adjusted proportional to the negative of the velocity divergence by

$$\delta p^{n}_{i,j} = - \beta \cdot \bar{D}^{n}_{i,j} \tag{3}$$

where n denotes the number of iterations and i,j are the cell indices in x,y direction. The idea, which stands behind this is that if $\bar{D}$ in a cell is positive there is a net mass outflow out of the cell. So the fluid pressure in the cell must decrease to reduce $\bar{D}$, while a negative $\bar{D}$, correspondinng to a net mass inflow, requires an increase in cell pressure [11].
Now the pressure in the cell (i,j) is changed by

$$p^{n+1}_{i,j} = p^{n}_{i,j} + \delta p^{n}_{i,j} \tag{4}$$

and the velocities at the cell faces are simultaneously changed by

$$\begin{aligned} u^{n+1}_{i+1/2,j} &= u^{n}_{i+1/2,j} + \frac{\delta t}{(\delta x_i + \delta x_{i+1}) / 2} \cdot \delta p \\ u^{n+1}_{i-1/2,j} &= u^{n}_{i-1/2,j} - \frac{\delta t}{(\delta x_i + \delta x_{i-1}) / 2} \cdot \delta p \, . \end{aligned} \tag{5}$$

Equations similar to (5) are used for the other velocity components. The factor β in eq. (3) is a relaxation factor which is defined by

$$\beta = \frac{\beta_0}{2 \, \delta t \, ((1/\delta x^2) + (1/\delta y^2))} \, . \tag{6}$$

While a change in one cell will affect neighboring cells, the pressure and velocity adjustments have to be performed iteratively until all cells have simultaneously achieved a zero mass change. Therefor a simple SOR-technique is chosen and the factor β_0 in eq. (6) refers to the overrelaxation parameter.
Normally β_0 is a constant between 1 and 2. Computational experiments have shown that it is not always necessary to keep β_0 constant during one iteration cycle. One can vary β_0 by the number of iteration sweeps, starting with a value for β_0 slightly larger than unity and rising up to a given maximum value $\beta_{0_{max}}$ during a prescribed number of iterations [7].
After the pressure and velocity adjustment has converged one can finally start a new cycle and calculate new velocities from the momentum equations (1).
For cases where steady state solutions exist a large saving of computer time can be achieved by restricting the number of iterations during a time step. At the beginning of a computational cycle it is not necessary to reach a velocity divergence in each cell less than ε, but only to filter out the highly oscillating errors, which is achieved after only a few iteration cycles, so that a restriction to the maximum number of iterations in the beginning of the calculation can be imposed in order to save computer time. A further saving of computer time can be achieved by the implementation of a Multi-grid technique [12].

RESULTS AND DISCUSSION

In the present work the effects of Reynolds number and blockage ratio D/B are investigated in order to study their influences on:

- the structure of the wake behind the cylinder
- the pressure loss along the channel walls and
- the friction factor along the channel walls.

The channel inlet profile in u - direction is either parabolic or uniform and v = 0.
Computations have first been performed for a blockage ratio D/B = 0.2 at low Reynolds numbers. From fluid flow in infinite medium around a circular cylinder it is known that the existence of a wake bubble (Fig. 4) behind the cylinder could be observed for Re > 5.

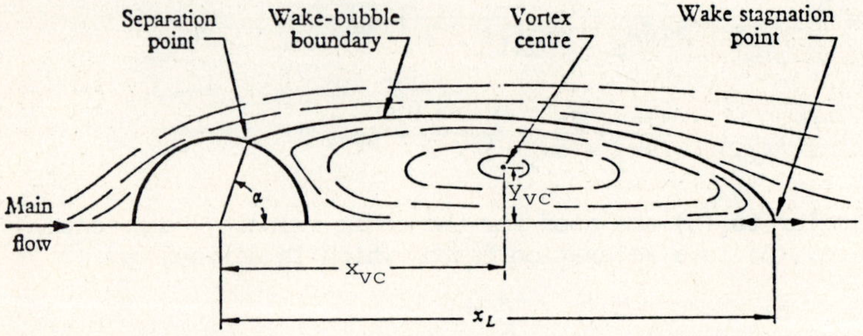

Fig. 4 : Steady separated flow past a circular cylinder

The length of the wake bubble x_L grows with increasing Reynolds numbers. Grove et al. [13] have measured the wake length x_L as a function of Reynolds number Re for different blockage ratios of the channel. In Fig. 5 a comparison has been made between the experimental results of Grove for D/B = 0.1 and 0.2 and our calculated results for D/B = 0.2 .

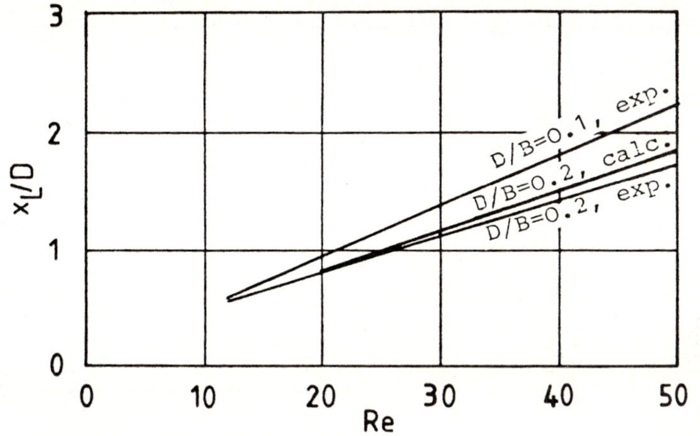

Fig. 5 : Effects of the confining walls on the length of the wake bubble (experimental results give the best line through data)

In the range of Reynolds number of computation a linear relationship between the wake length and Reynolds number is obvious. The wake length depends also on the blockage ratio D/B. Figure 5 shows good agreement between the experimental and calculated results for D/B = 0.2 .
In infinite medium the wake flow behind the cylinder becomes periodic when the Reynolds number reaches 40 and a Karman vortex street appears. Effects of confining walls on the periodicity of the wake flow behind the circular cylinder were experimentally investigated by Shair et al. [14] . They have found that the Reynolds number at which periodicity starts grows with increasing blockage ratio.
In Fig. 6a a computed steady solution for D/B = 0.2 and Reynolds number Re = 53 is shown. At Re = 67 the flow has become periodic in the near wake (Fig. 6b).

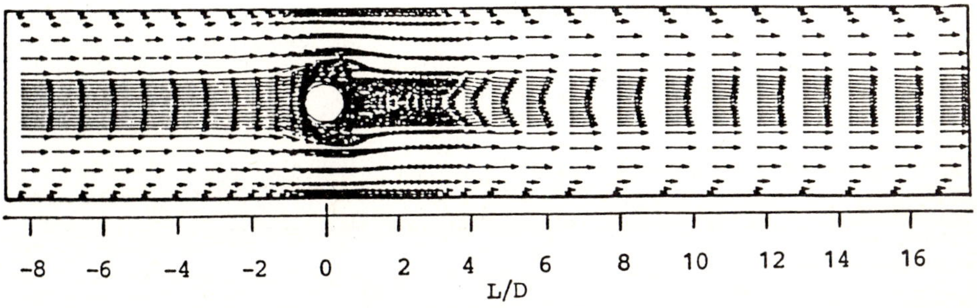

Fig. 6a : Re = 53

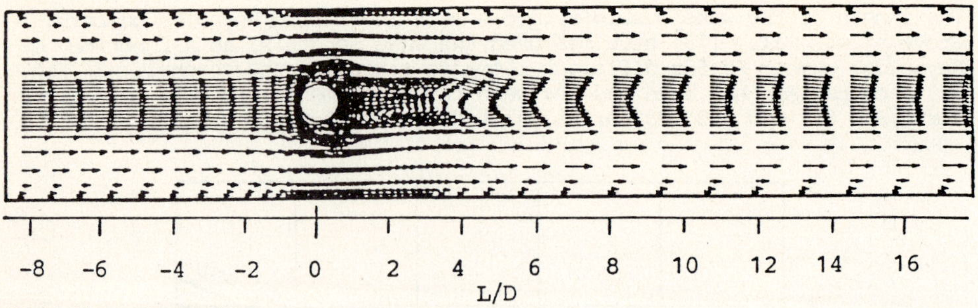

Fig. 6b : Re = 67

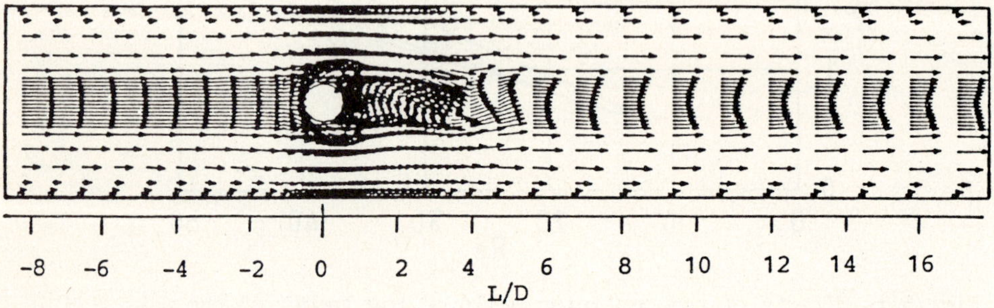

Fig. 6c : Re = 134

Fig. 6 : Velocity vectors in the channel for D/B = 0.2 , parabolic inlet profile

The periodicity is suppressed far downstream of the cylinder by the channel walls and the flow at the exit of the channel tends again to become steady parabolic flow. This is not really apparent from the figure, but can be seen from the computed data.
Shair's [14] experiments show a critical Reynolds number where periodicity starts of about 130 . Here the Reynolds number is defined with a maximum velocity u as to be the velocity which would exist at the same location as that of the centre of the cylinder under flow conditions identical with those of the experiment but in absence of the cylinder.
Our Reynolds number of beginning periodicity, based on a similar definition as that of Shair, is about 100 . This discrepancy between experimental and calculated results can be explained by noting that the beginning of oscillation could much earlier be observed at lower Reynolds numbers from computed data than only by flow visualization from experiments. Besides, the initial profile in the experiment of Shair et al. [14] is not clearly defined and may not be exactly parabolic as we have used in our computations.
With an increase in Reynolds number upto 134 - but now again based on the average of the channel inlet velocity - the periodicity of the wake becomes stronger (Fig. 6c), but is again strongly damped by the channel walls.
The flowfield in front of the cylinder seems to be nearly independent of the Reynolds number and the structure of the wake [8]. One may conjecture, that the influence of the position of the cylinder in the channel L_1/L is not of great significance for the wake flow behind the cylinder. This situation of course will change for a non - parabolic velocity profile at

the channel entrance.
The effects of confining walls become prominent with increasing blockage ratio. For D/B = 0.5 steady state solutions are obtained with Re = 67 and also 134 (Fig. 7a,b). Periodic wake flow appears first at Re > 170 (Fig. 7) but is again strongly damped downstream of the cylinder by the channel walls. For the case D/B = 0.5 again little influence of Reynolds number or the structure of the wake on the flowfield in front of the cylinder can be ascertained.

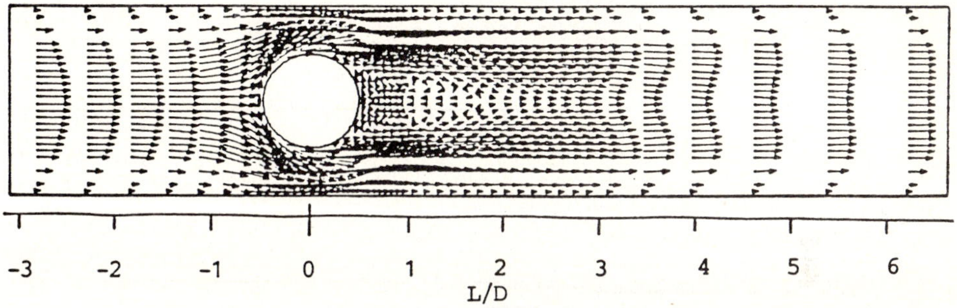

Fig. 7a : Re = 67

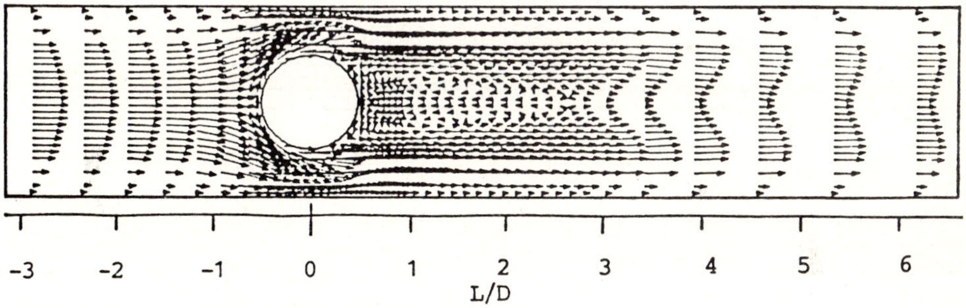

Fig. 7b : Re = 134

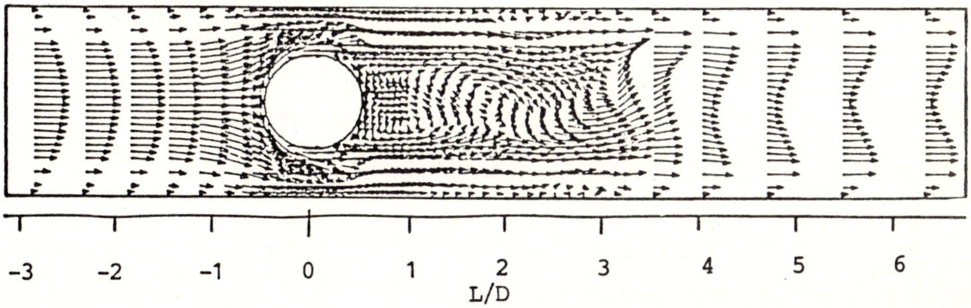

Fig. 7c : Re = 268

Fig. 7 : Velocity vectors , D/B = 0.5 , parabolic inlet profile

Although the structure of the periodic flow for the same Reynolds number at two different blockage ratios is nearly the same (Fig. 8), the damping effects of the channel walls on the periodical vortex flow are greater for the blockage ratio D/B = 0.5 (Fig. 8b) than for D/B = 0.33 (Fig. 8a). But the fluid flow tends earlier to become parabolic for D/B = 0.33 than for D/B = 0.5. Here the velocity values beside the cylinder are not as large as for D/B = 0.5, and so a parabolic profile will evolve earlier.

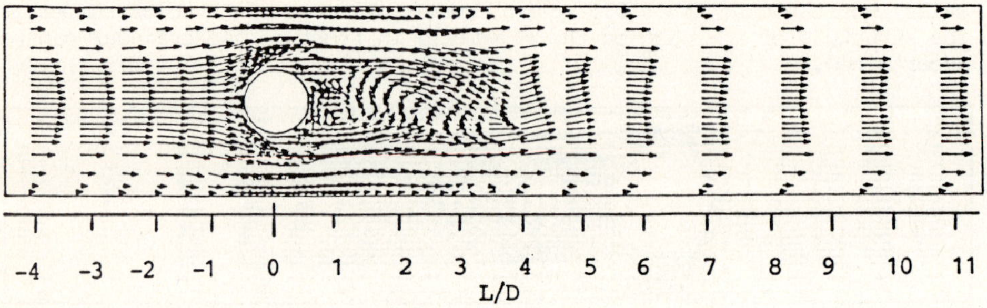

Fig. 8a : D/B = 0.33

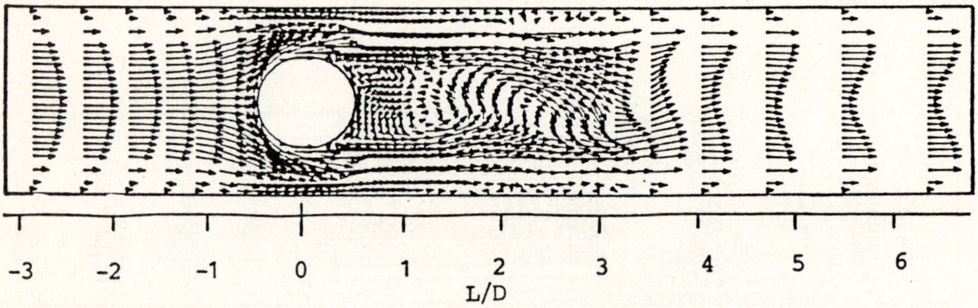

Fig. 8b : D/B = 0.2

Fig. 8 : Velocity vectors in the channel for Re = 268 , parabolic inlet profile

In Fig. 7c and 8 a,b also seperation areas at the channel walls behind the cylinder can be seen. In Fig. 9 a,b,c c_f Re_B along the channel walls are computed for blockage ratios D/B = 0.2, 0.33 and 0.5 at several different Reynolds numbers. Here the Reynolds number Re_B is defined by $Re_B = (u_{av} \cdot B/\nu)$.

The recirculation areas behind the cylinder at the channel walls are characterized by negative values for c_f Re_B. While for D/B = 0.2 no existence of a recirculation area could be seen at the computed Reynolds numbers, c_f Re_B reaches higher negative values with increasing blockage ratio for the same Reynolds number (Fig. 9 b,c - Re_D = 268). Also the location of maximum negative c_f Re_B moves nearer to the cylinder with increasing blockage ratio.

Further it can be seen that the maximum value of c_f Re_B grows of course with increasing Reynolds numbers at the same blockage ratio, and also with increasing blockage ratio at the same Reynolds number. Investigations of friction factor c_f show that c_f becomes lower with increasing Reynolds number, while the product of c_f Re_B grows.

In Fig. 9 it could be seen, that with increasing Reynolds number the point of maximum c_f Re_B moves slightly to the rear of the cylinder.

Figure 9 shows that c_f Re_B at the channel entrance is equal to 12 (the

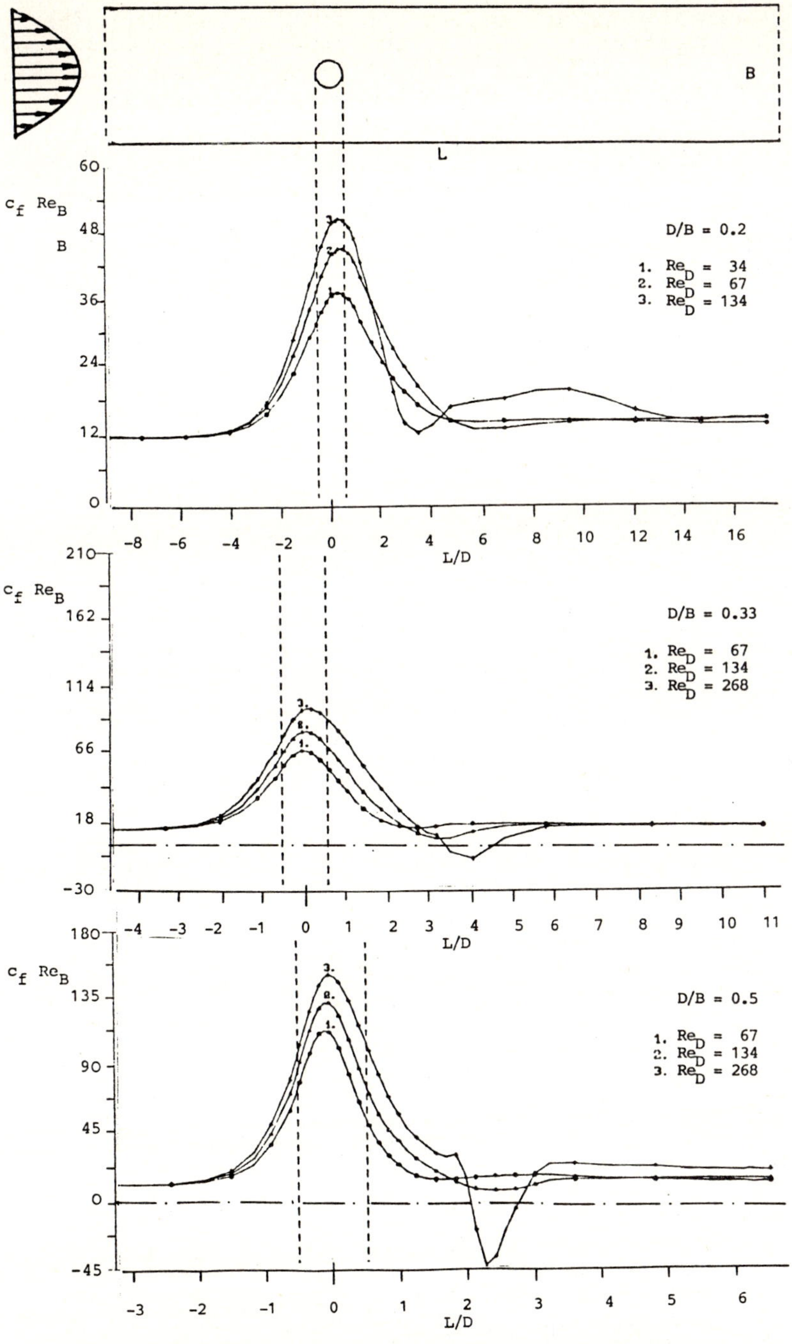

Fig. 9 : c_f Re_B along one channel wall, parabolic inlet profile (dotted lines - position of the cylinder)

$Re_B = u_{av} \cdot B / \nu$, $c_f = \tau_w / (\rho \cdot u_{av}^2 / 2)$

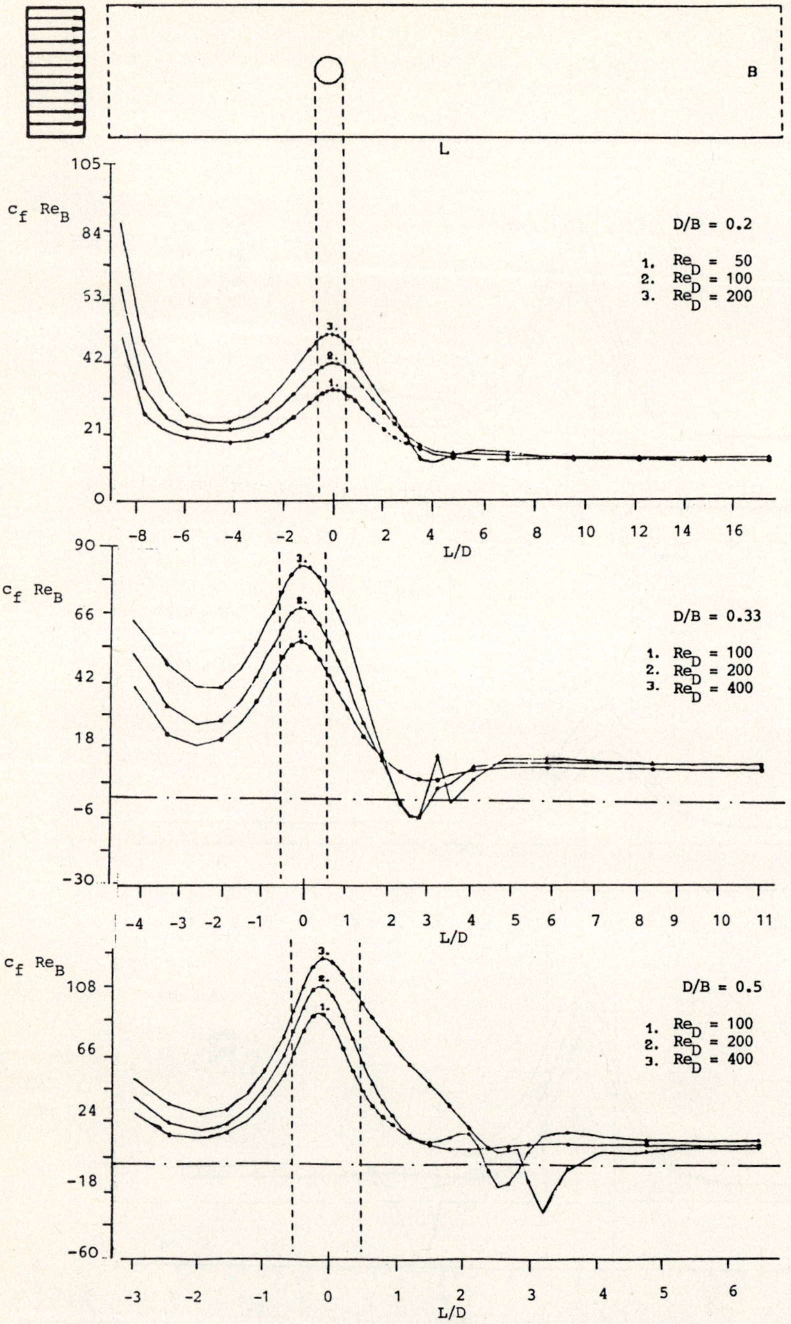

Fig. 10 : c_f Re_B along one channel wall, constant inlet profile (dotted lines - position of the cylinder)

$Re_B = u_{av} \cdot B / \nu$, $c_f = \tau_w / (\rho \cdot u_{av}^2 / 2)$

value of the fully developed laminar flow in a 2-D channel) and tends to this value far downstream from the cylinder. The deviation of $c_f\ Re_B$ from 12 in front of the cylinder upto a distance of 2·D shows the upstream influence of the cylinder.
When a uniform velocity instead of a parabolic profile at the channel entrance is used, the boundary layer theory on a flat plate gives

$$c_f = \frac{0.664}{\sqrt{Re_B \frac{x}{B}}} . \tag{7}$$

Now $c_f\ Re_B$ in front of the cylinder depends strongly on the Reynolds number, while far beind the cylinder, where again a parabolic flow profile will evolve, $c_f\ Re_B$ tends to 12 (Fig. 10).
The location of maximum $c_f\ Re_B$ again moves to the rear of the cylinder with increasing Reynolds number.
A comparison between $c_f\ Re_B$ on both walls is shown in Fig. 11 for D/B = 0.5 at a Reynolds number of 400 and a uniform velocity profile at the inlet.

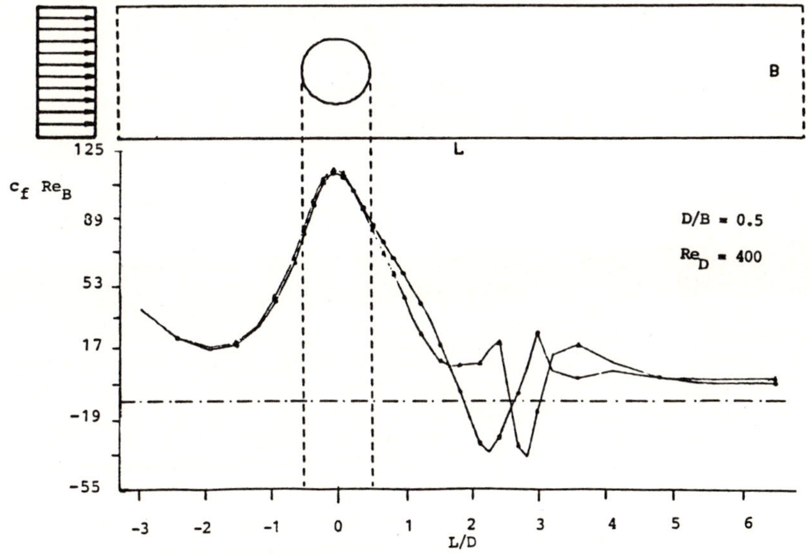

Fig. 11 : $c_f\ Re_B$ along top and bottom walls, D/B = 0.5, uniform channel inlet profile
$Re_B = u_{av} \cdot B / \nu$, $c_f = \tau_w / (\rho \cdot u_{av}^2 /2)$

While the structure of $c_f\ Re_B$, which is proportional to the shear stresses in front of the cylinder is nearly the same on the both channel walls, it becomes just the opposite on the rear side of the cylinder. When the top wall has maximum values of shear stress, on the bottom wall one will find a minimum, and vice-versa. This oscillating structure is damped far downstream of the cylinder. No oscillation in shear stress on both channel

walls means that the velocity profile tends again to become steady parabolic flow.
The same oscillating structure between the two walls can be seen in the pressure distribution along the channel walls. This is shown in Fig. 12 again for D/B = 0.5, Re = 400 and a uniform velocity profile at the channel inlet.
From Fig. 12 it becomes also apparent that behind the cylinder the pressure increase on both side walls is very large, compared to the minimum value immediately behind the cylinder.

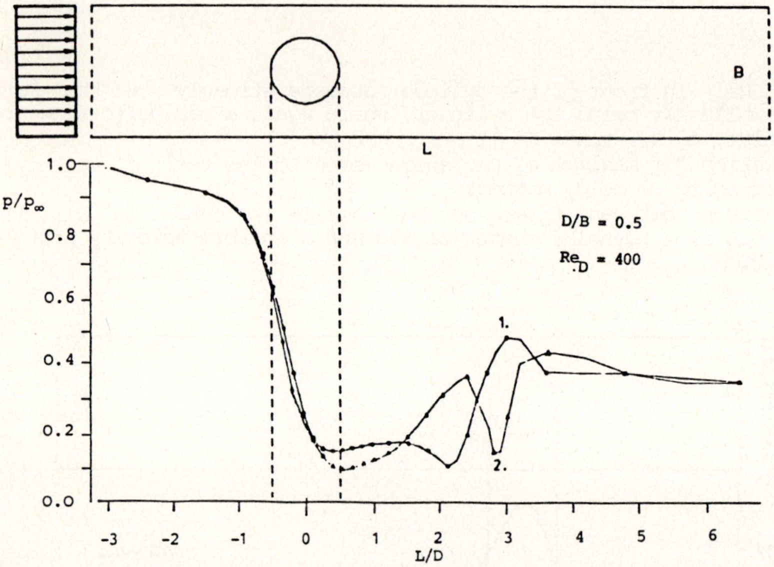

Fig. 12 : Pressure distribution along top and bottom walls, Re = 400, uniform channel inlet profile

This changes when the channel inlet profile is parabolic. Now the fluid flow in the beginning of the channel is fully developed and so no significant change in pressure along the channel wall is required as long as the fluid is far from the cylinder (Fig. 13).
Here one can also see, that the influence of the cylinder on the pressure loss in front of the cylinder decreases with increasing blockage ratio. Also the influence of Reynolds number on pressure loss upstream of the cylinder vanishes with increasing blockage ratio.

CONCLUSION

Channel walls have a strong damping influence on the wake flow behind a circular cylinder. The beginning of periodicity appears at higher Reynolds numbers with greater blockage ratios and the periodicity itself is suppressed far downstream of the cylinder.
The wake length and the critical Reynolds number at which periodicity starts for a blockage ratio of D/B = 0.2 show good agreement between experimental and computed results.

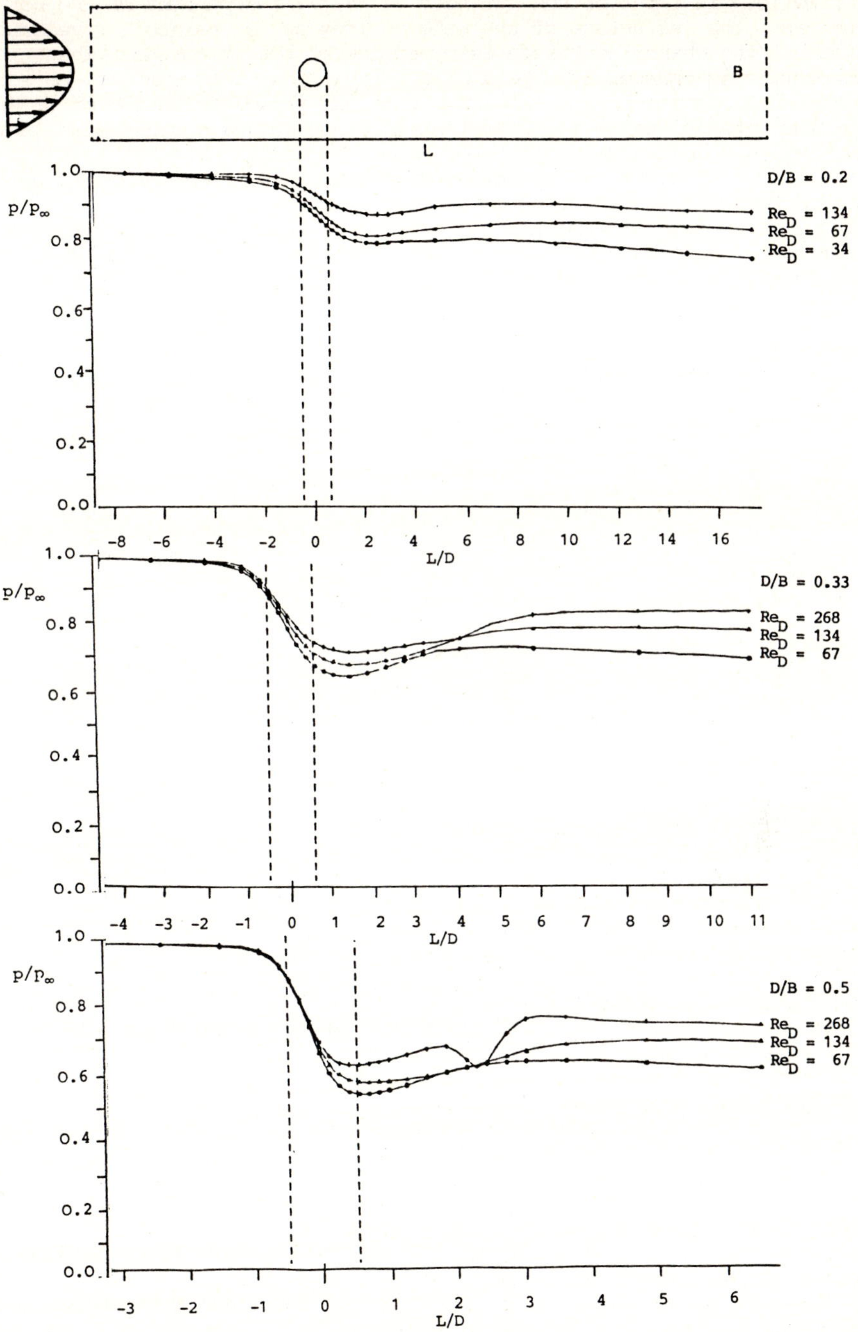

Fig. 13 : Pressure loss along one side wall, parabolic inlet profile
(dotted lines - position of the cylinder)

The flowfield in front of the cylinder is nearly independent of Reynolds number and the structure of the wake as long as a parabolic velocity profile at the channel entrance is prescribed, but it changes when the inlet profile is changed from parabolic to uniform. This can also be seen from computed results for the pressure loss and the distribution of $c_f\ Re_B$ along the channel walls for different blockage ratios and Reynolds numbers.
Investigations of $c_f\ Re_B$ and pressure distributions on both channel walls show an oscillating structure between minimum and maximum values on the walls for a periodic flow behind the cylinder. This oscillation decreases far behind the cylinder and vanishes when the flow profile tends again to become steady parabolic flow.

Acknowledgement: This work has been supported by the Deutsche Forschungsgemeinschaft.

REFERENCES

[1] Thoman, D. C. and Szewczyk, A. A.,
"Time dependent viscous flows over a circular cylinder",
Physics of fluids, 12, 3, 1969, pp. II 77 - II 86.

[2] Lin, C. L., Pepper, W. W. and Lee, S. D.,
"Numerical methods for separated flow solutions around a circular cylinder",
AIAA J., 14, 2, 1976, pp. 900 - 906.

[3] Fornberg, B.,
"A numerical study of steady viscous flows past a circular cylinder",
JFM, 98, 4, 1980, pp. 819 - 855.

[4] van Dyke, M.,
"An album of fluid motion",
published by the departement of Mech. Eng.,
Stanford University, Stanford, California, USA, 1982.

[5] Hirt, C. W., Nichols, B. D. and Romero, N. C.,
"SOLA - A numerical Solution Algorithm for transient Fluid Flow"
Los Alamos, Report LA - 5852, 1975.

[6] Wnek, W. J., Ramshaw, J. D., et. al.,
"Transient three-dimensional thermal-hydraulic Analysis of Nuclear Reactor Fuel Rod Arrays",
Aerojet Nuclear Company, ANCR - 1207, 1975.

[7] Kiehm, P., Mitra, N. K. and Fiebig, M.,
"Numerical study of confined 2-D and 3-D Laminar Flows around a Circular Cylinder",
To appear in Proc. 6. GAMM Conference on Numerical Methods in Fluid Mechanics, 1985.

[8] Kiehm, P., Mitra, N. K. and Fiebig, M.,
"Numerical Investigation of 2-D and 3-D confined Wakes behind a Circular Cylinder in a Channel",
To be presented in AIAA 24th Aerospace Sciences Meeting, 1986, in Reno, Nevada.

[9] Launder, B. E. and Massey, T. H.,
"The Numerical Prediction of Viscous Flow and Heat Transfer in Tube Banks",
Journal of Heat Transfer, vol. 100, 1978, pp. 565 - 571.

[10] Kiehm, P.
Doctoral - Dissertation at the Institute for Thermo- and Fluiddynamics,
Ruhr - University Bochum, in preparation.

[11] Nichols, B. D., Hirt, C. W.,
"Calculating Three-Dimensional Free Surface Flows in the Vicinity of Submerged and Exposed Structures",
J. of Comp.Phys., 12, 1973, pp. 234-246.

[12] Brockmeier, U. , Mitra, N. K. , Fiebig, M.
"Multigrid Marker and Cell (SOLA) Algorithm for three dimensional Flow Computation"
To appear in Proc. 6. GAMM Conference on Numerical Methods in Fluid Mechanics, 1985.

[13] Grove, A. S., Shair, F. H., Peterson, E. E. and Acrivos, A.,
"An experimental Investigation of the Steady Separated Flow past a Circular Cylinder",
J. Fluid Mech., 19, p. 60, 1964.

[14] Shair, F. H., Grove, A. S., Peterson, E. E. and Acrivos, A.,
"The effect of confining walls on the Stability of the Steady Wake behind a Circular Cylinder"
J. Fluid Mech., 17, p. 546, 1965.

EXPLICIT METHOD FOR SOLVING NAVIER STOKES EQUATIONS USING A FINITE ELEMENT FORMULATION

W. Koschel, M. Lötzerich, A. Vornberger
Institut für Strahlantriebe und Turboarbeitsmaschinen
RWTH Aachen, Templergraben 55, 5100 Aachen

SUMMARY

Using an explicit Taylor Galerkin finite element scheme in the solution of transport equations, the "Rotating Cone Problem" is solved to test numerical accuracy. The scheme was applied to solve viscous laminar flow problems on a flat plate and at the mean section of a turbine guide vane cascade. The results are compared with the respective solutions derived from boundary layer theory.

1. INTRODUCTION

The finite element method has become a powerful method in solving general engineering problems by the extension to other areas than structural mechanics. The solution of high speed aerodynamic problems took place in the last years. Some key papers for the understanding of the problem addressed in this paper are given in the following. Hughes [1] developed a tensor formulation of an artifical viscosity term acting in the streamwise direction. Donea [2] determined in his Taylor Galerkin method the amount of artifical viscosity by a Taylor series expansion in time, as in the well-known Lax-Wendroff scheme. The formulation of the second derivatives follows Hughes's approach. Löhner [3] changed to a corresponding two step scheme avoiding the evaluation of the second derivatives, and combined it with a Domain/Time Splitting Procedure [4], in order to increase the efficiency of the explicit scheme.

In section 2 the explicit Taylor Galerkin formulation for a 1-dimensional transport equation is given, in section 3 results on the "Rotating Cone Problem" with and without time-splitting are presented. Finally the laminar viscous flow over a flat plate and through the mean section of a turbine guide vane were computed by using a stream-function-vorticity formulation of the Navier-Stokes equations. The results were compared with solutions derived from boundary layer theory.

2. THE EXPLICIT TAYLOR GALERKIN METHOD

The explicit Taylor Galerkin method is illustrated for an one-dimensional scalar hyperbolic equation

$$\frac{\partial \omega}{\partial t} + \frac{\partial F}{\partial x} = 0 \qquad (1)$$

$$\frac{\partial \omega}{\partial t} + u \frac{\partial \omega}{\partial x} = 0 \qquad \text{with} \qquad u = \frac{\partial F}{\partial \omega}\,. \tag{2}$$

A Taylor series expansion in time gives:

$$\omega^{n+1} = \omega^n + \Delta t \left(\frac{\partial \omega}{\partial t}\right)^n + \frac{\Delta t^2}{2!} \frac{\partial^2 \omega}{\partial t^2} \tag{3}$$

$$\frac{\partial \omega^2}{\partial t^2} = - \frac{\partial}{\partial t} \left(u \frac{\partial \omega}{\partial x}\right) = u^2 \frac{\partial^2 \omega}{\partial x^2}\,. \tag{4}$$

Using (3) and (4) we get

$$\omega^{n+1} = \omega^n - \Delta t \; u \frac{\partial \omega}{\partial x} + \frac{\Delta t^2}{2!} u^2 \frac{\partial^2 \omega}{\partial x^2}\,. \tag{5}$$

This equals the well-known Lax-Wendroff-method. When extending to higher dimensions we get mixed derivatives in opposition to the Lax-Wendroff method of the kind:

$$\frac{\omega^{n+1} - \omega^n}{\Delta t} = - \bar{u} \, \nabla \, \omega^n + \frac{\Delta t}{2} \bar{u} \, (\bar{u}\nabla) \, \nabla \, \omega^n. \tag{6}$$

The second expression on the right hand side has the same structure as the artificial viscosity in the "Streamline Upwind Petrov-Galerkin Formulation" given by Brooks and Hughes [1].

By this the second term on the right hand side can be interpreted as part of the time derivatives or as an artificial viscosity in streamwise direction. Using a Galerkin weighted residual approach we get for the one-dimensional transport equation

$$\sum_j \left(\int_{\Omega e} N_i \, N_j \, d\Omega \right) \Delta\omega^{n+1} = - \sum_j \int_{\Omega e} \left(\tfrac{1}{2} \Delta t \; u \; N_i \frac{\partial N_j}{\partial x} - \frac{1}{2} \Delta t^2 \, u^2 \frac{\partial N_i}{\partial x} \frac{\partial N_j}{\partial x} \right) d\Omega \;\; \omega^n + \sum_j \int_{Se} \frac{1}{2} \Delta t^2 \, u^2 \, N_i \frac{\partial \omega}{\partial n} \, dS \tag{7}$$

which can be written as a system of equations

$$C \Delta\omega^{n+1} = B_\Omega + B_S \qquad \text{and} \qquad \Delta\omega^{n+1} = \omega^{n+1} - \omega^n \tag{8}$$

where C is the consistent mass matrix, $\Delta\omega$ is the vector of unkowns, B_Ω and B_S are the area integral and line integral respectively.

As the C-matrix is dominant on the diagonal, the system of equations can be solved iteratively by using a scheme proposed by Donea and Giuliani [5]:

$$C_{LM} \left(\Delta\omega_K^{n+1} - \Delta\omega_{K-1}^{n+1}\right) = B_\Omega + B_S - C\Delta\omega_{K-1}^{n+1}; \qquad 1 \le K \le \text{niter}$$
$$C_{LMi,i} = \sum_j C_{i,j}; \quad C_{LMi,j \; i\neq j} = 0; \quad \Delta\omega_o^{n+1} = 0\,. \tag{9}$$

C_{LM} is the diagonalized mass matrix. The results which are presented in the paper were computed with niter = 4. In the case that niter is set equal to 1 we get the Lax-Wendroff-scheme.

Due to the Euler time stepping the stability is bounded by the local Courant number C < 1.

3. THE "ROTATING CONE PROBLEM"

To test the numerical accuracy of the formulation, the "Rotating Cone Problem" was solved. This problem is settled by an unsteady twodimensional transport equation. Figure 1 shows the computational domain, the mesh and the boundary conditions. The computational domain is a square with 31x31 mesh points or 900 bilinear elements. A concentration of sinusoidal shape -"the cone" - is placed in a rotating flow field as depicted in fig. 2 at t = 0. A view of the cone after one revolution or the equivalent 160 time steps is given in fig. 3. After this full rotation the cone is reduced to 98.3% of its original height, at the same time having a maximum undershoot of 0.5%.

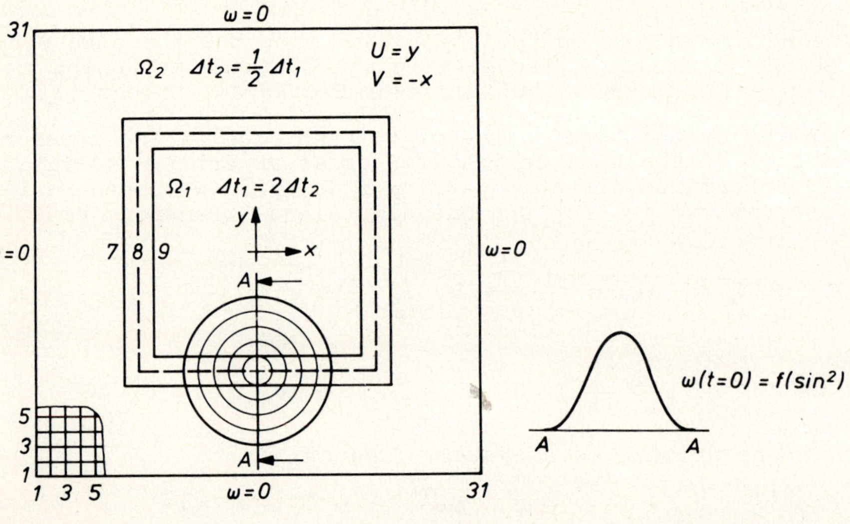

Fig. 1: The "Rotating Cone Problem"

As explicit solution schemes are bounded by the CFL-criteria, we get a restriction in the time step due to the associated local Courant-number. Local mesh refinements can lead to highly uneconomic time steps.

In the case, that the computations are performed with the pertinent local time steps, the solution is advanced on a wharped time surface, leading to an uncorrect unsteady solution. By subdividing the computational domain, we can achieve the unsteady solution. In each of this subdomains the computations are done with the pertinent time steps. Such a procedure is given by Löhner et.al. in [4].

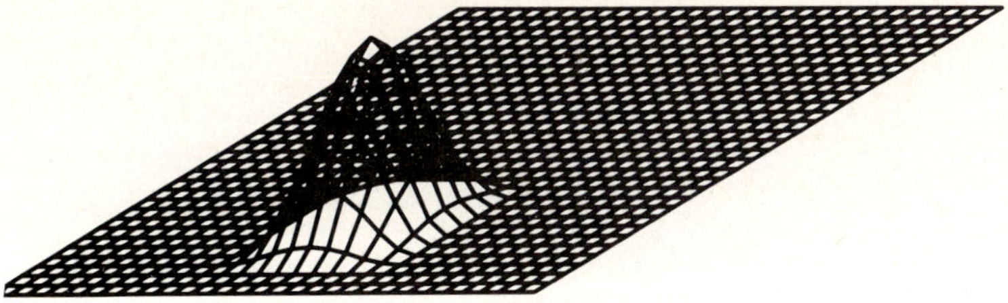

Fig. 2: Rotating Cone at t = 0

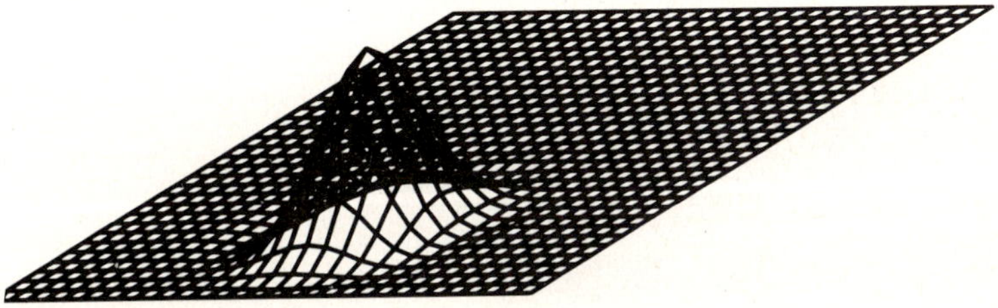

Fig. 3: Rotating Cone after one revolution (160 time steps)

By using this procedure, it is possible to achieve economic solutions of the unsteady equations with explicit methods even for domains with local mesh refinements. Using the subdomains Ω_1 and Ω_2 as depicted in figure 1, the cone is reduced after a full rotation to 97% of its original height, at the same time having a maximum undershoot of 1%.

4. CALCULATION OF TWO-DIMENSIONAL VISCOUS FLOWS

The numerical procedures outlined in chapter 2 were applied to a two-dimensional, blade to blade flow using the Navier-Stokes equations in streamfunction-vorticity formulation. They are given for a surface of revolution on an orthogonal m,ϕ-coordinate system:

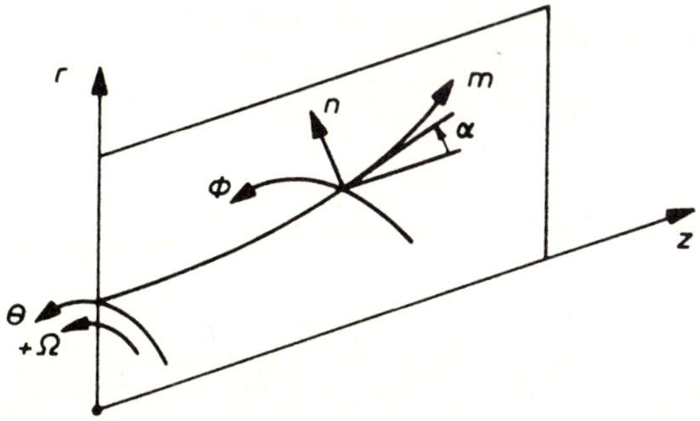

Continuity:

$$\frac{\partial}{\partial m} (\rho\ br\ W_m) + \frac{\partial}{\partial \phi} (\rho\ b\ W_\phi) = 0 \qquad (10)$$

Meridional momentum:

$$\rho \left\{ W_m \frac{\partial W_m}{\partial m} + \frac{W_\phi}{r} \frac{\partial W_m}{\partial \phi} - \frac{W_\phi 2}{r} \sin\alpha - 2\Omega W_\phi \sin\alpha - \Omega^2 r \sin\alpha \right\}$$

$$= - \frac{\partial p}{\partial m} - \frac{2}{3} \frac{\partial}{\partial m} (\mu\ \nabla \cdot \bar{w}) + \frac{1}{r} \frac{\partial}{\partial m} (2\ \mu r \frac{\partial W_m}{\partial m})$$

$$+ \frac{1}{r} \frac{\partial}{\partial \phi} \{\mu\ (r \frac{\partial}{\partial m} (\frac{W_\phi}{r}) + \frac{1}{r} \frac{\partial W_m}{\partial \phi})\}$$

$$- \frac{2\mu}{r} \left\{\frac{1}{r} \frac{\partial W_\phi}{\partial \phi} + \frac{W_m}{r} \sin\alpha \right\} \sin\alpha \qquad (11)$$

Tangential momentum:

$$\rho \left\{ W_m \frac{\partial W_\phi}{\partial m} + \frac{W_\phi}{r} \frac{\partial W_\phi}{\partial \phi} + \frac{W_m\ W_\phi}{r} \sin\alpha + 2\ \Omega\ W_m \sin\alpha \right\}$$

$$= - \frac{1}{r} \frac{\partial p}{\partial \phi} - \frac{2}{3} \frac{1}{r} \frac{\partial}{\partial \phi} \{\mu\ \nabla \cdot \bar{w}\}$$

$$+ \frac{1}{r} \frac{\partial}{\partial m} \left\{ r\mu\ (r \frac{\partial}{\partial m} (\frac{W_\phi}{r}) + \frac{1}{r} \frac{\partial}{\partial \phi} (W_m)) \right\}$$

$$+ \frac{1}{r} \frac{\partial}{\partial \phi} \left\{ 2\mu\ (\frac{1}{r} \frac{\partial W_\phi}{\partial \phi} + \frac{W_m}{r} \sin\alpha) \right\}$$

$$+ \frac{\mu}{r} \left\{ r \frac{\partial}{\partial m} (\frac{W_\phi}{r}) + \frac{1}{r} \frac{\partial}{\partial \phi} (W_m) \right\} \sin\alpha . \qquad (12)$$

The continuity equation is satisfied by introducing a stream-function which is related to the velocity components by:

$$W_m = \frac{1}{\rho\ br} \frac{\partial \psi}{\partial \phi} \qquad W_\phi = - \frac{1}{\rho\ b} \frac{\partial \psi}{\partial m} . \qquad (13)$$

If the pressure terms are eliminated from the momentum equations by cross differentiation and the vorticity ω is introduced one obtains the vorticity transport equation:

$$\frac{\partial}{\partial m} (\frac{1}{b} \frac{\partial \psi}{\partial \phi}\ \omega) - \frac{\partial}{\partial \phi} (\frac{1}{b} \frac{\partial \psi}{\partial m}\ \omega) + S = \frac{\partial}{\partial m} (r \frac{\partial}{\partial m} (\mu\omega))$$

$$+ \frac{\partial}{\partial \phi} (\frac{1}{r} \frac{\partial}{\partial \phi} (\mu\omega)) \qquad (14)$$

where $S = 2\ \Omega\ (\frac{\partial}{\partial m} (\frac{1}{b} \sin\alpha \frac{\partial \psi}{\partial \phi}) - \frac{\partial}{\partial \phi} (\frac{1}{b} \sin\alpha \frac{\partial \psi}{\partial m}))$

$$+ \frac{\partial \rho}{\partial m} \frac{\partial\ W2/2}{\partial \phi} - \frac{\partial \rho}{\partial \phi} \frac{\partial\ W2/2}{\partial m} + \Omega^2 r \sin\alpha \frac{\partial \rho}{\partial \phi}$$

$$W^2 = W_m{}^2 + W_\phi{}^2$$

and ω is defined:

$$\omega = \frac{1}{r} \left(\frac{\partial}{\partial m} (W_\phi\, r) - \frac{\partial W_m}{\partial \phi} \right) \tag{15}$$

or in terms of the stream function:

$$\frac{\partial}{\partial m} \left(\frac{r}{b\rho} \frac{\partial \psi}{\partial m} \right) + \frac{\partial}{\partial \phi} \left(\frac{1}{\rho\, br} \frac{\partial \psi}{\partial \phi} \right) = - \omega \cdot r . \tag{16}$$

The non-slip boundary condition is implemented in the vorticity transport equation:

$$\frac{\partial \psi}{\partial n} = 0 \rightarrow \omega_W = \frac{3\, (\psi_{W+1} - \psi_W)}{\Delta n^2} - \frac{1}{2}\, \omega_{W+1} + O(\Delta n^2). \tag{17}$$

The impermeability of the wall requires:

$$\frac{\partial \psi}{\partial S} = 0 \rightarrow \psi = \text{const.}$$

Therefore the blade surface is treated as a streamline. Further details are given by Khahil [6].

The pressure field can be calculated from a Poisson equation, which is derived by differentiation of the momentum equations:

$$\frac{\partial}{\partial m} \left(r \frac{\partial P}{\partial m} \right) + \frac{\partial}{\partial \phi} \left(\frac{1}{r} \frac{\partial P}{\partial \phi} \right) = - 2\rho \left\{ \frac{\partial W_m}{\partial \phi} \frac{\partial W_\phi}{\partial m} - \frac{\partial W_m}{\partial m} \frac{\partial W_\phi}{\partial \phi} \right.$$

$$\left. + \sin\alpha \quad \Omega \left(\left(\frac{\partial W_m}{\partial \phi} + \frac{\partial W_\phi\, r}{\partial m} - \Omega r \sin\alpha \right) - W_\phi \frac{\partial W_\phi}{\partial m} - W_m \frac{\partial W_m}{\partial m} \right) \right\}. \tag{18}$$

For the energy equation the assumption of constant total enthalpy of the gas is made. The enthalpy is defined as:

$$H = c_p\, T + W^2/2 + \Omega\, W_\phi\, r + \frac{1}{2}\, \Omega^2 r^2. \tag{19}$$

4.1 FLOW OVER A FLAT PLATE WITHOUT PRESSURE GRADIENT

In a first test we calculated the laminar flow over a flat plate without pressure gradient. Due to the fact, that the procedure is set up for flow calculations in turbomachines, we calculate the flow through a cascade of plates instead of the flow around a single plate. The mesh consists of 3000 bilinear elements, the plate length is 7 m and the distance between two plates is 5 m.

In figure 4 the non-dimensionalized boundary layer profiles at two different locations, corresponding to the local Reynolds numbers of 400 and 1000, are shown in comparison with the Blasius profile. In figures 5 and 6 the calculated displacement- and momentum thicknesses are compared with an analytic solution.

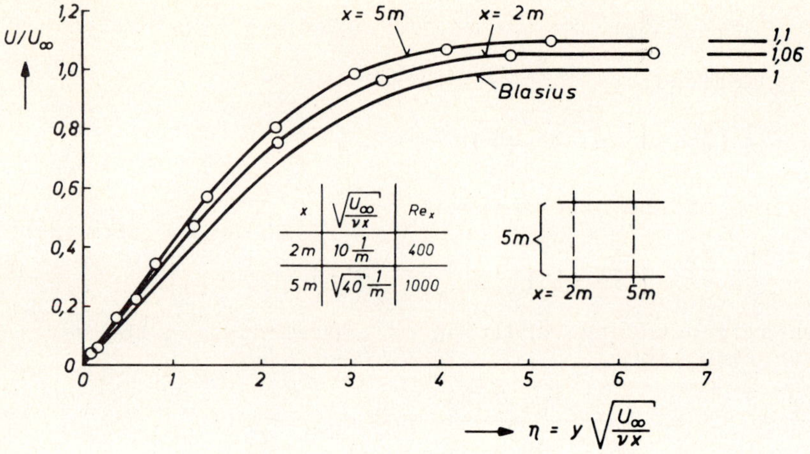

Fig. 4: Non-dimensionalized boundary layer profiles

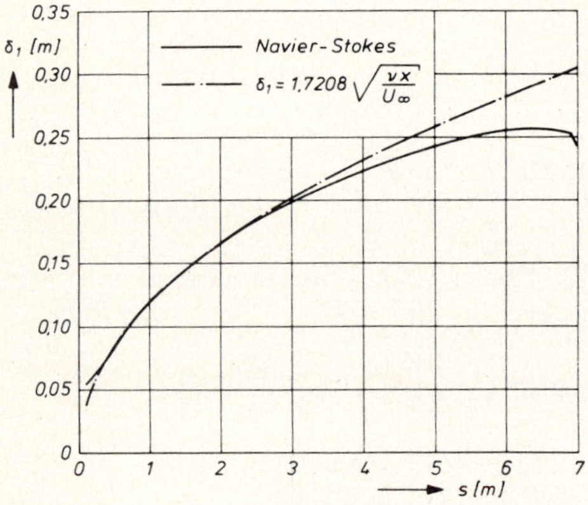

Fig. 5: Displacement thickness

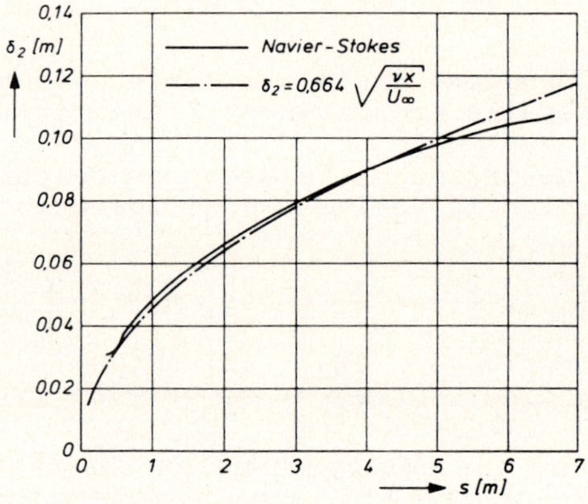

Fig. 6: Momentum thickness

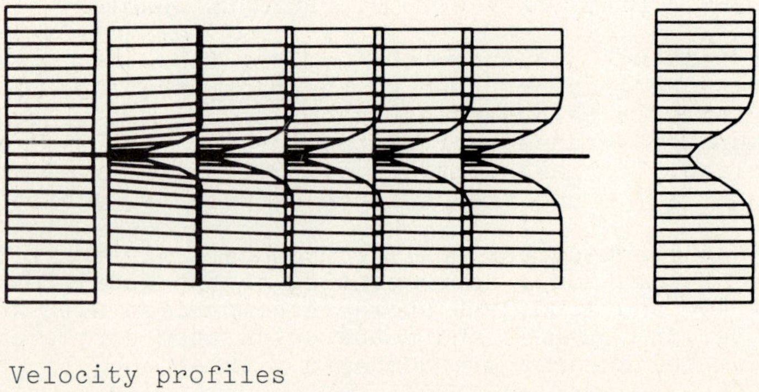

Fig. 7: Velocity profiles

The increasing difference towards the end of the plate is thought to be caused by the small favourable pressure gradient and the trailing edge. This interpretation is confirmed by calculations varying the plate length and the channel height. In fig. 7 boundary layer profiles at different locations are shown.

4.2 FLOW ON A S1-STREAM SURFACE

The laminar flow was calculated at the mean section of a turbine guide vane cascade. A view of the guide vane in the meridional plane is given in fig. 8. The computational domain (fig. 8) is generated by rotating the corresponding streamline around the center axis.

Density field, mass flow, meridional shape and the local thickness of the stream surface are provided by a 3-dimensional inviscid flow computation on S1-S2 stream-surface [7].

For the flow field computation a near design point was chosen, this corresponds to a downstream Machnumber of 0.7. The viscous flow field was calculated for two different Reynolds numbers of 3700 and 37000, the Reynolds number defined as:

$$Re = \frac{\rho(w_1+w_2)\ S}{2\mu} \qquad \text{where } S = \text{mean camber line.}$$

Fig. 9 shows the velocity distribution around the blade for the small Reynolds number. Due to the adverse pressure gradient in the rear part of the suction side laminar separation occurs. The same separation point was detected in experimental studies. In the case of the high Reynolds number a kind of vortex street is developing at the trailing edge (fig. 10).

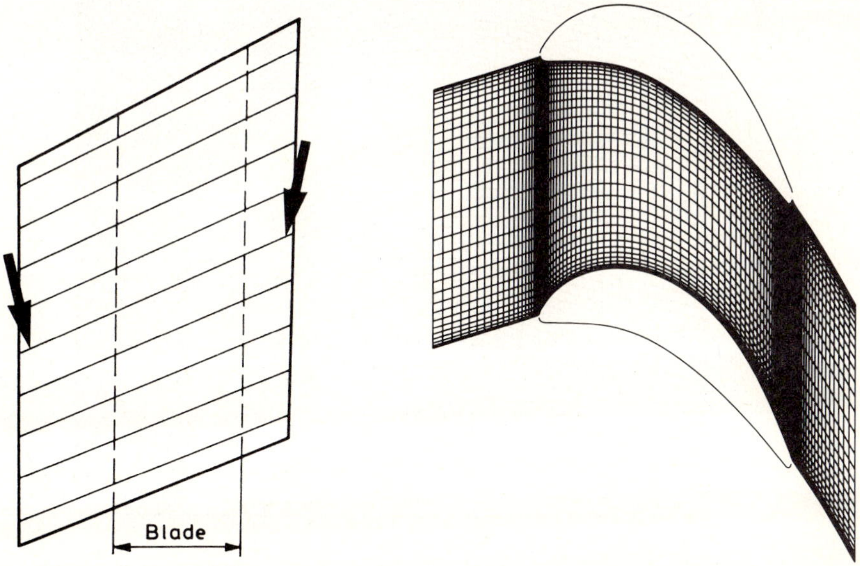

Fig. 8: Computational domain

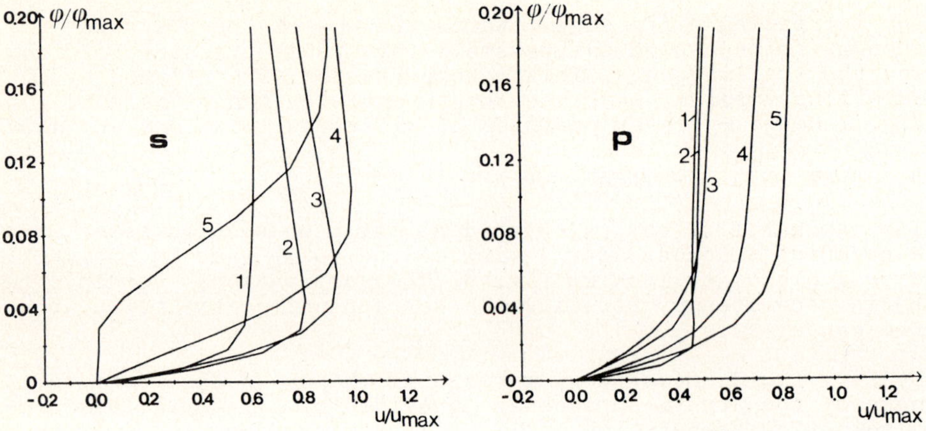

s = Suction side

p = Pressure side

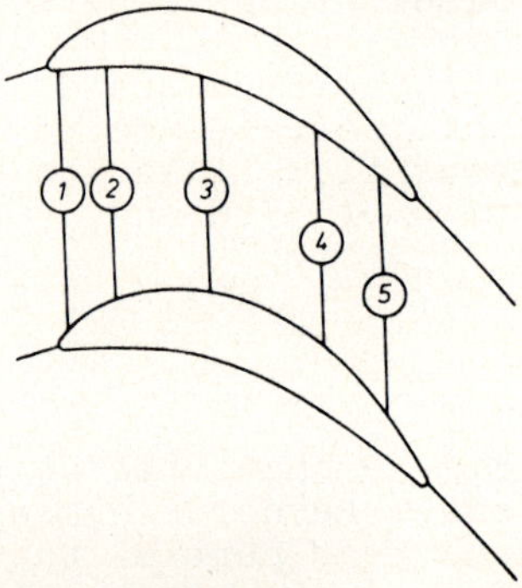

Fig. 9: Velocity distribution at Re=3700

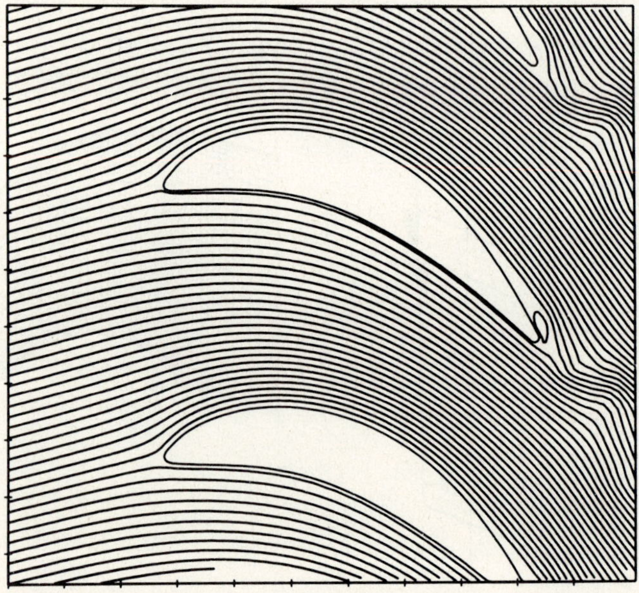

Fig. 10: Streamlines at Re=37000

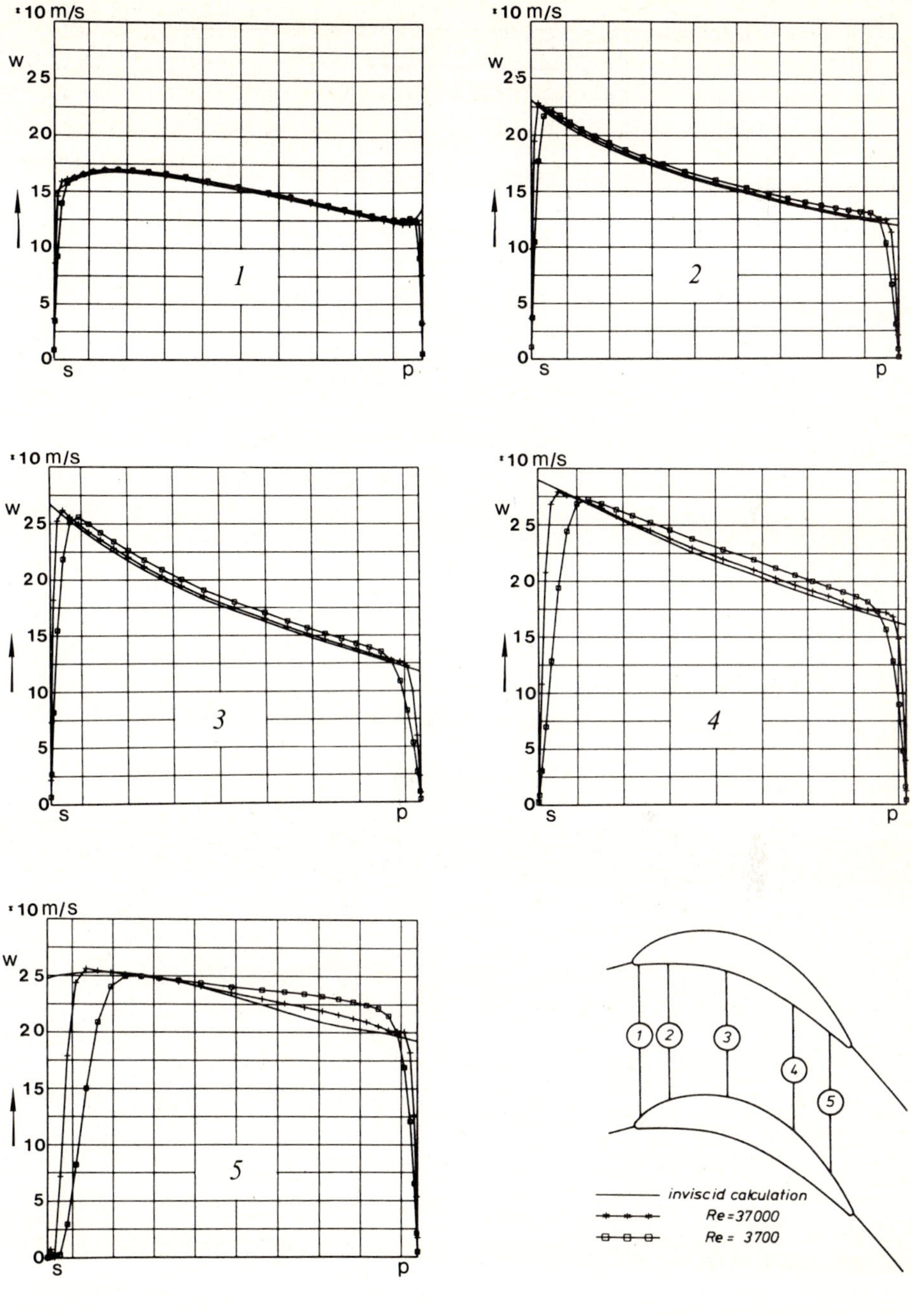

p= Pressure side

s= Suction side

Fig. 11: Velocity profiles at different locations

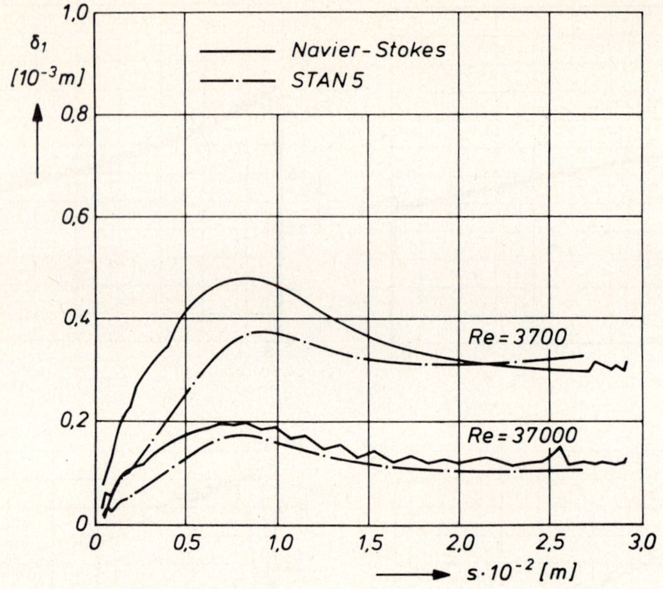

Fig. 12: Displacement thickness on pressure side

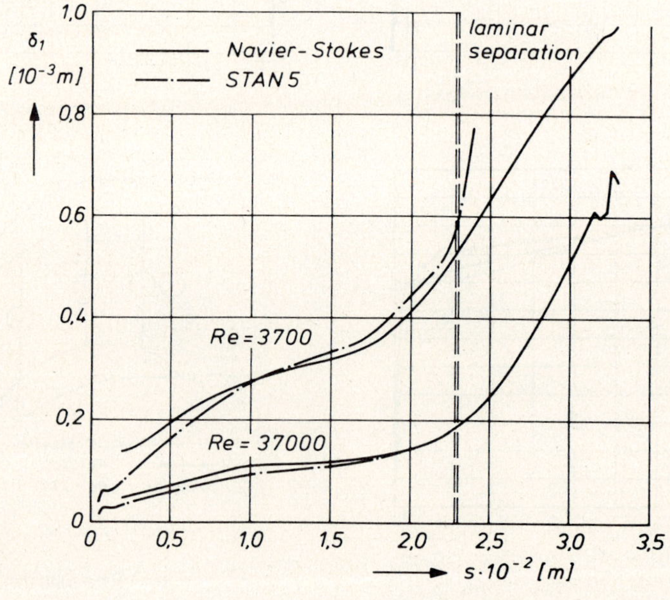

Fig. 13: Displacement thickness on suction side

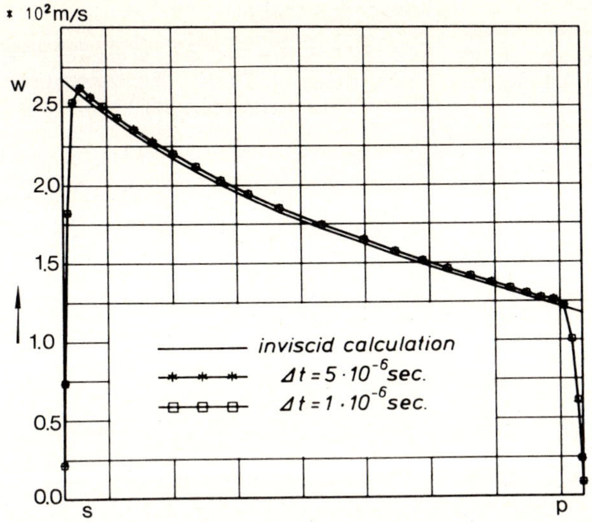

Fig. 14: Influence of time step size on velocity profiles.

Fig. 11 shows the velocity profiles in circumferential direction for both Reynolds numbers at different locations in comparison with the inviscid solution. Even for the high Reynolds number the boundary layer is well resolved with 3 to 4 mesh points. Due to the turning of the blade surface, the boundary layer thickness cannot be extracted directly from fig. 11.

In order to check the computed boundary layers the integral quantities displacement and momentum thickness were calculated from the Navier-Stokes solutions. These were compared with results from the boundary layer code STAN5 [8], based on the pressure field from the inviscid solution. Fig. 12 shows the comparison for the pressure side. On the suction side we have a separation region where boundary layer theory is not valid anymore - nevertheless, we calculated also a displacement thickness from the Navier Stokes solution in this region (fig. 13).

The discrepancies in the results are presumably caused by the coarse mesh and the determination of the integral quantities.

Several authors mentioned a dependency between the final result and the time step size when using a Lax-Wendroff formulation. A check on this problem was done rerunning the calculation for the high Reynolds number-case with a time step reduced by a factor of five. Figure 14 shows the velocity profiles from both calculations in comparison with the inviscid solution. There is no visible influence of the time step size on the resulting velocity profiles.

5. CONCLUSIONS

The Taylor-Galerkin-Method is shown to be a practical solution procedure for high Reynolds number flow. The results are encouraging for further studies with finite element formulations.

REFERENCES

[1] Brooks, A., and T. Hughes: Streamline Upwind/Petrov Galerkin Formulations for Convection Dominated Flows with Particular Emphasis on the Incompressible Navier Stokes Equations.
Comp. Method in Applied Mech. and Engrg., Vol. 32, 1982, pp. 199-259.

[2] Donea, J.: A Taylor Galerkin Method for Convective Transport Problems.
Internat. J. Numer. Meths. Eng. 20, 1984, pp. 101-120.

[3] Löhner, R., Morgan, K. and O.C. Zienkiewicz: The Solution of Nonlinear Hyperbolic Equation Systems by the Finite Element Method.
J. Num. Meth. Fluids, 1984.

[4] Löhner, R., Morgan, K. and O.C. Zienkiewicz: The Use of Domain Splitting with an Explicit Hyperbolic Solver.
Comp. Meths. Appl. Mech. Engrg., Vol. 45, 1984, pp. 313-319.

[5] Donea, J. and S. Giuliani: A Simple Method to Generate High Order Accurate Convection Operators for Explicit Schemes Based on Linear Finite Elements.
Internat. J. Numerical Methods in Engrg. 1, 1981, pp. 63-79.

[6] Khahil, J.M. and W. Tabakoff: Viscous Flow Analysis in Mixed Flow Rotors.
J. of Engineering for Power, Vol. 102, 1980, pp. 193-201.

[7] Lötzerich, M., und A. Vornberger: Die Berechnung der reibungsfreien dreidimensionalen Strömung in Turbomaschinen. Institut für Strahlantriebe und Turboarbeitsmaschinen der RWTH Aachen, Institutsbericht 05-85.

[8] Crawford, M.E. and W.M. Kays: STAN5 - A Program for Numerical Computation of two-dimensional Internal and External Boundary Layer Flows.
NASA CR 2742, 1976.

[9] Roache, J.P.: Computational Fluid Dynamics.
Hermosa Publishers Albuquerque, New Mexico.

[10] Lax, P.D. and B. Wendroff: Systems of Conservation Laws.
Communications on Pure and Applied Mechanics, Vol. 13, pp. 217-237.

[11] Zienkiewicz, O.C.: The Finite Element Method.
McGraw-Hill.

FINITE ELEMENT SCHEMES FOR AN IMPROVED COMPUTATION OF CONVECTIVE TRANSPORT IN FLUIDS

H.M. Leismann, B. Herrling
Institute of Hydromechanics, University of Karlsruhe
Kaiserstr. 12, D-7500 Karlsruhe, Germany

SUMMARY

The paper deals with the numerical solution of the differential equation for one-dimensional convective transport by the finite element method. It is demonstrated, that without inducing numerical diffusion at the same time, a remarkable reduction of oscillations can only be obtained by increasing the order of consistence. This is achieved using linear shape functions only. The corresponding schemes are derived from Taylor series expansions. An additional smoothing is obtained introducing an unsymmetrical scheme. The respective weighting functions are presented.

INTRODUCTION

The numerical computation of the differential equation describing the convective transport in fluids has been an active research topic for several years. The crux of the problem is that the numerical results show more or less strong oscillations which can render the computation useless. In recent years several proposals have been made to overcome this problem, but with doubtful success. In this paper, finite element schemes are presented with a higher order of consistence, which improves the results considerably. Only the one-dimensional differential equation is considered because of its better representation. However all results presented can be adapted for two-dimensional computations.

BASIC SOLUTION AND DEMONSTRATION OF THE PROBLEM

The one-dimensional convection equation is generally presented as

$$\frac{\partial u}{\partial t} + \frac{\partial (vu)}{\partial x} = q \qquad (1)$$

with u as the quantity transported (e.g. concentration of a pollution in a fluid), v the velocity, x and t the coordinates in space and time and q a sink and source term. For the analysis, equation (1) is considered with constant coefficients. This is admissible, as in many applications the change of the velocity is small from element to element. Further, the sink and source term is omitted causing no problems in the numerical computation. In this way equation (1) changes to

$$\frac{\partial u}{\partial t} + v \frac{\partial u}{\partial x} = 0 \qquad \text{or } u,_t + vu,_x = 0 \tag{2}$$

with $,_t$ and $,_x$ denoting partial differentiations with respect to time and space. At the influx boundary Dirichlet-conditions are prescribed.

Solving equation (2) the finite element method (FEM) is used. Applying the method of weighted residuals, equation (2) results in

$$\int_L W\, u,_t\, dx + v \int_L W\, u,_x\, dx = 0 \tag{3}$$

where W is a weighting function and L is the length of the one-dimensional domain. In the following, only linear shape functions are used for the unknown u. These can be handled simply and give sufficiently accurate results when the solution scheme is skilfully chosen. Thereby equation (4) follows (written for only one element of length l):

$$\int_l W_F\, N_E (u,_t)_E\, dx + v \int_l W_F\, N_{E,x}\, u_E\, dx = 0 \tag{4}$$

where F,E = 1,2 are indices for the nodes and N_E are linear shape functions. The time derivative is realised using a simple differential quotient between two time levels as usual:

$$u,_t = \frac{u^1 - u^0}{\Delta t} \quad . \tag{5}$$

When the weighting function W_F is equal to the shape function, the Bubnov-Galerkin method or short the Galerkin method is applied, otherwise the Petrov-Galerkin method is used.

Equation (4) describes the integral across one element. For the integration across the entire domain L the integrals of each element have to be summed, see [1]. This results in a set of linear equations which can be solved in the usual manner. Each equation of the set is attached to one node. This is comparable to an equation derived from the finite difference method (FDM) for a central and neighboring nodes. In case of the Galerkin method, equation (6) attached to one node results from equations (4) and (5):

$$\frac{1}{k}\left\{\left(\frac{1}{6}u^1_{-1} + \frac{2}{3}u^1_0 + \frac{1}{6}u^1_1\right) - \left(\frac{1}{6}u^0_{-1} + \frac{2}{3}u^0_0 + \frac{1}{6}u^0_1\right)\right\}$$

$$+ \Theta \frac{v}{2h}\left(u^1_1 - u^1_{-1}\right) + (1-\Theta)\frac{v}{2h}\left(u^0_1 - u^0_{-1}\right) = 0 \quad . \tag{6}$$

The notation are illustrated in figure 1, upper indices mark the time, lower indices the position. In equation (6), Δx is denoted by h and Δt by k. Θ is a parameter describing the time weighting of terms without temporal derivatives. For $\Theta = 0.5$ the Crank-Nicolson method is presented.

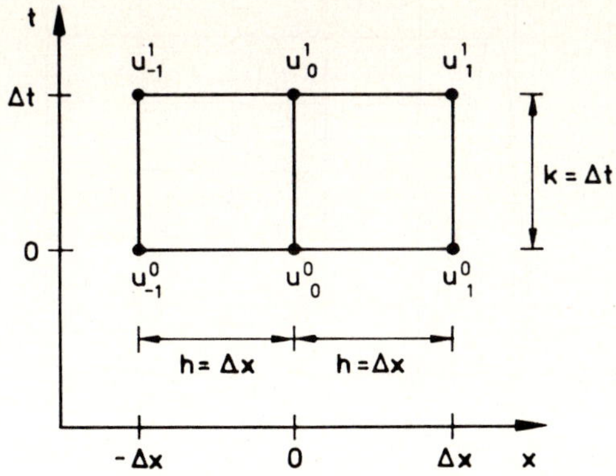

Fig. 1 Notations in a one-dimensional patch

A comparison of equation (6) with the conventional central finite difference approximation (equation (7)) demonstrates that no fundamental difference exists between the FEM and FDM. In both cases there are differential quotients. In equation (6) the spatial averaging of the term $u,_t$ results from the Galerkin weighting function. In case of the FDM (equation (7)), only the differential quotient of the middle node is used in general:

$$\frac{1}{k}\,(u^1_0 - u^0_0) + \theta\,\frac{v}{2h}(u^1_1 - u^1_{-1}) + (1-\theta)\,\frac{v}{2h}\,(u^0_1 - u^0_{-1}) = 0\,. \qquad (7)$$

In the following, a given test situation will be used to compare the different numerical schemes: A one-dimensional domain of length 10 m is discretized into 100 elements, each 0.1 m long. At the left part of the model a sinusoidal peak is prescribed as the initial condition which is transported by the flow moving with a velocity of v = 0.001 m/s. the boundary condition at the left boundary is specified by $u \equiv 0.0$. For lack of space results are presented only for time steps of 80 s plotted at different times. This results in a Courant-number of $Cr = vk/h = 0.8$. The dotted lines represent the exact solution.

Figure 2 demonstrates results computed with equation (6) and figure 3 those of equation (7), each by application of the Crank-Nicolson scheme ($\theta = 0.5$). Both cases demonstrate considerable oscillations, the FEM applying the Galerkin method is better than the FDM.

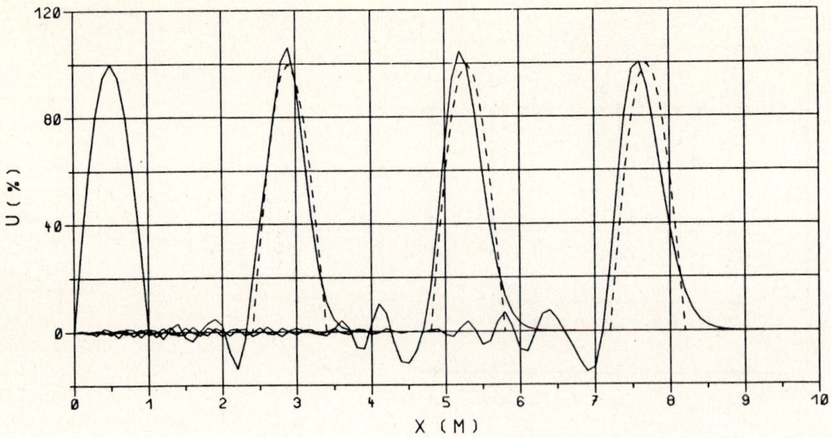

Fig. 2 Galerkin scheme, $\Theta = 0.5$ and Cr = 0.8

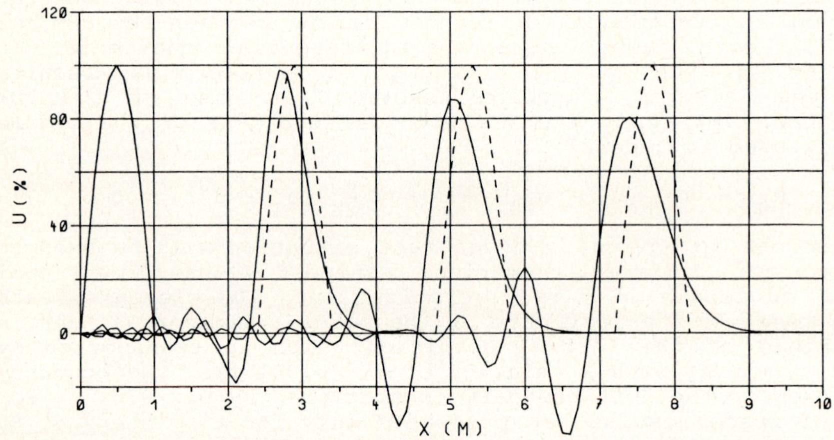

Fig. 3 Central finite difference scheme, $\Theta = 0.5$ and Cr =0.8

DEVELOPMENT OF IMPROVED NUMERICAL SCHEMES

To investigate the reasons for the poor results in figures 2 and 3, the truncation errors of the respective schemes are determined. This is obtained by expanding equations (6) and (7) in Taylor series and replacing time derivatives by space derivatives using equation (2). The Galerkin scheme in combination with Crank-Nicolson results in

$$u,_t + vu,_x + (v^3 k^2/12)u,_{xxx} + 0(h^3,k^3) = 0 \quad \text{or}$$
$$u,_t + vu,_x + 0(h^3,k^2) = 0. \qquad (8)$$

The central finite difference scheme combined with Crank-Nicolson yields

$$u,_t + vu,_x + (vh^2/6 + v^3k^2/12)u,_{xxx} + 0(h^3,k^3) = 0 \quad \text{or}$$
$$u,_t + vu,_x + 0(h^2,k^2) = 0 \quad . \tag{9}$$

Thus the Galerkin scheme presents the higher order of spatial consistence. The only difference between equations (6) and (7) lies in the spatial averaging of the storage term (time derivative term). The coefficients in the FEM depend on the weighting function. So far equation (6) represents a special case. The application of other weighting functions (Petrov-Galerkin method) results in different coefficients. The advantage of the Galerkin scheme against the FDM consists of the elemination of the second order error due to the space discretization. According to this, it seems promising to eliminate the error of second order due to the time discretization choosing suitable weighting functions. The weighting function itself is only of theoretical importance. In fact, only the coefficients of the element matrix are required, resulting from the integration over one element. For illustration, the weighting functions will, however, be presented.

In the first instance, only the Crank-Nicolson procedure is considered. The respective Petrov-Galerkin scheme runs in generalized form using a symmetrical weighting function as follows:

$$\frac{1}{k}\{(a\,u^1_{-1} + b\,u^1_0 + a\,u^1_1) - (a\,u^0_{-1} + b\,u^0_0 + a\,u^0_1)\}$$
$$+ \frac{v}{4h}(u^1_1 - u^1_{-1}) + \frac{v}{4h}(u^0_1 - u^0_{-1}) = 0 \quad . \tag{10}$$

a and b are coefficients having to be fixed. The Taylor series expansion yields the effectively solved differential equation:

$$(2a + b)u,_t + vu,_x + \{vh^2(1/6-a) + v^3k^2/12\}u,_{xxx}$$
$$+ 0(h^3,k^3) = 0 \quad . \tag{11}$$

A complete scheme of third order of consistence is present when satisfying the following equation:

$$2a + b = 1 \tag{12}$$

$$vh^2(1/6 - a) + v^3k^2/12 = 0 \quad . \tag{13}$$

These result in

$$a = 1/6 + Cr^2/12 \qquad \text{and} \tag{14}$$

$$b = 2/3 - Cr^2/6 \quad . \tag{15}$$

From the conditions (14) and (15) the coefficients of a cubic polynomial as weighting function, dependent on the Courant number, can be determined. Such a weighting function for Cr =

0.8, and the Galerkin weighting function for comparison, are given in figure 4.

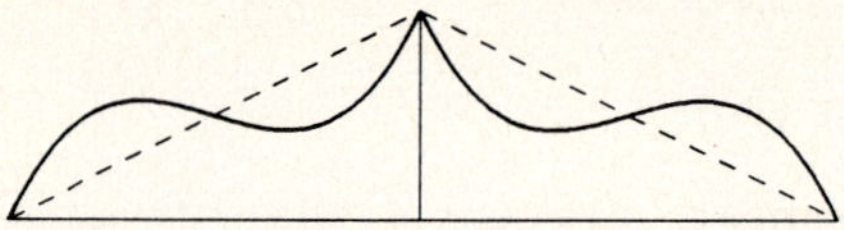

Fig. 4 Cubic weighting function for Cr = 0.8 and Galerkin weighting function; scheme of third order and Θ = 0.5

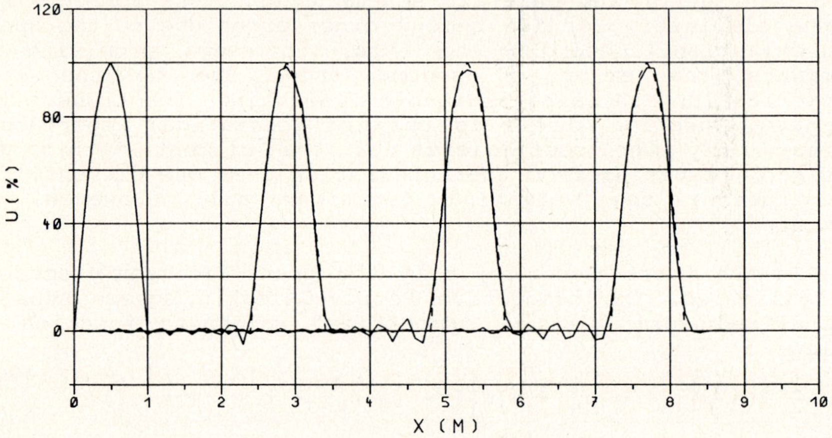

Fig. 5 Petrov-Galerkin scheme of third order, Θ = 0.5 and Cr = 0.8

Figure 5 represents the numerical results for the same test situation as before. Compared with the preceeding findings, the amplitudes of the oscillations are distinctly lower and the phase accuracies are considerably improved.

The upwind-technique is an often used method for the computation of convective propagation problems. The scheme deviates from the central differential quotient for the approximation of the convective term in that a backward difference is used (e.g. [2]). The same effects are obtained in the FEM by [3] and [4] by applying unsymmetrical weighting functions. In generalized form the Petrov-Galerkin scheme with an unsymmetrical weighting and the Crank-Nicolson procedure is presented in the following for flow in the positiv direction of x:

$$\begin{aligned}&\frac{1}{k}\{(a\,u^1_{-1} + b\,u^1_0 + c\,u^1_1) - (a\,u^0_{-1} + b\,u^0_0 + c\,u^0_1)\}\\&+ \frac{v\gamma}{h}\{\frac{1}{2}(u^1_0 + u^0_0) - \frac{1}{2}(u^1_{-1} + u^0_{-1})\}\\&+ \frac{v(1-\gamma)}{h}\{\frac{1}{2}(u^1_1 + u^0_1) - \frac{1}{2}(u^1_0 + u^0_0)\} = 0 \quad . \qquad (16)\end{aligned}$$

In this case the coefficients a, b and c have to be fixed. γ is the upwind-parameter ($0.5 \leq \gamma \leq 1$) whereby $\gamma = 1$ denotes complete upwinding and $\gamma = 0.5$ no upwinding.

The Taylor series expansion yields the effectively solved differential equation

$$(a+b+c)u,_t + vu,_x + \{vh(1/2-\gamma) + vh(a-c)\}u,_{xx}$$

$$+ \{vh^2[1/6 - (a+c)/2] + v^3k^2/12\} u,_{xxx} + 0(h^3,k^3) = 0 \ . \quad (17)$$

From equation (17) there follows of course

$$a + b + c = 1 \ . \quad (18)$$

Further, it is evident that the Galerkin method with a = c = 1/6 or the FDM with a = c = 0, each combined with upwinding, merely result in a scheme of first order of consistence. Accordingly, the upwind technique in combination with the FDM yields exaggerated numerical diffusion with possible solutions being of no use.

A scheme of second order of consistence arised if

$$a - c = \gamma - 1/2 \ . \quad (19)$$

A scheme of third order is achieved with the additional condition

$$a + c = 1/3 + Cr^2/6 \ . \quad (20)$$

Using equation (18) to (20), the coefficients a, b and c are finally fixed by the definition equations (21)

$$a = -1/12 + \gamma/2 + Cr^2/12 \ ,$$

$$b = 2/3 - Cr^2/6 \qquad \text{and} \quad (21)$$

$$c = 5/12 - \gamma/2 + Cr^2/12 \ .$$

The upwind parameter γ can, in principle, be chosen independently, but for reasons of stability the condition

$$\gamma \geq 0.5 \quad (22)$$

must be observed.

The corresponding weighting function, being determined using the above mentioned method, is demonstrated in figure 6 for the case of Cr = 0.8 and γ = 1.0. The corresponding numerical results of the above test study are shown in figure 7 using the Petrov-Galerkin scheme of equation (16) and the coefficients of equation (21). The finding is better than that computed without upwinding, as the oscillations vanish exept for the zone in the immediate neighbourhood of the peak.

In principle, it is conceivable to use the upwind parameter γ to improve the order of consistence. However, an analysis indicates that a scheme of fourth order of consistence exists for the case γ = 0.5, without upwinding. The result is shown in figure 5, which indicates that the fourth-order scheme obviously has poorer properties.

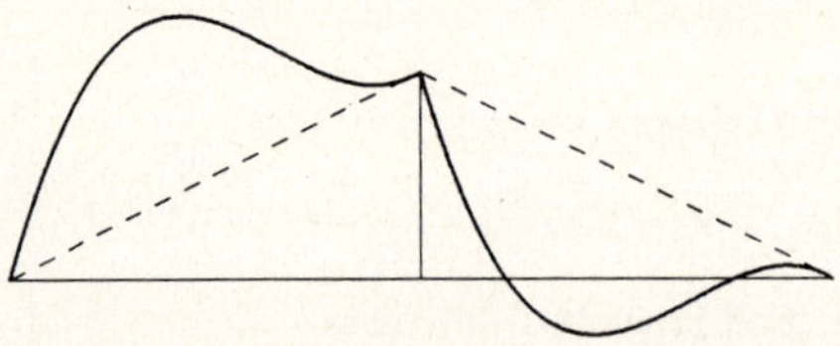

Fig. 6 Cubic Petrov-Galerkin weighting function for Cr = 0.8 and γ = 1.0 and Galerkin weighting function; scheme of third order and Θ = 0.5

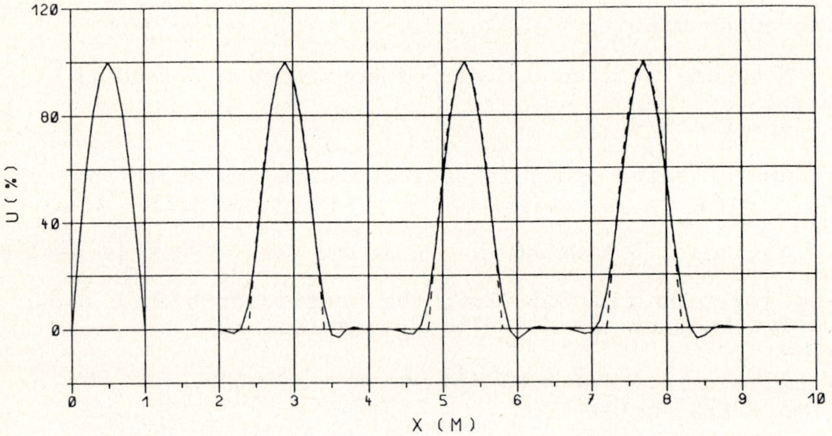

Fig. 7 Petrov-Galerkin scheme of third order, Cr = 0.8, γ = 1.0 and Θ = 0.5

At this stage it should be mentioned that in [4], the weighting function according to equation (21) with Cr = 0 is used; that is, only a scheme of second order of consistence is obtained. The respective result of the test study is demonstrated in figure 8. The solution is indeed smoother than that of the Galerkin scheme (figure 2) but the remarkable effect of the scheme of third order of consistence is not obtained by the method presented in [4].

An additional smoothing was obtained simply by using an unsymmetrical scheme. In [4] it is demonstrated that, for the two-dimensional case, the upwinding has to take place in the direction of the streamlines. This results in additional matrix operations. In case, an unsymmetry is not used in the spatial but in the time direction, supplementary calculations are not necessary to regard the flow directions. In the following, such a scheme without upwinding (γ = 0.5) is studied.

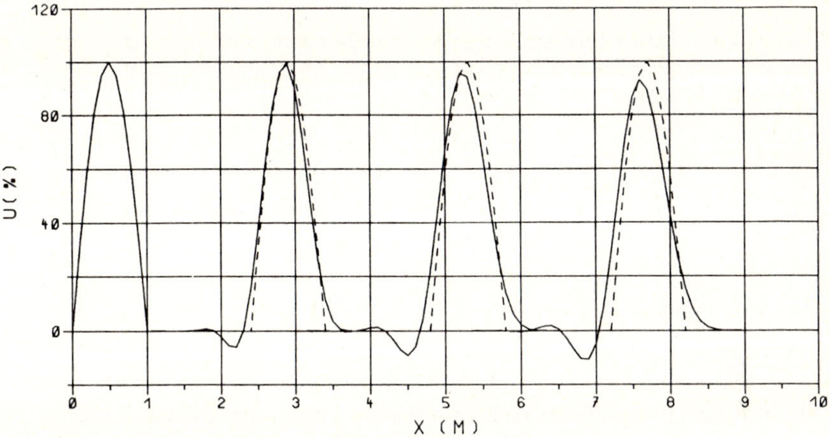

Fig. 8 Petrov-Galerkin scheme of second order according to [4], Cr = 0.8, γ = 1.0 and Θ = 0.5

Schemes of second or third order of consistence are only stable in the range of $\Theta \leq 0.5$. Here the special case $\Theta = 0$ shall be regarded as this scheme produces advantages to be described later. Corresponding to the unsymmetry between the time levels an unsymmetrical weighting in the time direction is permitted. The Petrov-Galerkin scheme runs in generalized form

$$\frac{1}{k}\{(a\,u^1_{-1} + b\,u^1_0 + a\,u^1_1) - (c\,u^0_{-1} + d\,u^0_0 + c\,u^0_1)\}$$
$$+ \frac{v}{2h}(u^0_1 - u^0_{-1}) = 0 \qquad (23)$$

with

$$2a + b = 1 \qquad \text{and} \qquad (24)$$

$$2c + d = 1 \quad . \qquad (25)$$

Here a, b, c and d are the coefficients having to be determined. The Taylor series expansion yields the effectively solved differential equation

$$u,_t + vu,_x + \{(a-c)h^2/k + v^2k/2\}u,_{xx}$$
$$+ \{vh^2(1/6-a) - v^3k^2/6\}\,u,_{xxx} + O(h^3,k^3) = 0 \quad . \qquad (26)$$

To obtain a scheme of third order of consistence the following conditions must hold:

$$a = 1/6 - Cr^2/6 \qquad \text{and} \qquad (27)$$

$$a - c = - Cr^2/2 \quad . \qquad (28)$$

This results in the coefficients

$$a = 1/6 - Cr^2/6 , \qquad b = 2/3 + Cr^2/3 ,$$
$$c = 1/6 + Cr^2/3 \quad \text{and} \quad d = 2/3 - 2Cr^2/3 . \tag{29}$$

The weighting function now being different for the old and the new time level is presented together with the Galerkin weighting function, for comparison, in figure 9.

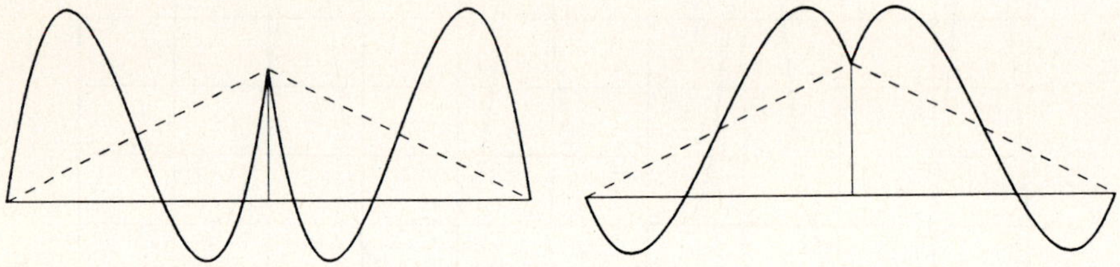

Fig. 9 Cubic Petrov-Galerkin weighting functions, different in each time level (left: old, right: new time level) for Cr = 0.8; and Galerkin weighting functions.Scheme of third order and $\theta = 0.0$

It can be easily ascertained that the resulting coefficient matrix of the set of equations for the whole domain is symmetric and positive definite when using the coefficients of equation (29). The reason for this is that the convective term is only estimated at the old time level. In addition, the elements on the principal diagonal of this matrix are dominant, allowing the use of iterative procedures to solve the equation set. The numerical results of the test study corresponding to equations (23) and (29) are shown in figure 10. The results are practically identical to those obtained using upwinding (figure 7).

A reason for the smoothing effect of an unsymmetrical scheme is yielded by the spectral analysis (see e.g. [5]), in which the transport of separate Fourier components can be investigated with respect to damping and phase error. Figure 11 shows the relative phase error and the damping after one time step as functions of the wave length for the symmetrical scheme (Crank-Nicolson, no upwinding) of third order of consistence; figure 12 for the unsymmetrical scheme with full upwinding ($\gamma = 1.0$); and figure 13 for the scheme which is unsymmetrical in time ($\theta = 0.0$). All schemes propagate the high frequencies with a phase error causing the oscillations. While the symmetrical scheme transports all frequencies undamped, the unsymmetrical schemes damp the high frequencies with their phase errors, and thus have a smoothing effect.

As described above the setting-up of the weighting functions is of theoretical meaning only. From the practical point of view, only the weighting coefficients are of interest. A

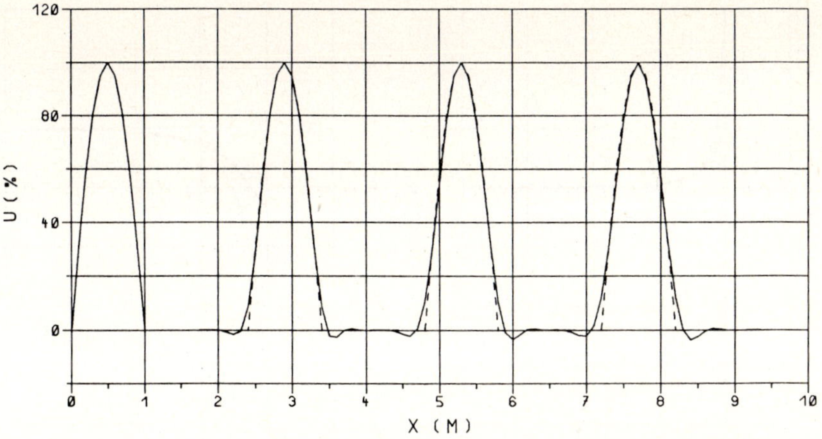

Fig. 10 Petrov-Galerkin scheme of third order, Cr = 0.8 and $\Theta = 0.0$.

simple way to determine these is as follows: After choosing any weighting function, a respective nodal equation is set up and is expanded in a Taylor series. The truncation errors are converted into a configuration suitable for a numerical approximation and are employed with opposite sign in the finite element formulation. The schemes thus obtained are identical to those shown above.

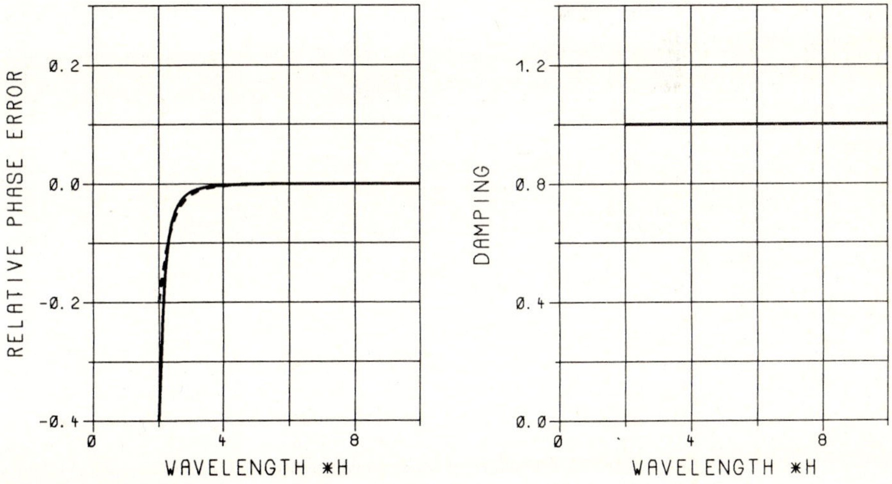

Fig. 11 Relative phase error and damping for the symmetrical scheme of third order (equations (10), (14) and (15)), solid line: Cr = 0.8, dashed line: Cr = 0.4

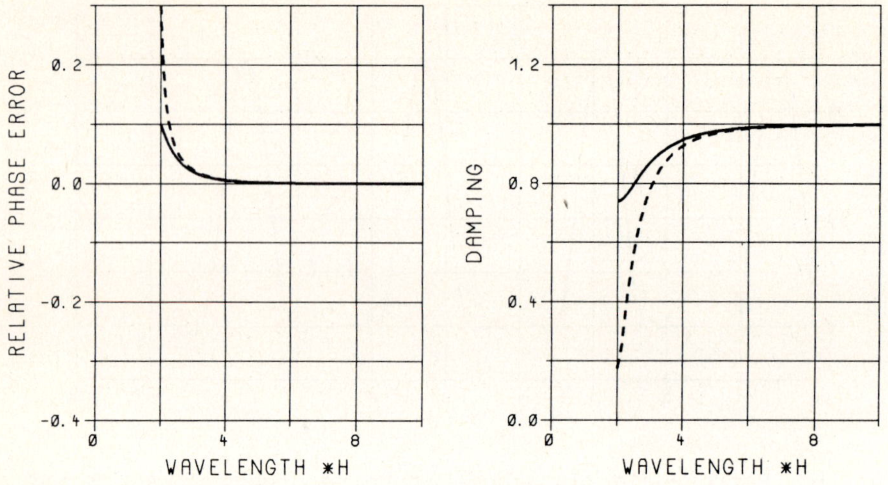

Fig. 12 Relative phase error and damping for the unsymmetrical scheme of third order with upwinding $\gamma = 1.0$ (equations (16) and (21)), solid line: Cr = 0.8, dashed line: Cr = 0.4

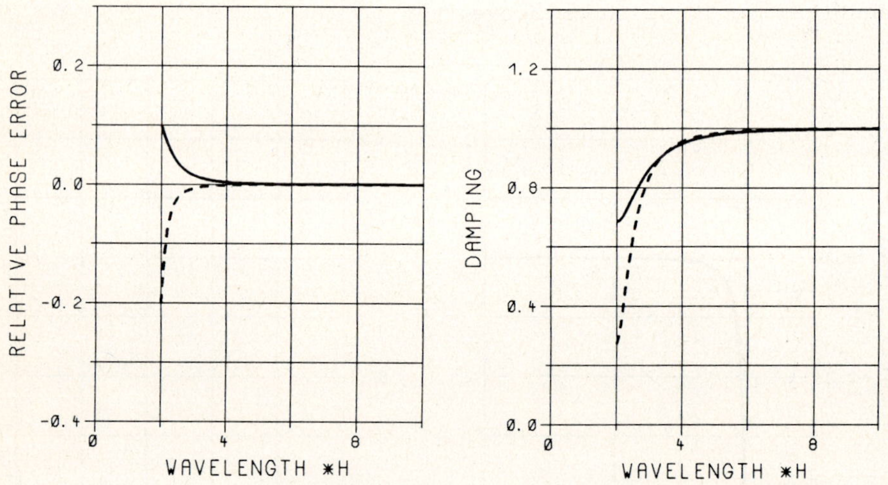

Fig. 13 Relative phase error and damping for the in time unsymmetrical scheme of third order with $\Theta = 0.0$ (equations (23) and (29)), solid line: Cr = 0.8, dashed line: Cr = 0.4

This method is employed in [6] but only a semi-discretization in time and the usual Galerkin weighting function are used. A reason for the latter is not given in [6]. As it can be easily demonstrated, this method fails when other weighting functions are used.

Finally it should be mentioned that all schemes of third order of consistence are only stable for

$$Cr = vk/h \leq 1 \quad . \tag{30}$$

Further, they yield exact solutions in the case of Cr = 1, with uniform discretization and constant coefficients. The direct use of the developed schemes for irregular discretizations and variable coefficients presents no problems; the schemes are well-conditioned.

CONCLUSION

The paper presents finite element schemes based on the Petrov-Galerkin weighting for the computation of convective transport in fluids. The schemes achieve third order consistence and yield results with minimal oscillations and minimal numerical diffusion. An additional smoothing, not effecting the third order of consistence, is attained by introducing an unsymmetry, either by upwinding or by deviation from the Crank-Nicolson time weighting. The latter is finally prefered: firstly, the matrix operations of the upwind method to regard the flow direction are unnecessary for two-dimensional flow; and secondly for the special case of $\Theta = 0.0$ there results a symmetrical coefficient matrix for the equation set, with dominant coefficients on the principal diagonal being suitable for iterative procedures.

ACKNOWLEDGEMENTS

The investigation was supported by the German Research Community (DFG), grant no. Th 159/14. The authors express their appreciation for their subsidy. Further they wish to thank Dr. S. Gärtner, Fluid Mechanics Institute, University of Hannover, for his valuable advice during several discussions.

REFERENCES

[1] ZIENKIEWICZ, O.C.: "The Finite Element Method", 3rd ed., McGraw-Hill, London, 1977.

[2] SPALDING, D.B.: "A Novel Finite Difference Formulation for Differential Expressions Involving Both First and Second Derivatives", Int. J. for Num. Meth. in Eng., Vol. 4 (1972), pp. 551 - 559.

[3] HEINRICH, J.C., HUYAKORN, P.S., ZIENKIEWICZ, O.C.: "An 'Upwind' Finite Element Scheme for Twodimensional Convective Transport Equation", Int. J. for Num. Meth. in Eng., Vol. 11 (1977), pp. 131 - 143.

[4] BROOKS, A.N., HUGHES, T.J.R.: "Streamline Upwind/Petrov-Galerkin Formulations for Convection Dominated Flows with Particular Emphasis on the Incompressible Navier- Stokes Equations", Comp. Meth. in Applied Mech. and Eng., 32 (1982), pp. 199 - 259.

[5] LAPIDUS, L., PINDER, G.F.: "Numerical Solution of Partial Differential Equations in Science and Engineering", John Wiley and Sons, New York, 1982.

[6] DONEA, J.: "A Taylor-Galerkin Method for Convective Transport Problems", Int. J. for Num. Meth. in Eng., Vol. 20 (1984), pp. 101 - 119.

COMPARING FINITE DIFFERENCES WITH A PARTICLE-IN-CELL METHOD ON SHOCKED UNSTEADY FLOW PAST A RECTANGLE

Thomas Mietzner

Fachbereich Mathematik, Universität Kaiserslautern

Erwin-Schrödinger-Str., D-6750 Kaiserslautern, Germany

INTRODUCTION

This paper presents numerical results obtained with GAP (grid and particles) by B. M. Marder [1] and multidimensional FCT (flux corrected transport) by S. T. Zalesak [2, 3]. They are compared to experimental data of H.-U. Hassenpflug [4]. He investigated a flow with a plane shock wave running through a square channel and interacting with a cube that lies on the bottom of the channel filling all its width. The incident shock wave is reflected and diffracted, and a region of very small density is produced right above the sharp front edge of the cube. Similar things happen at the back side of the cube when the incident shock wave passes it. In particular, Hassenpflug found a strong vortex close to the (upper) downstream edge. He suggests that this vortex is produced by shock wave diffraction, not by boundary layer effects. For that reason, we model the flow by Euler's equations

$$\frac{\partial \rho}{\partial t} + \nabla \cdot (\rho v) = 0 \quad , \tag{1a}$$

$$\frac{Dv}{Dt} = -\frac{1}{\rho} \nabla p \quad , \tag{1b}$$

$$\frac{De}{Dt} = -\frac{p}{\rho} \nabla \cdot v \quad , \tag{1c}$$

where ρ denotes density, v velocity, p pressure and e specific internal energy. $\frac{D}{Dt}$ is an abbreviation for the substantial derivative $\frac{\partial}{\partial t} + v \cdot \nabla$. Following Hassenpflug, we suppose that the state of the gas only depends on the length and height coordinates (x and y) of the channel. The equation of state

$$p = (\gamma - 1) e \rho \tag{1d}$$

of an ideal gas with $\gamma = 1.4$ for air closes the system. Accordingly, we demand impenetrable but free slip walls.

NONREFLECTING BOUNDARY CONDITIONS

To obtain a well posed problem, we need conditions at the

$l_j(u,\omega) = l_j^1(u) + l_j^2(u)\omega_2 + O(|\omega-\bar{\omega}|^2)$, we deduce that

$$b_j^o = l_j^1, \quad b_j^2 = -l_j^2 \cdot A_1 \tag{4}$$

is a possible choice. Different ones can easily be derived from (3) along the same lines. If system (2a) is linear, then, of course, our boundary conditions correspond to the second order approximation of Engquist & Majda's absorbing boundary conditions.

In our computations, we also take into account the natural boundary conditions

$$l_j^1(u)\,\frac{\partial u}{\partial t} = -l_j^1(u)\,(A_1(u)\,\frac{\partial u}{\partial x} + A_2(u)\,\frac{\partial u}{\partial y}) \quad \text{if } \lambda_j < 0.$$

GRID AND PARTICLES IN TWO SPACE DIMENSIONS

For this numerical method, the Euler equations (1) are cast in Lagrangian coordinates, yielding

$$\dot{X} = v \tag{5}$$

$$\dot{v} = -u_s \nabla p \tag{6}$$

$$\dot{e} = -p\dot{u}_s \tag{7}$$

where $X = (x,y)$ denotes position, v velocity, u_s specific volume, p pressure and e specific internal energy. The method uses particles that have

a mass	M_n
a position	$X_n = (x_n, y_n)$
a velocity	$V_n = (v_{xn}, v_{yn})$
a volume	U_n
an internal energy	I_n .

The grid they flow through is necessary to compute

the density	ρ_{jk}
the grid velocity	V_{jk}
the grid internal energy	I_{jk}
the pressure	p_{jk} .

This grid consists of rectangular cells C_{jk} with center X_{jk} and side lengths Δx and Δy. In addition, the particles are given a shape function inflow and outflow boundary of our simulation region. To this end, we derived multidimensional absorbing boundary conditions in the spirit of Engquist & Majda [5] and Hedstrom [6]. To be specific, consider the hyperbolic system

$$u_t + A_1(u)u_x + A_2(u)u_y = 0 \tag{2a}$$

for the unknown vector $u(t,x,y)$. Suppose that the matrices

$$A(u,\omega) = A_1(u)\omega_1 + A_2(u)\omega_2$$

can be diagonalized for all $\omega = (\omega_1,\omega_2)^T$ and all u in a suitable open set G. Suppose further that we are given initial values

$$u(0,x,y) = u_o(x,y) \quad , \tag{2b}$$

u_o being constant for $x < 0$. We try to obtain the solution of the problem (2) for $x > 0$ by introducing an artificial boundary at $x = 0$ and giving nonreflecting boundary conditions on it. For this purpose, we look for boundary conditions that suppress simple plane waves coming in from the artificial boundary. If we denote the typical eigenvalue and corresponding left and right eigenvector of $A(u,\omega)$ by λ_j, l_j and v_j, then a simple plane wave solves (2a) and satisfies

$$\frac{\partial u}{\partial x} = \alpha \cdot \omega_1 \cdot r_k(u,\omega) \quad ,$$

$$\frac{\partial u}{\partial y} = \alpha \cdot \omega_2 \cdot r_k(u,\omega) \quad ,$$

$$\frac{\partial u}{\partial t} = -\alpha \cdot \lambda_k(u,\omega) \cdot r_k(u,\omega)$$

for some index k, unit vector ω and scalar function α.
We consider first order differential operators

$$B_j = b_j^o(u)\frac{\partial}{\partial t} + b_j^2(u)\frac{\partial}{\partial y}$$

and try to achieve

$$B_j u = -\lambda_k(u,\omega) \cdot \alpha \cdot \delta_{jk} + O(|\omega-\bar{\omega}|^2) \tag{3}$$

for all such simple plane waves, where $\bar{\omega} = (1,0)^T$. Then, demanding

$$B_j u = 0 \qquad \text{if} \qquad \lambda_j > 0$$

produces simple plane wave reflections of the order $O(|\omega-\bar{\omega}|^2)$ and a well posed problem. We may suppose $l_j \cdot r_k = \delta_{jk}$. Expanding l_j, λ_k and r_k into powers of $\omega-\bar{\omega}$ and setting

Here, h abbreviates $c^a\Delta t$; c^a and ρ^a are evaluated at $X_n = (x,y)$. As only grid quantities occur on the right side of (6^+), there is no need for velocity smoothing any more. Accordingly, we used $F_n \equiv 1$ for smoothing the internal energy.

To treat the two-dimensional problems, one has to determine initial points for the particles. We generated an approximation to the uniform distribution using Hammersley sequences with the bases 2, 3 or 5 or Fibonacci sequences (see [7]), and transformed these distributions to the initial density. The results of the flow simulation depended only weakly on the sequence chosen. However, for this purpose, these sequences are known to be superior to random distributions.

FLUX CORRECTED TRANSPORT

This method is described in [2]. We used Zalesak's leapfrog-trapezoidal scheme as higher order method and applied ZIP-differencing as proposed in [3]. In addition, we employed the fourth order artificial viscosity that Kreiss & Oliger [8] used with the leapfrog part of this scheme. Rusanov's scheme [9] served as low order method. We used Zalesak's scheme up to the boundary without modification, defining flow variables outside the computational domain by symmetric extrapolation. In both simulation methods, we treated the grid points on the edges of the cube like interior points. No other point on its surface knows of these edges.

VECTORIZATION

It was a straightforward job to vectorize flux corrected transport on the CYBER 205. WHERE, the vector-IF, is the key ingredient of CYBER-FORTRAN for that task. Grid and particles are more complicated as they require frequent conversion of particle and grid quantities into each other. A data mix using the vector routine Q8VGATHR suffices to compute particle from grid quantities. To do the other direction, we first number the particles (with the lower left corner) in a cell consecutively. This step was not vectorized, but has to be done only once per timestep. In addition, we use an array divided into equal parts of grid structure. Into this array we rearrange the particle masses, e.g., in such a way that the mass of the j-th particle in a cell is assigned to the corresponding storage location of the j-th part of the array. This step was vectorized using the mixing routine Q8VSCATR. From here, the cell masses are calculated simultaneously for all cells. This technique reduced the CPU-time to one third compared to a program that computed cell from particle quantities sequentially, but everything else in parallel.

COMPARISON OF NUMERICS AND EXPERIMENTS

Let's first state some parameters. Hassenpflug's channel is 140 mm high. We got along with a fairly small part of its length, 280 mm. The cube in it is 60 mm high, with its front 110 mm down stream of our inflow boundary. The shock starts at

80 mm, hitting the cube at $t = 0$. Hassenpflug used air at room conditions in front of the shock, which we translated into a density of 1205 g/m^3 and a pressure of 10^5 Pa.

With GAP, we used 54000 particles and 84×42 square cells. Our timestep was calculated from

$$\frac{\Delta t}{\Delta x} (|v| + c)_{max} = q/\sqrt{2} \; , \tag{10}$$

using $q = 0.6$ with equation (6^+). The original scheme, with eq. (6*), required half the timestep with slightly worse results and almost twice as much CPU-time. Therefore, we will only show results of the modified scheme.

The FCT-results were produced using 98×49 square cells and the same value of q in (10).

Fig. 1 shows FCT-densities (in g/m^3) for a shock wave with Mach number 1.12 and a cube thickness of 30 mm. The corresponding GAP-results are displayed in fig. 2. Experimental pictures for this setup can be found in [4], fig. 13III. Observe the sharp shocks of the FCT-calculation, showing no overshoots behind them. The GAP-shocks are much thicker. The theoretical density of 1733 g/m^3 behind the reflected shock is reproduced well by both schemes. Furthermore, both show spots of very low density at the front edge of the cube. However, next to it, FCT places a strong unphysical density maximum. Both phenomena are induced by the "sharp edge" boundary condition and are greatly influenced by the method of extrapolation. A closer look at the edge singularity will be necessary to get reasonable results in this region. In contrast, the sharp front edge is no problem for GAP. Its local density minimum is comparable to, but somewhat higher than the experimental value. Hassenpflug's results show the minimum at some distance to the cube, and moving, in contrast to the GAP prediction. This holds at the rear edge, too. Here, FCT gives higher density minima than GAP, but at $t = 611$ µs it is detached from the cube and compares very well to Hassenpflug's experimental picture.

Especially at 311 µs, the FCT-plot displays unphysical density oscillations behind the cube. A comparable phenomenon occurs in the GAP-results at 610 µs.

Have a look now at the velocity plots in fig. 3. In fact, they show some multiple of $\frac{v}{|v|} \cdot \log(|v|/3+1)$, thus magnifying small velocities compared to large ones. FCT gives a pronounced vortex, GAP just predicts a little bit of back flow behind the cube. The pertinent recirculation velocities are about 100 $\frac{m}{s}$ in the experiment (for a cube of 12 mm thickness), 40 $\frac{m}{s}$ calculated by FCT and 1.3 $\frac{m}{s}$ in the GAP simulation.

Finally, we should point out that the GAP computation took about 4.5 times longer than the FCT one, comparing CPU-time.

SUMMARY

We improved "grid and particles" (GAP) and compared it to "flux corrected transport" (FCT), an elaborate finite difference scheme for the solution of hyperbolic systems of conservation laws, implementing both on a vector computer. Experiments on vortex generation by shock wave diffraction served as a reference. To keep the simulated part of the flow as small as possible, we derived nonlinear absorbing boundary conditions of second order in two space dimensions. They were used to treat the inflow and outflow boundary of our problem.

The results show that FCT is much faster than GAP, gives steeper shocks, and predicts much stronger, realistic vortices. Problems arise at the front edge of the obstacle in FCT, but not in GAP. Excluding the shocks, neither FCT nor GAP produce as smooth density contours as the experiment. The FCT results confirm that the vortices observed by Hassenpflug essentially are produced by shock wave diffraction.

It should be said that the particle-in-cell methods originally were designed to give a rough idea of the events in supersonic flows. Detailed comparisons with experiment were rarely published. And for our problem with a free stream Mach number of about 1/4, these methods were said to be inappropriate. Bearing this in mind, GAP turns out to be better than expected. Comparing the amount of work that has been invested in particle and in difference methods, we believe that a lot can be done to improve the former ones. In particular, they should not only take into account the v-characteristic, which is their great advantage in supersonic flows, but also the whole set $\{v + c\cdot\omega: |\omega| = 1\}$ of characteristic directions. Moreover, smoothing the density will be more than a beauty aid.

Finally, let us add that a detailed report of this work has been published in [10].

ACKNOWLEDGEMENTS

I would like to thank Prof. H. Neunzert. He introduced me to fluid dynamics and supported me scientifically and organizationally. Without the frequent help of the co-operative staffs at the computer centers of the universities of Karlsruhe and Kaiserslautern, I would never have finished this work. It was done as part of the program "Finite Approximationen in der Strömungsdynamik" of the Deutsche Forschungsgemeinschaft.

BIBLIOGRAPHY

[1] MARDER, B. M.: GAP - A PIC-Type Fluid Code, Mathematics of Computation 29 (1975), 434-446.

[2] ZALESAK, S. T.: Fully Multidimensional Flux-Corrected Transport Algorithms for Fluids. Journal of Computational Physics 31 (1979), 335-362.

[3] ZALESAK, S. T.: High Order "ZIP"-Differencing for Convective Terms (Note). Journal of Computational Physics 40 (1980), 497-508.

[4] HASSENPFLUG, H.-U.: Untersuchungen zur Wirbelbildung durch Beugung einer Stoßwelle (Dissertation). RWTH Aachen, Fakultät für Maschinenwesen, 1976.

[5] ENGQUIST, B., MAJDA, A.: Absorbing Boundary Conditions for the Numerical Simulation of Waves. Mathematics of Computation 31 (1977), 629-651.

[6] HEDSTROM, G. W.: Nonreflecting Boundary Conditions for Nonlinear Hyperbolic Systems. Journal of Computational Physics 30 (1979), 222-237.

[7] WICK, J.: Numerical Aspects of Particle Simulation in the Plasma-Physical Case. In: Applications of Mathematics in Technology, V. Boffi/H. Neunzert (Eds.), Stuttgart: Teubner 1984.

[8] KREISS, H.-O., OLIGER, J.: Comparison of Accurate Methods for the Integration of Hyperbolic Equations. Tellus 24 (1972), 199-215.

[9] RUSANOV, V. V.: The Calculation of the Interaction of Non-Stationary Shock Waves and Obstacles. U.S.S.R. Computational Mathematics and Mathematical Physics 1 (1962), 304-320.

[10] MIETZNER, Th.: Umströmung von Ecken und Kanten, Teil 2. Bericht Nr. 10 der Arbeitsgruppe Technomathematik, Universität Kaiserslautern.

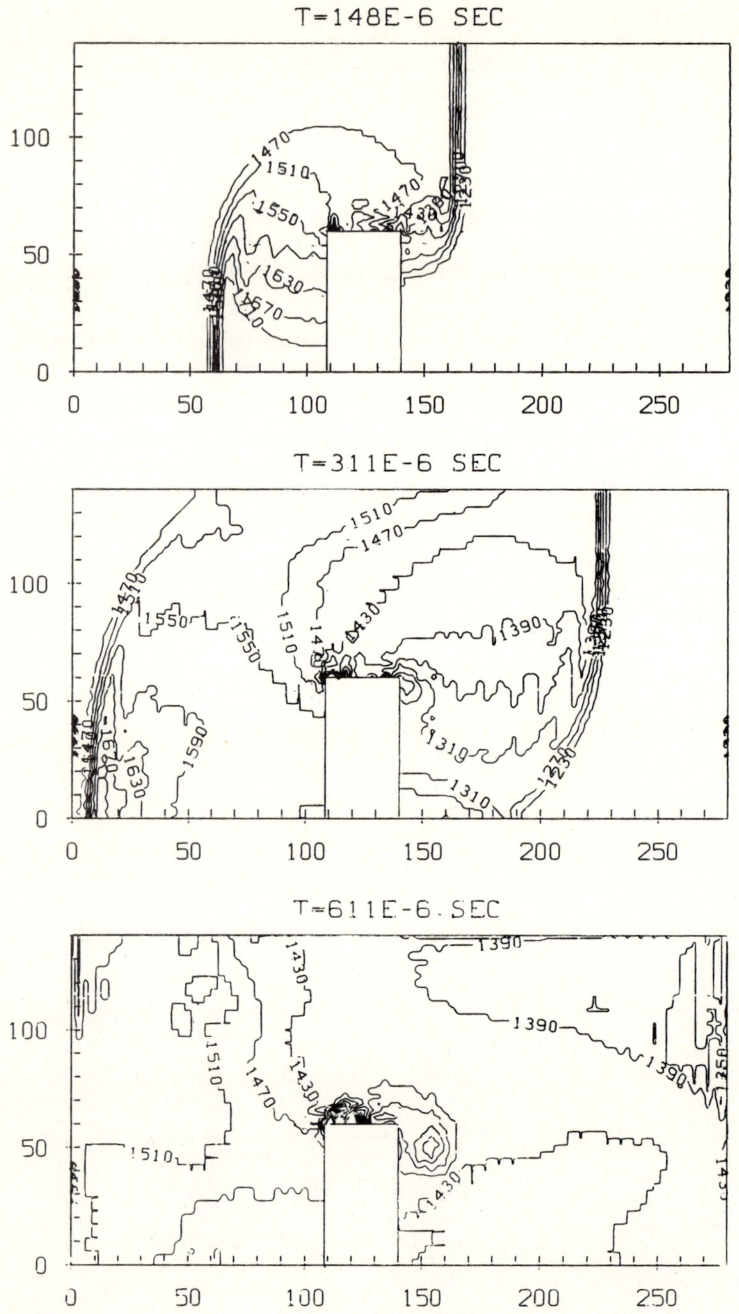
T=148E-6 SEC
1470
1510
1550
1630
1670
1230
0
50
100
150
200
250
T=311E-6 SEC
1510
1470
1550
1590
1630
1430
1390
1310
1230
T=611E-6 SEC
1390
1430
1510
1470
1350

Fig. 1

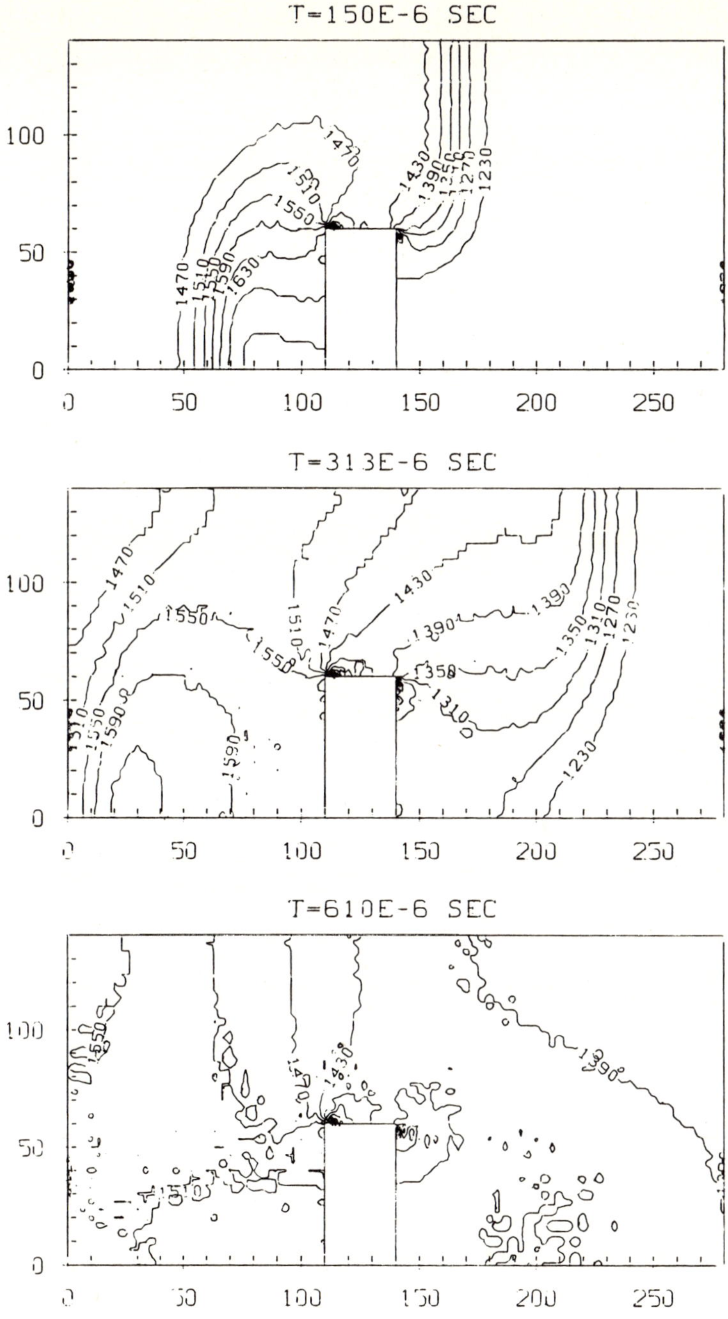

Fig. 2

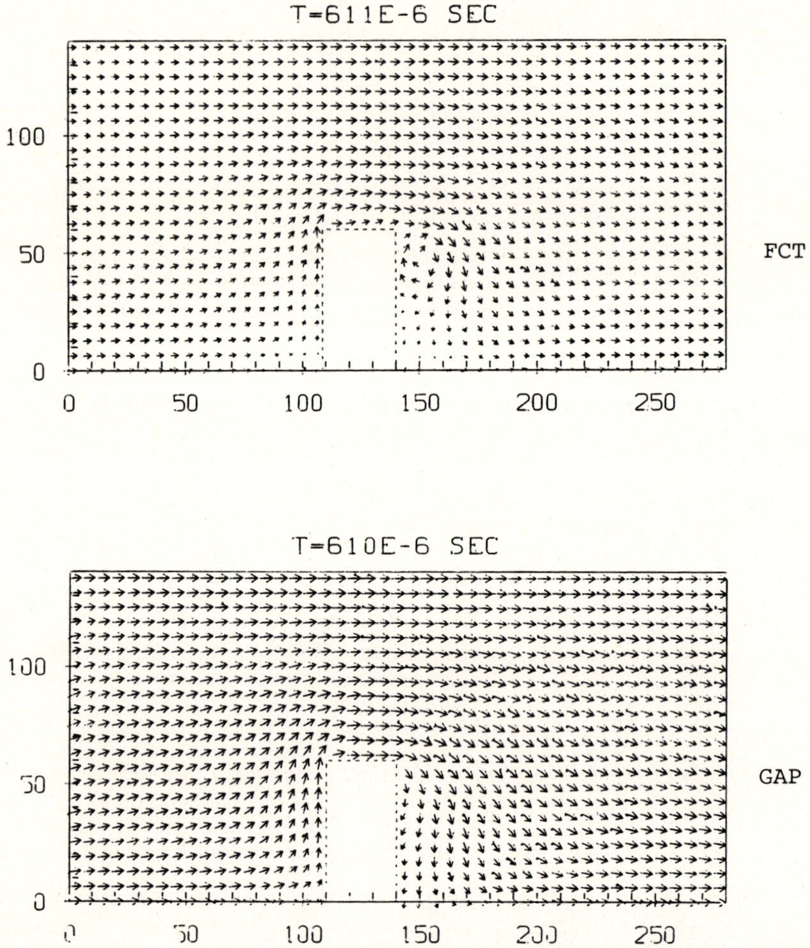
T=611E-6 SEC
100
50
0
0
50
100
150
200
250
FCT
T=610E-6 SEC
100
50
0
0
50
100
150
200
250
GAP

Fig. 3

IMPLICIT FINITE-DIFFERENCE SIMULATION OF SEPARATED HYPERSONIC FLOW OVER AN INDENTED NOSETIP

B. Müller and D. Rues
DFVLR, Institut für Theoretische Strömungsmechanik
Bunsenstr. 10, D-3400 Göttingen, FRG

SUMMARY

Laminar hypersonic flow over a severely indented blunt nosetip at zero angle of attack is simulated by solving the axisymmetric Navier-Stokes equations using an implicit finite-difference method. New forms of convergence acceleration to the steady state are presented. The validity of the thin-layer approximation is confirmed for flow with a large separation region at high Reynolds numbers.

INTRODUCTION

The indented nosetip problem arises in the design of reentry vehicles, because the effects of ablation on the aerodynamic behaviour and on the thermal protection system of a reentry vehicle have to be determined accurately. Hypersonic flow over severely indented nosetips was investigated numerically by Kutler et al. [1] and Hsieh [2] using the thin-layer approximation of the Navier-Stokes equations, and experimentally by Morrison et al. [3].

The purpose of the present paper is to demonstrate the efficient use of an implicit finite-difference method to solve the Navier-Stokes equations on a vector computer, and to check the validity of the thin-layer approximation for flows with large separation bubbles, as they occur in axisymmetric hypersonic flow over severely indented nosetips.

In the present approach, the axisymmetric Navier-Stokes equations of a compressible fluid are solved by means of the implicit scheme of Beam and Warming [4]. A factorized implicit treatment of the fourth-order damping terms, applied in the near wall-tangential ξ-direction only, accelerates the convergence to the steady state. The solution of the unfactored finite-difference equation using a zebra line relaxation method with the lines in the near wall-normal η-direction proves to be even more efficient. The bow shock is treated by the shock fitting approach, similar to Kutler et al. [1]. The computer code is vectorized on the CRAY-1S vector computer of the DFVLR.

The numerical simulation of axisymmetric laminar hypersonic flow over a severely indented nosetip at high Reynolds numbers indicates separated flow in the concave region. An embedded shock causes the deflection of the detached bow shock near the second expansion shoulder. Thus, a contact discontinuity is generated. The investigation of the thin-layer approximation at medium Reynolds numbers reveals local deviations near the separation and reattachment points compared with the Navier-Stokes results and agreement at high Reynolds numbers.

GOVERNING EQUATIONS

For axisymmetric Newtonian fluid flow of a perfect gas with constant specific heats, without external forces and without heat sources, the time-dependent Navier-Stokes equations may be written in dimensionless conservation-law form as [5]:

$$\frac{\partial q}{\partial \tau} + \frac{\partial E^{(\xi)}}{\partial \xi} + \frac{\partial E^{(\eta)}}{\partial \eta} + H = Re_{\infty,L}^{-1} \left[\frac{\partial E_v^{(\xi)}}{\partial \xi} + \frac{\partial E_v^{(\eta)}}{\partial \eta} + H_v\right] \qquad (1)$$

where the independent variables are transformed (Fig. 1) by $\tau = t$, $\xi = \xi\ (t,x,y)$, $\eta = \eta\ (t,x,y)$. $q=J^{-1}(\rho,\rho u,\rho v,e)^T$ is the vector of the conservative variables scaled by the Jacobian of the transformation. $E^{(\xi)}$, $E^{(\eta)}$ and H, $E_v^{(\xi)}$, $E_v^{(\eta)}$ and H_v are the inviscid and viscous fluxes in ξ- and η-directions and axisymmetric source terms, resp. [6]. The Reynolds number $Re_{\infty,L}$, the Prandtl number Pr, and the Mach number M_∞ are defined in terms of the freestream values and a characteristic length L, which is used to make the axial and radial coordinates x and y,resp., non-dimensional. Stokes's hypothesis is employed, and Pr is assumed to be constant. The dependence of the viscosity coefficient μ on the temperature T is obtained from the Sutherland law with the Sutherland constant $S_1 = 110°K/T_\infty$ for air.

If the given body conforms to a line of constant η and if among the viscous terms only the gradients in the near wall-normal η-direction need to be resolved, the thin-layer approximation may be employed by neglecting all viscous terms containing ξ-derivatives [7]. Ordering the viscous terms according to their derivatives

$$
\begin{aligned}
E_v^{(\xi)} &= F_0(q) + F_1(q,\tfrac{\partial q}{\partial \xi}) + F_2(q,\tfrac{\partial q}{\partial \eta}) \\
E_v^{(\eta)} &= G_0(q) + G_1(q,\tfrac{\partial q}{\partial \xi}) + G_2(q,\tfrac{\partial q}{\partial \eta}) \qquad (2) \\
H_v &= H_0(q) + H_1(q,\tfrac{\partial q}{\partial \xi}) + H_2(q,\tfrac{\partial q}{\partial \eta})
\end{aligned}
$$

the thin-layer approximation amounts to neglecting all viscous terms, except G_0, G_2, H_0 and H_2. Thus, the thin-layer equations contain the Euler equations and the most important boundary-layer terms [8].

ALGORITHM

The implicit factorized finite-difference scheme of Beam and Warming [4] is employed to solve the Navier-Stokes equations (1):

ξ-sweep:

$$
\begin{aligned}
&[I+\tfrac{\vartheta\Delta\tau}{1+\psi}(\mu_\xi\delta_\xi A^n - Re_{\infty,L}^{-1}(\delta_\xi L_1^n + \mu_\xi\delta_\xi L_0^n + N_1^n)) \\
&-\varepsilon_I(J^{-1}\delta_\xi{}^2 J)^n]\ \Delta q^{*n} = \\
&= -\tfrac{\Delta\tau}{1+\psi}[\mu_\xi\delta_\xi E^{(\xi)n} + \mu_\eta\delta_\eta E^{(\eta)n} + H^n \\
&\qquad -Re_{\infty,L}^{-1}(\delta_\xi F_1^n + \delta_\eta G_2^n + \delta_\xi F_2(\tilde{q}^n) + \delta_\eta G_1(\tilde{q}^n) \\
&\qquad\qquad +\mu_\xi\delta_\xi F_0{}^n + \mu_\eta\delta_\eta G_0^n + H_v^n)] \\
&\quad +\tfrac{\psi}{1+\psi}\Delta q^{n-1} - \varepsilon_E(J^n)^{-1}[\delta_\xi^4 + \delta_\eta^4](Jq)^n \qquad (3a)
\end{aligned}
$$

η-sweep:

$$
\begin{aligned}
&[I+\tfrac{\vartheta\Delta\tau}{1+\psi}(\mu_\eta\delta_\eta B^n + C^n - Re_{\infty,L}^{-1}(\delta_\eta M_2^n + \mu_\eta\delta_\eta M_0^n + N_2^n + N_0^n)) \\
&-\varepsilon_I(J^{-1}\delta_\eta{}^2 J)^n]\ \Delta q^n = \Delta q^{*n} \qquad (3b)
\end{aligned}
$$

$$q^{n+1} = q^n + \Delta q^n \tag{3c}$$

where $\tilde{q}^n = q^n + \vartheta \Delta q^{n-1}$.

A, B, C, L_0, L_1, M_0, M_2, N_0, N_1, N_2 are the Jacobian matrices of $E^{(\xi)}$, $E^{(\eta)}$, H, F_0, F_1, G_0, G_2, H_0, H_1, H_2, resp. (cf. [1],[8]), and I is the identity matrix. The classical finite-difference operators are defined by

$$\delta_\xi a_{i,j} = a_{i+1/2,j} - a_{i-1/2,j} \tag{4a}$$
$$\delta_\eta a_{i,j} = a_{i,j+1/2} - a_{i,j-1/2} \tag{4b}$$
$$\mu_\xi a_{i,j} = (a_{i+1/2,j} + a_{i-1/2,j})/2 \tag{4c}$$
$$\mu_\eta a_{i,j} = (a_{i,j+1/2} + a_{i,j-1/2})/2. \tag{4d}$$

The viscous terms F_2 and G_1 leading to mixed derivatives have to be treated explicitly to allow for the factorization. For efficiency reasons, a one-leg method is used instead of the corresponding linear multistep method [4].

The three-point-backward time differencing formula, i.e. $\psi=1/2$ and $\vartheta=1$ in (3), is preferred rather than the Euler implicit formula, i.e. $\psi=0$ and $\vartheta=1$ in (3), because the former is second-order accurate and less dissipative. With $\varepsilon_E = \Delta\tau$ and $\varepsilon_I = 2\varepsilon_E$ chosen for consistency and stability reasons, resp., the effects of the numerical damping coefficients on accuracy and convergence are found to be optimum for the inviscid two-dimensional Prandtl-Meyer expansion [8].

The ξ- and η-sweeps lead to block-tridiagonal linear systems, the formation of which is easily vectorizable. On a serial computer, these systems are efficiently solved by the Richmyer algorithm [9]. On a vector computer, the inherent recursions preclude the vectorization of the algebraic systems solver. However, if the Richmyer algorithm is applied to the mutually independent linear systems of the ξ-sweep and the η-sweep, resp., simultaneously instead of separately [9], the maximum vector performance is obtained, although at the cost of increased storage requirements (cf. Table 1).

CONVERGENCE ACCELERATION

The stability analysis of the implicit Beam and Warming scheme applied to the one-dimensional linear wave equation indicates, that the modulus of the amplification factor can be decreased to accelerate the convergence to the steady-state by treating the explicit fourth-order damping terms implicitly and skipping the implicit second-order smoothing terms [10]. By factorizing the implicit fourth-order damping terms and applying them in the near wall-tangential ξ-direction only, a new form of convergence acceleration is obtained, which takes only about 10 % more CPU-time per time step than the standard scheme (3):

$$L_\xi \, \Delta q^{*n} = RHS \tag{5a}$$
$$D_\xi \, \Delta q^{**n} = \Delta q^{*n} \tag{5b}$$
$$L_\eta \, \Delta q^n = \Delta q^{**n} \tag{5c}$$
$$q^{n+1} = q^n + \Delta q^n \tag{5d}$$

where $D_\xi = I + \varepsilon_E J^{-1} \delta_\xi^4 JI$.

RHS denotes the right hand side of (3a). L_ξ and L_η are the implicit operators of (3a) and (3b), resp., with $\varepsilon_I = 0$. Applied to the linear wave equation, the modified scheme (5) is unconditionally stable for $\vartheta \geq \frac{1}{2}$ and $\varepsilon_E \geq 0$ [8].

The efficiency of the modified scheme (5) is demonstrated for supersonic flow over an adiabatic hemisphere-cylinder at $M_\infty = 2.94$, $Re_{\infty,L} = 2.2\ 10^5$ using a 26*31 mesh, $\Delta\tau = 0.06$ and 800 time levels (cf. Fig. 2).

For the two-dimensional linear wave equation, the stability analysis of the Beam and Warming scheme shows, that the high frequency modes are the less damped the higher the time steps are due to the factorization error [11]. The unfactored finite-difference equation is solved by a zebra line relaxation method with the lines in the near wall-normal η-direction, because this non-Gauss-Seidel method is suitable for vectorization and the Courant numbers in the near wall-normal η-direction CFL_η are in many high Reynolds number applications much higher than CFL_ξ. Since only the steady-state solution is of interest, only one iteration per time step is performed. Similar to the treatment of the mixed derivatives in the Beam and Warming scheme, a one-leg method is employed for the linearization of the fluxes in ξ-direction. Thus the following scheme is obtained for the axisymmetric Euler equations:

$$[I+\Delta\tau(\mu_\eta\delta_\eta B^n+C^n)-\varepsilon_I J^{-1}\delta_\eta{}^2 J+2\varepsilon_I]\ \Delta q^n_{i,j} =$$

$$= -\Delta\tau[\mu_\xi\delta_\xi E^{(\xi)}(q^n + \begin{matrix}\Delta q^{n-1}\\ \Delta q^n\end{matrix}) + \mu_\eta\delta_\eta E^{(\eta)n} + H^n] \tag{6}$$

$$+\varepsilon_I J^{-1}[\delta_\xi^2 J \begin{matrix}\Delta q^{n-1}\\ \Delta q^n\end{matrix} + 2 \begin{matrix}\Delta q^{n-1}\\ \Delta q^n\end{matrix}] - \varepsilon_E J^{-1}[\delta_\xi^4 J\bar{q} + \delta_\eta^4 J q^n]$$

for $n+i$ $\begin{matrix}\text{odd}\\ \text{even}\end{matrix}$,

where $\bar{q}$ is taken at the latest time level.

The Euler implicit time differencing is used, because the three-point-backward formula does not converge for the test problem. Numerical experiments indicate a stability condition of $CFL_\xi < 1$, but the capability of using $CFL_\eta = O(100)$. With the one-leg method, the scheme is easily extented to the thin-layer Navier-Stokes equations and to three dimensions. For the full Navier-Stokes equations, the treatment of $E^{(\xi)n+1}$ in the even sweep has to be modified. (6) needs the same amount of storage as the factorized Beam and Warming scheme, but for the axisymmetric thin-layer equations about 20% less CPU-time per time step, since the ξ-sweep is dropped. The unfactored scheme (6) is closely related to the hybrid explicit-implicit procedure of Rizk and Chaussee [12], which does not consider the linearization of the fluxes in ξ-direction. Scheme (6) needs 4 two-dimensional arrays more storage, because the delta-variables are stored, and for the axisymmetric thin-layer equations about 4% more CPU-time per time-step than the Rizk and Chaussee method. For the previously used test case, the good convergence properties of (6) are demonstrated by Table 2.

GRID GENERATION, INITIALIZATION, BOUNDARY TREATMENT

To calculate supersonic flow over a blunt body, the grid is generated algebraically. The lines of constant ξ are kept fixed, whereas the lines of constant η are clustered close to the body surface [1], [8] and moved according to the shock position.

The initial shape and position of the shock are used to initialize the shock layer on the lines of constant ξ. On the body surface, the initial surface pressure is prescribed according to a modified Newtonian theory. For an adiabatic wall, the temperature is initialized by the stagnation point temperature. In addition, the no-slip condition is imposed. For the first 100 time levels, the time step is linearly increased to its final value [8]

The physical region, bounded by the body surface, the detached bow shock, a symmetry and an outflow boundary, is mapped into a rectangular computational domain using a transformation of the independent variables (cf. (1) and Fig. 1). The pressure behind the shock is calculated explicitly from the admissible compatibility equation. Then the other flow variables and the shock velocity are evaluated from the Rankine-Hugoniot relations [13], [1]. The shock velocity determines the shock position at the new time level. Besides the no-slip condition and a temperature condition, the wall-normal momentum equation is used at the body surface. The reflection principle and extrapolation conditions are imposed at the symmetry and outflow boundary, resp., and treated implicitly.

RESULTS

Hypersonic flow over the severely indented nosetip considered here was investigated experimentally by Morrison et al. [3] and numerically by Hsieh [2] at $M_\infty = 5$, $Re_{\infty,L} = 2*10^6$, $T_w/T_\infty = 5.4$. Here the base radius is taken as characteristic length L. In the present numerical simulation, the flow is calculated at $\gamma = 1.4$, $Pr = 0.723$, and $T_\infty = 64°K$. Starting from a sphere-cone, the body is gradually deformed into the severely indented nosetip [8]. The final results are obtained on a 75*41 mesh (cf. Fig. 3) with the grid points condensed near the expansion shoulders to resolve the primary separation and reattachment points similar to [2]. $\varepsilon_E = 5\Delta\tau$ is used to improve the convergence rate of scheme (5). At the end of the calculation, $\|\Delta Jq/\Delta\tau\|_{max}=0.11$ with $\Delta\tau=0.005$, and the maximum modulus of the shock velocity v_S is 0.0010. Thus, the convergence acceleration (5) is insufficient in the present case. The same is expected for (6).

In Fig. 4, the shock location of the present thin-layer and Navier-Stokes results are seen to be in good agreement with [2]. The measured data [3] are corrected in [2] and agree well with the laminar computed results. The kink in the bow shock is predicted a little downstream compared with [2]. Whereas Hsieh found a secondary separation, the present results yield a secondary and a tertiary separation. The latter can hardly be visualized in Fig. 5, and is very sensitive to the spatial resolution and the numerical damping.

There is no difference in surface pressure between the Navier-Stokes and thin-layer results (cf. Fig. 6, where s denotes the arclength from the stagnation point) and very good agreement with Hsieh's thin-layer solution. Up to the first expansion shoulder, the numerical and experimental results agree well. Whereas in the experiment the surface pressure is almost constant between the expansion shoulders [3], the numerically determined surface pressure reflects separation and reattachment points by local minima and maxima, resp.. For the attached flow behind the primary reattachment point, the data agree well. It is noted that the resolution of the embedded shock and the secondary separation point is crucial to the accurate prediction of the surface pressure distribution. Deviations between the calculations and the experiment may be expected, since the flow is likely not to be laminar, as assumed in the numerical simulations, but transitional in reality.

The heat flux proves to be one of the most sensitive quantities. The differences of the Stanton number calculated from the thin-layer and the full Navier-Stokes equations are so small that they cannot be seen in Fig. 7. The present results agree rather well with Hsieh's numerical solution [2], except for the peak values near the expansion shoulders. No experimental data are available for comparison.

At $Re_{\infty,L} = 2*10^4$, the thin-layer and Navier-Stokes solutions are in good agreement, except for small deviations near the separation and reattachment points. The secondary reattachment point is predicted by the thin-layer result about 9% further downstream than by the Navier-Stokes result (cf. Fig. 8). Similar results are obtained at $Re_{\infty,L} = 2*10^3$, where only a primary separation bubble is found.

CONCLUSIONS

An implicit finite-difference procedure to solve either the Euler equations or the thin-layer or full Navier-Stokes equations is described. The mixed derivatives are efficiently treated. New forms of convergence acceleration to the steady state are presented. The vectorization of the scheme takes considerable advantage of the high performance of vector computers. The computer code is applied to calculate axisymmetric laminar hypersonic flow over a severely indented nosetip. For the meshes used, there is essentially no difference between the thin-layer and full Navier-Stokes solutions. The present results are in very good agreement with other numerical results, and show at least the same trend as the experimental data.

ACKNOWLEDGEMENTS

This research was partially supported by the Deutsche Forschungsgemeinschaft under the program "Finite Approximationen in der Strömungsmechanik". The authors thank Prof. E. Krause, Ph.D., RWTH Aachen, for his support of this work.

REFERENCES

[1] KUTLER,P., CHAKRAVARTHY,S.R. and LOMBARD,C.P., "Supersonic Flow over Ablated Nosetips Using an Unsteady Implicit Procedure", AIAA Paper 78-213, 1978.

[2] HSIEH,T., "Calculation of Viscous Hypersonic Flow over a Severely Indented Nosetip", AIAA J., Vol.22, No.7, 1984, pp. 935-941.

[3] MORRISON,A.M., YANTA,W.J. and VOISINET,R.L.P., "The Hypersonic Flow Field Over a Re-entry Vehicle Indented Nose Configuration", AIAA Paper 81-1060, 1981.

[4] BEAM,R.M. and WARMING,R.F., "Implicit Numerical Methods for the Compressible Navier-Stokes and Euler Equations", VKI Lecture Series 1982-04, 1982.

[5] PEYRET,R. and VIVIAND,H., "Computation of Viscous Compressible Flows Based on the Navier-Stokes Equations", AGARD - AG - 212, 1975.

[6] MÜLLER,B., "Navier-Stokes Solution for Hypersonic Flow over an Indented Nosetip", AIAA Paper 85-1504, 1985.

[7] BALDWIN,B.S. and LOMAX,H., "Thin-Layer Approximation and Algebraic Model for Separated Turbulent Flows", AIAA Paper 78-257, 1978.

[8] MÜLLER,B., "Berechnung abgelöster laminarer Überschallströmungen um nichtangestellte stumpfe Rotationskörper", Dissertation, RWTH Aachen, 1985, and DFVLR-FB 85-30, 1985.

[9] MÜLLER,B., "Vectorization of the Implicit Beam and Warming Scheme", in: Gentzsch,W., "Vectorization of Computer Programs with Applications to Computational Fluid Dynamics", Vieweg, Braunschweig, 1984, pp. 172-194.

[10] PULLIAM,T.H., "Euler And Thin Layer NavierStokes Codes: ARC2D,ARC3D", in: Reddy,K.C. and Steinhoff,J.S. (eds.), Computational Fluid Dynamics, UTSI Publication No. E02-4005-023-84, 1984, pp. 15.1-15.85.

[11] STEGER,J.L., "Implicit Finite-Difference Simulation of Flow About Arbitrary Two-Dimensional Geometries", AIAA Paper 77-665, 1977.
[12] RIZK,Y.M. and CHAUSSEE,D.S., "Three-Dimensional Viscous-Flow Computations Using a Directionally Hybrid Implicit-Explicit Procedure", AIAA Paper 83-1910, 1983.
[13] RUES,D., "Der Einfluß einfallender Stoßwellen auf ebene Überschallströmungen um stumpfe Körper", DLR-FB 72-68, 1972.

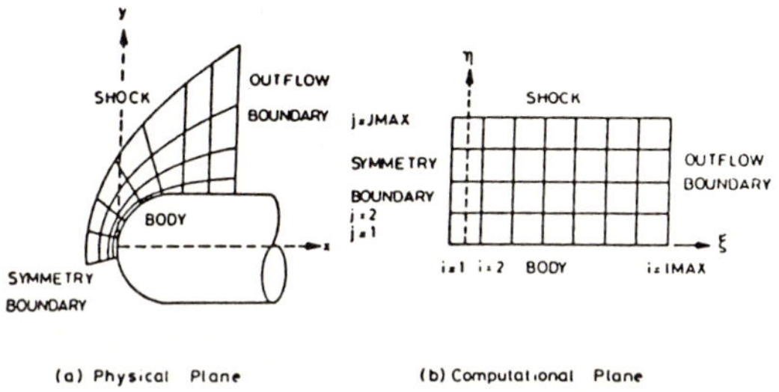

Fig. 1 Transformation of physical to computational plane and indexing of computational mesh for supersonic blunt body flow

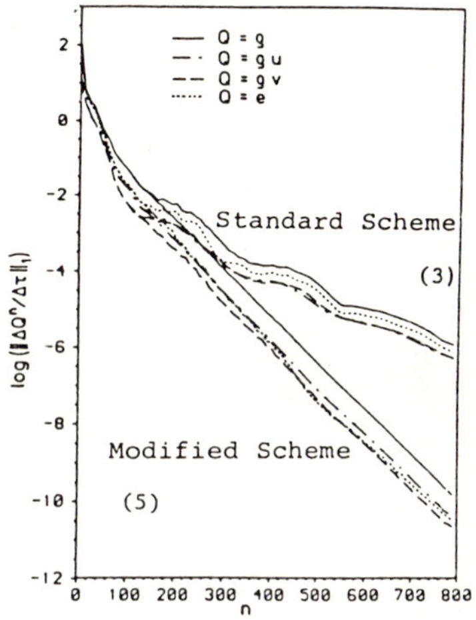

Fig. 2 Convergence history of thin-layer computation for adiabatic hemisphere-cylinder at $M_\infty=2.94$, $Re_{\infty,L}=2.2*10^5$

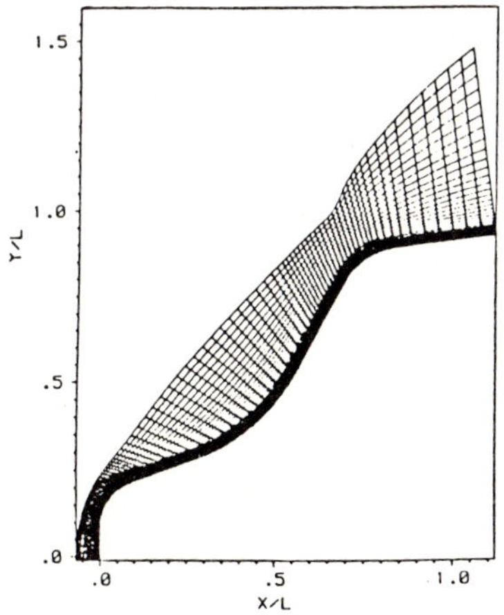

Fig. 3 75 * 41 mesh

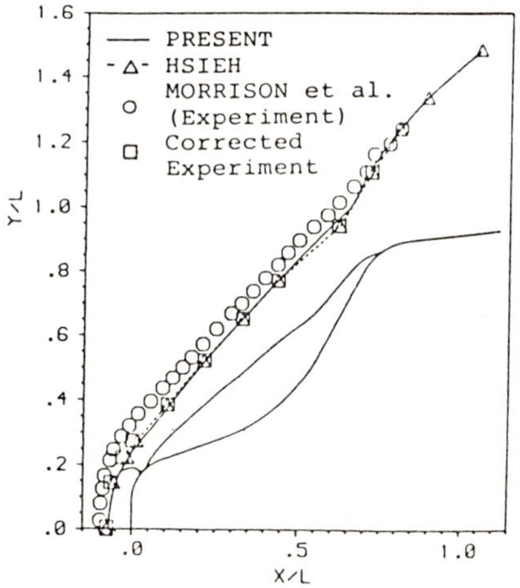

Fig. 4 Sonic line location and comparison of shock locations at $M_\infty=5$, $Re_{\infty,L}=2*10^6$, $T_w/T_\infty=5.4$

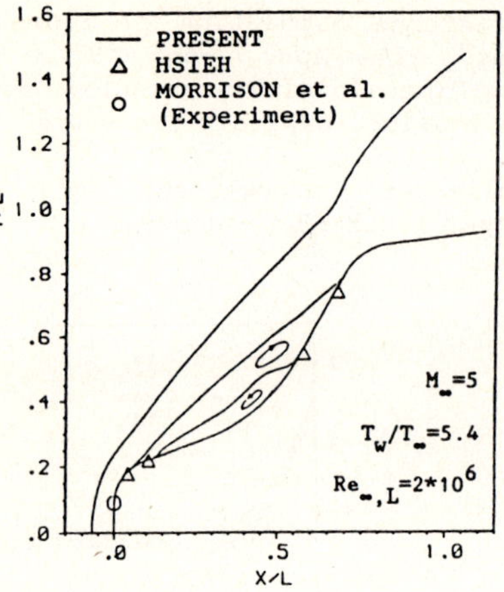

Fig. 5 Dividing streamlines and comparison of separation and reattachment points

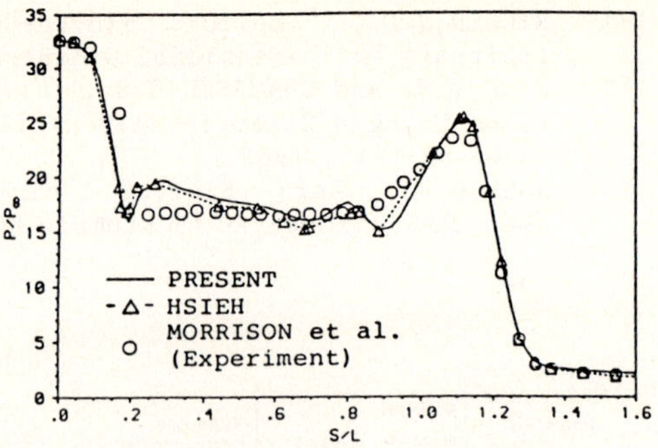

Fig. 6 Comparison of surface pressure distribution for indented nosetip at $M_\infty=5$, $Re_{\infty,L}=2*10^6$, $T_w/T_\infty=5.4$

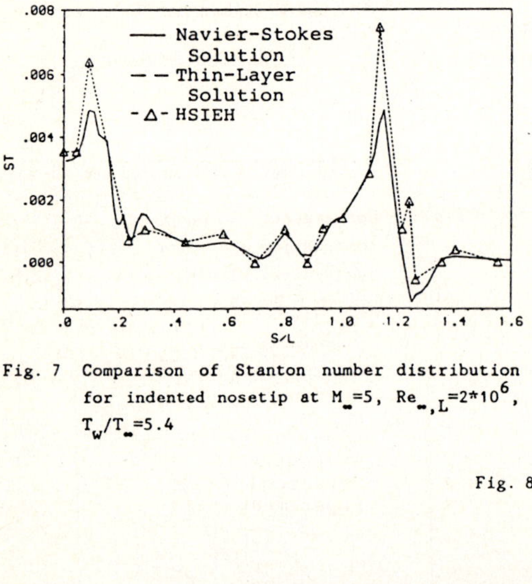

Fig. 7 Comparison of Stanton number distribution for indented nosetip at $M_\infty=5$, $Re_{\infty,L}=2*10^6$, $T_w/T_\infty=5.4$

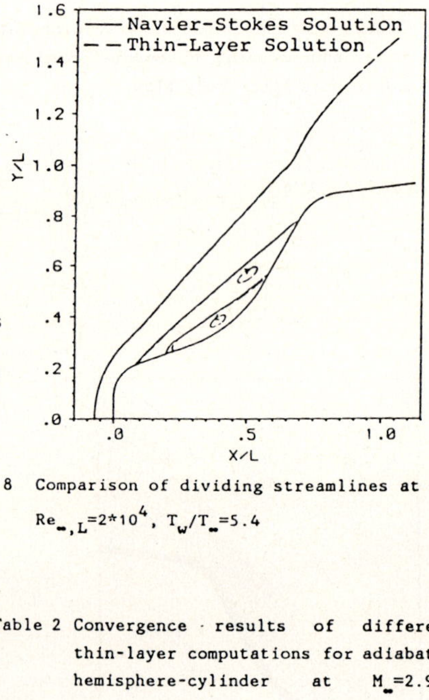

Fig. 8 Comparison of dividing streamlines at $M_\infty=5$, $Re_{\infty,L}=2*10^4$, $T_w/T_\infty=5.4$

Table 1 CPU-times in seconds per time level and per grid point for blunt body code on CRAY-1S

Equations	Richtmyer Version	Simultaneous Richtmyer Version
Euler	$6.2\ 10^{-5}$	$3.3\ 10^{-5}$
Thin-Layer	$7.8\ 10^{-5}$	$4.9\ 10^{-5}$
Navier-Stokes	$10.1\ 10^{-5}$	$7.2\ 10^{-5}$

Table 2 Convergence results of different thin-layer computations for adiabatic hemisphere-cylinder at $M_\infty=2.94$, $Re_{\infty,L}=2.2*10^5$

SCHEME	$\|\Delta Jq/\Delta\tau\|_{max}$	$\|v_s\|_{max}$
Beam & Warming (3)	$0.7\ 10^{-5}$	$0.7\ 10^{-6}$
Rizk & Chaussee [12]	$0.1\ 10^{-6}$	$0.9\ 10^{-8}$
Factorized (5)	$0.9\ 10^{-9}$	$0.2\ 10^{-10}$
Unfactored (6)	$0.3\ 10^{-9}$	$0.3\ 10^{-11}$

ON THE COMPARISON AND CONSTRUCTION OF TWO-STEP SCHEMES FOR THE EULER EQUATIONS

C.-D. Munz
Mathematisches Institut II der Universität Karlsruhe (TH)
Englerstr. 2, D-7500 Karlsruhe, Fed. Rep. of Germany

SUMMARY

This paper presents a survey and a comparison of two-step schemes for the one-dimensional Euler equations of compressible gas dynamics. These schemes are generalizations of van Leer's MUSCL scheme - a second-order extension to Godunov's upwind scheme. Within this context it is easy to convert every low order upwind difference scheme to a higher order scheme which does not generate spurious oscillations across shock waves and contact discontinuities, but capturesshocks within one or two mesh intervals. The advantages and disadvantages of various approaches are discussed and numerical results are presented.

1. INTRODUCTION

In recent years a number of new shock capturing finite difference approximations have been constructed. These schemes are second- or higher order accurate in regions where the solutions are smooth and they sharply resolve discontinuities, but do not exhibit spurious oscillations. Such finite difference schemes, called "high resolution schemes" after Harten [12], do not need any artificial viscosity contrary to classical second- or higher order methods. The starting point for the construction of such schemes are monotone or E-schemes for the scalar conservation law. These schemes and especially their generalization to systems are often called upwind schemes because they incorporate into the numerical solution the direction of nonlinear wave propagation given by the direction of the characteristics. In the scalar case the monotone or E-schemes satisfy a discrete entropy condition and possess the TVD-property (total variation diminishing) which guarantees that no spurious oscillations are generated. However, monotone and E-schemes are only first-order accurate and spread discontinuities over many grid points.

There are essentially three basic approaches to reduce this great amount of numerical dissipation. The high resolution schemes using flux limiters add an "antidiffusive" term to the numerical flux of an upwind scheme. This term is limited in such a way that the desirable properties of the upwind schemes, e.g., the TVD property or a discrete entropy condition are conserved, but the scheme is second-order accurate on smooth solutions. This concept was introduced by Boris and Book [3] and van Leer [21]. To get higher resolution Harten [12] applied in his approach an upwind scheme to a modified flux of the conservation law. The third way was indicated by the MUSCL scheme of van Leer [20] - a second-order extension to Godunov's upwind method. Here the higher order solution is achieved by using in every time step a more accurate representation of the initial distribution and then applying Godunov's method to these data.

The two-step schemes are generalizations of the MUSCL approach formulated in a two-step algorithm. With this concept it is easy to convert every upwind scheme to higher order accuracy. The aim of this article is to give a survey and a comparison of various two-step schemes for the one-dimensional Euler equations. The one-dimensional schemes may be easily extended to two or three dimensions by operator splitting ([35]). Recently a comparison of two-step schemes based on the flux-vector splitting schemes was given by Anderson et al. [2]. There were also discussed several advantages of the two-step approach. We will present in addition numerical results with the upwind schemes of Roe and Godunov and we will also consider various representations of initial distributions.

This paper is divided into five sections. Section 2 describes the construction of high resolution schemes and especially the two-step schemes in the scalar case. In section 3 we shall briefly review several upwind schemes for the Euler equations and we shall describe the generalization of the two-step schemes to this equation. Numerical results for two test problems will be presented in section 4, our conclusions in section 5.

2. THE SCALAR CASE

In this section we shall consider numerical approximations to weak solutions of the scalar Cauchy problem

$$u_t + f(u)_x = 0 \quad \text{in } \mathbb{R} \times \mathbb{R}^+ \quad , \tag{2.1}$$

$$u(x,0) = q(x) \quad , \quad x \in \mathbb{R} \quad , \tag{2.2}$$

where q is assumed to be a function of bounded total variation. It is well known that weak solutions of this Cauchy problem are non-unique; so an additional principle is needed to select the unique physical solution. This principle is given by the validity of an entropy inequality (see e.g.[16]).

2.1 Monotone, E-, TVD-Schemes

We shall consider explicit finite difference schemes for the scalar conservation law (2.1) written in conservation form

$$u_i^{n+1} = u_i^n - \lambda(g_{i+1/2}^n - g_{i-1/2}^n) \quad . \tag{2.3}$$

Here λ is the mesh ratio $\Delta t/\Delta x$ and u_i^n is an approximation of the average value of u in the grid zone $[x_{i-1/2}, x_{i+1/2}]$ at time $n\Delta t$. As usual, we shall use the notations

$$t_n = n\Delta t \ , \ x_i = i\Delta x \ , \ x_{i+1/2} = (x_i + x_{i+1/2})/2, \ i \in Z \ , \ n \in \mathbb{N}_o \ , \tag{2.4}$$

and we shall restrict us to uniform grids with $\Delta x = \text{constant}$.

We require the numerical flux g, which is a function of 2k variables

$$g_{i+1/2}^n = g(u_{i-k+1}^n, \dots, u_{i+k}^n) \quad , \tag{2.5}$$

to be Lipschitz continuous in all arguments and consistent with the physical flux in the following sense:

$$g(u,u,\dots,u) = f(u). \tag{2.6}$$

The initial values (2.2) are discretized by taking cell averages

$$u_i^o = \frac{1}{\Delta x} \int_{x_{i-1/2}}^{x_{i+1/2}} q(x)\,dx \quad . \tag{2.7}$$

The physically relevant weak solution has a monotonicity property which means that no local extrema in x can be created and the absolute values of local extrema do not increase. It follows from this that the total variation of u(x,t) in x is a nonincreasing function of t:

$$TV(\cdot,t_2) \leqq TV\,u(\cdot,t_1) \quad \text{for all } t_2 \geqq t_1 \quad . \tag{2.8}$$

Of course, it is desirable that difference schemes preserve such a property and satisfy a discrete version of (2.8). The scheme (2.3) is called total variation diminishing (TVD) if u_Δ with

$$u_\Delta(x,t) = u_i^n \quad , \quad (x,t) \in (x_{i-1/2}, x_{i+1/2}] \times (t_{n-1}, t_n] \quad , \tag{2.9}$$

satisfies (2.8) ([11]). A TVD-scheme especially guarantees that no unphysical oscillation will occur near discontinuities or steep gradients.

Harten [11] has given a sufficient condition for the scheme (2.3) to be TVD. If the scheme in conservation form is rewritten in the incremental form

$$u_i^{n+1} = u_i^n + C_{i+1/2}^n (u_{i+1}^n - u_i^n) - D_{i-1/2}^n (u_i^n - u_{i-1}^n) \quad , \tag{2.10}$$

then the coefficients must satisfy the inequalities

$$C_{i+1/2}^n \geqq 0 \quad , \quad D_{i-1/2}^n \geqq 0 \quad , \tag{2.11}$$

$$C_{i+1/2}^n + D_{i+1/2}^n \leqq 1 \tag{2.12}$$

for all values of i.

An important class of TVD-schemes are the "monotone schemes", which are three point schemes (k = 1) with a numerical flux being monotone in the following sense:

$$\begin{aligned} &h(v,w) \quad \text{is monotone nondecreasing in } v \quad , \\ &h(v,w) \quad \text{is monotone nonincreasing in } w \quad . \end{aligned} \tag{2.13}$$

By monotone schemes the conservation law is discretized by using differences biased in the direction of the characteristics. These schemes may be considered as an extension of the Courant-Isaacson-Rees scheme ([8]) to nonlinear problems. They are often called upwind-differencing or upwind schemes and in subsonic and supersonic regions all of them are identical to the first-order scheme using one-sided differences.

In [24] Osher introduced a class of schemes where the numerical flux satisfies the inequality

$$\operatorname{sgn}(u_{i+1} - u_i)[h_{i+1/2} - f(u)] \leqq 0 \tag{2.14}$$

for all values of u between u_i and u_{i+1} and called them E-schemes. By (2.13) it is easy to show that the monotone schemes belong to this class. The inequality (2.14) guarantees the TVD property and a discrete entropy inequality to be valid. Therefore, E- as well as monotone schemes exclude the approximation of nonphysical solution. They are very robust, but spread discontinuities over many grid points and are at most first-order accurate.

The class of TVD-schemes consequently contains the E- and monotone schemes. However, a TVD-scheme does not, in general, satisfy a discrete entropy condition. It is well known that the method of Roe [28] which is TVD may approximate stable expansion shocks. But with an entropy fix this method can be modified to be an E-scheme (e.g.,[12],[18]). A TVD scheme can be of second-order global accuracy; however, it necessarily degenerates at local extrema to first-order accuracy in the sense of local truncation error ([24],[11]). The TVD property plays also a significant role in the convergence theory; from stability in L^∞-norm and in total variation follows the compactness of the approximate solutions. For a detailed description of the theory of finite difference schemes approximating weak solutions of a scalar conservation law, see, e.g., [11],[24],[34].

2.2 High Resolution Schemes

The so-called high resolution schemes have been constructed by converting the first-order E- or monotone three point schemes to five point second- or higher order schemes. This conversion can be done in a way such that the favorable properties of the E- or monotone schemes as TVD property may be conserved. The first attempt in this direction was the FCT algorithm of Boris and Book [3] and the scheme of van Leer [21]. The concept of flux limiters was introduced by these authors. Recently Sweby [33] succeeded in giving a general framework for the construction of these schemes. We will briefly outline this approach and then describe the two-step algorithm.

First we consider the second-order Taylor series of u in the time direction where the t-derivatives are replaced by x-derivatives by means of the conservation law (2.1)

$$u(x,t+\Delta t) = u(x,t) - \Delta t\, f(u(x,t))_x + \frac{\Delta t^2}{2}(a^2(u(x,t))u_x(x,t))_x + O(\Delta t^3) \quad (2.15)$$

where $a(u) := f'(u)$. If the space derivatives are then approximated by central difference quotients, we get the Lax-Wendroff scheme [17] which is second-order in space and time and has the numerical flux

$$g^n_{i+1/2} = \frac{1}{2}(f^n_{i+1} + f^n_i) - \frac{\lambda}{2} a^n_{i+1/2}(f^n_{i+1} - f^n_i) \quad . \quad (2.16)$$

Here we used the standard notations $f^n_i = f(u^n_i)$, $a^n_{i+1/2} = a((u^n_{i+1} + u^n_i)/2)$.

Instead of central differences we can also use upwind differences. Let $h \in C^1$ be a numerical flux of a monotone or E-scheme then we can write (2.16) in the form

$$g^n_{i+1/2} = h^n_{i+1/2} + \frac{1}{2}(h_v - h_w)^n_{i+1/2}(u^n_{i+1} - u^n_i) - \frac{\lambda}{2} a^n_{i+1/2}(f^n_{i+1} - f^n_i) + O(\Delta x^2) \quad (2.17)$$

where h_v, h_w denote the partial derivatives in the first and second argument, respectively. The flux (2.17) can be interpreted as the sum of the E-flux and a correction for second-order accuracy by an "antidiffusive flux". To avoid the spurious oscillation of the Lax-Wendroff scheme before and after steep gradients, the antidiffusive flux has to be limited by a flux limiter function φ. This limiting process can be performed in such a way that the resulting scheme is TVD and second-order accurate because it is locally a weighted average of the Lax-Wendroff scheme and the second-order one-sided difference scheme of Beam and Warming ([33]).

The two-step schemes start instead of (2.15) from the Taylor series

$$u(x,t+\frac{1}{2}\Delta t) = u(x,t) - \frac{\Delta t}{2}\, f(u(x,t))_x + O(\Delta t^2)$$
$$u(x,t+\Delta t) = u(x,t) - \Delta t\, f(u(x,t+\frac{\Delta t}{2}))_x + O(\Delta t^3) \;. \tag{2.18}$$

Approximating the space derivatives by central difference quotients gives the two-step Lax-Wendroff procedure of Richtmyer [27].

An upwind scheme of second-order accuracy can be obtained by defining in each mesh interval boundary values

$$u^n_{i\pm} = u^n_i \pm \frac{\Delta x}{2} s^n_i \tag{2.19}$$

with s^n_i as described below. These values $u^n_{i\pm}$ are advanced to $t_{n+1/2}$ by

$$u^{n+1/2}_{i\pm} = u^n_{i\pm} - \frac{\lambda}{2}(f^n_{i+} - f^n_{i-}) \tag{2.20}$$

and then an E-scheme is applied

$$u^{n+1}_i = u^n_i - \lambda(h^{n+1/2}_{i+1/2} - h^{n+1/2}_{i-1/2}) \tag{2.21}$$

where

$$h^{n+1/2}_{i+1/2} = h(u^{n+1/2}_{i+}, u^{n+1/2}_{(i+1)-}) \;.$$

The boundary values (2.19) are defined by interpolating slopes in each grid zone which means that the initial values, instead of a piecewise constant, form a piecewise linear distribution (see Fig.2.1)

$$u^n(x) = u^n_i + (x - x_i)\, s^n_i \;, \quad x_{i-1/2} < x < x_{i+1/2} \;. \tag{2.22}$$

The slopes are weighted averages of right- and left-hand difference quotients and satisfy

$$s^n_i = u_x(x_i, t_n) + O(\Delta) \;, \tag{2.23}$$

where Δ denotes the discretization parameter.

A more general class of slopes is achieved by defining instead of (2.19) the boundary values as

$$u^n_{i\pm} = u^n_i \pm \delta^n_{i\pm} \;. \tag{2.24}$$

In the case $\delta^n_{i\pm} = \Delta x s^n_i/2$ this reduces to (2.19). Various slopes will be discussed in the next section.

The first scheme, in which such a piecewise linear distribution is used to get higher resolution, was the MUSCL-scheme of van Leer [20] for the Euler equations. The class of schemes in the form (2.19)-(2.21) was proposed in [18].

2.3 Construction of Two-step Schemes

In the semi-discrete case (method of lines) Osher [33] proved the convergence of such schemes and gave some criteria for these schemes to be second-order and TVD. Using the two-step approach to get a fully discrete

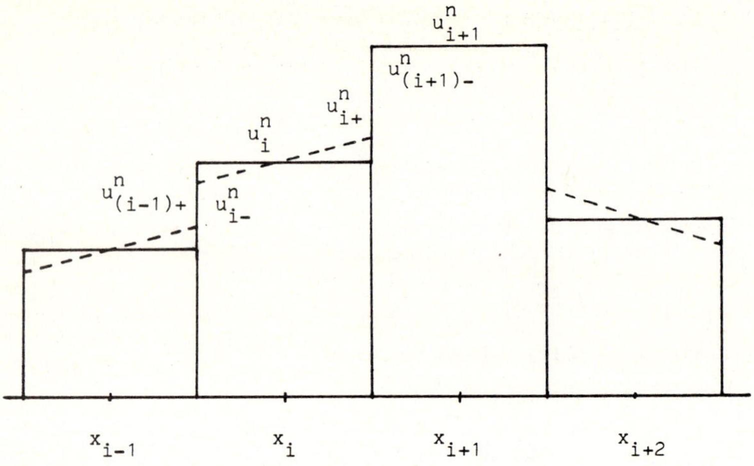

Fig.2.1 Piecewise linear distribution

second-order scheme similar results are valid. A necessary condition to be second-order accurate is

$$g^n_{i+1/2} - h^{n+1/2}_{i+1/2} = O(\Delta^2) \tag{2.25}$$

where $g^n_{i+1/2}$ denotes the numerical flux of the Lax-Wendroff scheme (2.16) and Δ the discretization parameter. Relation (2.2) is a sufficient condition if the coefficients in the $O(\Delta^2)$ term are differentiable (see [12]).

A two-step scheme with a Lipschitz continuous flux satisfies (2.25) in smooth regions of the solution, if

$$\frac{\Delta x s^n_i}{u^n_i - u^n_{i-1}} = 1 + O(\Delta) \quad , \quad \frac{\Delta x s^n_i}{u^n_{i+1} - u^n_i} = 1 + O(\Delta) \tag{2.26}$$

which becomes also a sufficient condition for second-order accuracy if the numerical flux h is C^2. Especially it follows from (2.26) that the slopes satisfy (2.23).

Next we will investigate conditions for the slopes which make the two-step scheme TVD. For this we rewrite the two-step scheme in the incremental form (2.10); a possible choice of the coefficients is

$$C^n_{i+1/2} = \frac{\lambda}{u^n_{i+1} - u^n_i}(h^{n+1/2}_{i+1/2} - h(u^{n+1/2}_{i+}, u^{n+1/2}_{i-})) \quad , \tag{2.27}$$

$$D^n_{i-1/2} = \frac{\lambda}{u^n_i - u^n_{i-1}} (h(u^{n+1/2}_{i+}, u^{n+1/2}_{i-}) - h^{n+1/2}_{i-1/2}) \quad . \tag{2.28}$$

If h is the numerical flux corresponding to a monotone scheme, then the first conditions in Harten's TVD-criterion (2.11) are fulfilled if

$$\frac{u^{n+1/2}_{(i+1)-} - u^{n+1/2}_{i-}}{u^n_{i+1} - u^n_i} \geqq 0 \quad , \quad \frac{u^{n+1/2}_{i+} - u^{n+1/2}_{(i-1)+}}{u^n_i - u^n_{i-1}} \geqq 0 \quad . \tag{2.29}$$

Using the definitions (2.19),(2.20) and the mean value theorem we get the inequalities

$$0 \leqq \frac{\Delta x s^n_i}{u^n_i - u^n_{i-1}} , \frac{\Delta x s^n_i}{u^n_{i+1} - u^n_i} \leqq \Psi \tag{2.30}$$

where $\Psi \leqq 2/(1+\mu)$ and μ denotes the CFL-number $\mu = \lambda \max f'(u)$.

The conditions (2.30) can be weakened in several cases. If we are away from sonic points $(f'(u) = 0)$ one of the coefficients (2.27),(2.28) equals zero. Instead of (2.30) we get in supersonic regions $(C^n_{i+1/2} = 0,$ $f'(u) > 0)$

$$0 \leqq \frac{\Delta x s^n_i}{u^n_i - u^n_{i-1}} \quad , \quad 0 \leqq \frac{\Delta x s^n_i}{u^n_{i+1} - u^n_i} \leqq 2 \tag{2.31}$$

and in subsonic regions $(D^n_{i-1/2} = 0,\ f'(u) > 0)$ the analogous inequalities; only the denominators in (2.31) have to be changed. So away from sonic points we can switch from (2.30) to these weaker conditions.

The CFL-like condition (2.12) of Harten's TVD-criterion causes a restriction on the mesh ratio. Because of the implicit nature of the two-step schemes and using the triangle inequality this leads to a very pessimistic condition; for $\Psi = 1$ we obtained $\lambda \max f'(u) < 0{,}3$. In practice, we have never observed such cruel restrictions.

In the linear case $f(u) = au$, $a \in \mathbb{R}$, a sufficient condition to be TVD is

$$0 \leqq \frac{\Delta x s^n_i}{u^n_i - u^n_{i-1}} , \frac{\Delta x s^n_i}{u^n_{i+1} - u^n_i} \leqq 2 \tag{2.32}$$

under the original CFL-condition $\lambda a \leqq 1$ of the monotone scheme.

For linear problems the Lax-Wendroff scheme and its two-step formulation are identical. The analogy is obtained between two-step schemes and flux limiter schemes. For this, the slopes are defined by

$$s^n_i = \frac{u^n_{i+1} - u^n_i}{\Delta x} \phi^n_i \quad , \quad \phi^n_i := \phi(r^n_i) \quad , \quad r^n_i := \frac{u^n_i - u^n_{i-1}}{u^n_{i+1} - u^n_i} \tag{2.33}$$

where the function ϕ switches between right- and left-hand difference quotients. Then the two-step scheme defined in this way agrees with a flux limiter scheme using the limiter function ϕ. In the context of two-step schemes we will call the function ϕ rather a switching or transition function than a limiter function.

Hence, in the linear case we can use the theory of schemes with flux limiters and every flux limiter defines an appropriate slope according to (2.33). The second-order TVD conditions (2.26),(2.32), expressed in terms of the switching function, agree with that of Sweby [33] for the flux limiters and we get his famous second-order TVD-region in the (r,ϕ)-plane (see Fig.2.2). In Table 2.1 we have listed the various switching functions and the associated slopes. The functions ϕ_k, $1 \leqq k \leqq 2$, ϕ_{CO}, ϕ_{VL} are the class of limiter functions of Sweby [33] including the limiters of Roe [30], and these functions proposed by Chakravarthy, Osher [25] and van Leer [21], respectively (see also [33]). The functions ϕ_M, ϕ_{VA} are switching functions associated with the slopes proposed in [20],[1]. The minmod function is defined by

$$\text{minmod}(a,b) = \begin{cases} a & \text{for} \quad |a| \leq |b| \quad , \; ab > 0 \\ b & \text{for} \quad |a| > |b| \quad , \; ab > 0 \\ 0 & \text{otherwise} \end{cases} \quad . \tag{2.34}$$

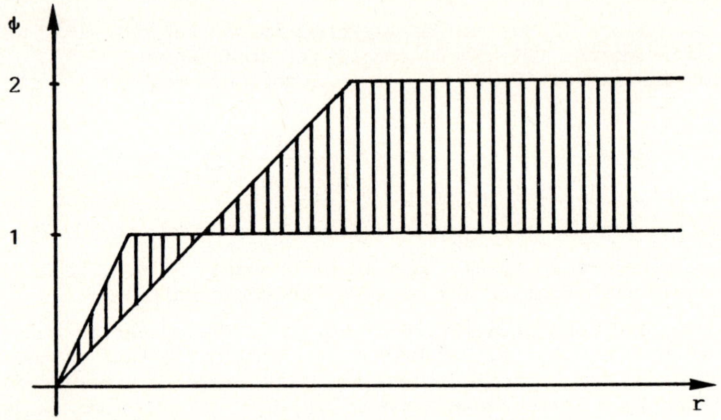

Fig.2.2 Second-order TVD region ([33])

All slopes except of ϕ_{CO} satisfy the symmetric condition

$$s(a,b) = s(b,a) \tag{2.35}$$

which ensures that backward and forward gradients are treated in the same fashion and so symmetric properties of the solution are maintained.

In [4] Chakravarthy and Osher proposed a new class of flux limiter TVD-schemes including a third-order accurate scheme. By the corresponding switching functions a class of two-step schemes is obtained where the boundary values in the grid zones are defined by

$$\begin{aligned} \delta_{i+} &= \frac{1}{4}\,[(1-\theta)\tilde{\tilde{s}}_i + (1+\theta)\tilde{s}_i] \\ \delta_{i-} &= \frac{1}{4}\,[(1-\theta)\tilde{s}_i + (1+\theta)\tilde{\tilde{s}}_i] \end{aligned} \tag{2.36}$$

where

$$\tilde{s}_i = \text{minmod}(u_{i+1} - u_i,\; \beta(u_i - u_{i-1}))$$

$$\tilde{\tilde{s}}_i = \text{minmod}(u_i - u_{i-1},\; \beta(u_{i+1} - u_i))$$

and $-1 \leq \theta \leq 1$. The compression parameter is given within the range

$$1 \leq \beta \leq \frac{3-\theta}{1-\theta} \; . \tag{2.37}$$

The value $\theta = 1/3$ leads to a third-order scheme, $\theta = 1$ to a fully one-sided and $\theta = 1/2$ to a central second-order approximation (see [4]). None of these schemes satisfies the symmetry property (2.35). A variation in which the continuous slope s_{VA} was used was given in [1]. For $\tilde{\tilde{s}} = \tilde{s}$ the schemes are identical to these defined by (2.19).

From TVD criterion (2.30) follows that at local extrema the slope is zero. Although a TVD-scheme can be made second-order accurate in the sense of global error in the L^1 norm, they necessarily degenerate to first-order accuracy in the sense of local truncation error (see [12],[25]). This means that local extrema of the numerical solution are stronger damped. Usually this is called the clipping phenomenon. Except for the scheme corresponding to the slope s_{VA} all schemes are TVD and the switching functions lie in the TVD region plotted in Fig.2.2.

To overcome this drawback of TVD-schemes at first van Albada et al. [1] used the above mentioned slope for which the transition function does not vanish for negative values of r. Recently Harten and Osher [14] have introduced the class of so-called uniformly non-oscillating (UNO) schemes. These schemes are required to diminish not the total variation but only the number of local extrema. In [14] these authors defined slopes which are second-order approximations of u_x. The associated scheme was constructed as a direct approximation of the conservation law and these piecewise linear initial data. This definition of slopes can also be used in a two-step algorithm. The slopes in [14] are given by

$$s_i = \text{minmod}(u_{i+1} - u_i - \frac{1}{2} d_{i+1/2},\ u_i - u_{i-1} + \frac{1}{2} d_{i-1/2}) \tag{2.38}$$

where d denotes the limited "second-order" terms

$$d_{i+1/2} = \text{minmod}(u_{i+2} - 2u_{i+1} + u_i,\ u_{i+1} - 2u_i + u_{i-1}) \quad . \tag{2.39}$$

Table 2.1 Switching functions, slopes

Switching Function	Slope
$\phi_1(r) = \max\{0,\min(r,1)\}$	$s_1(a,b) = \frac{1}{\Delta x}\,\text{minmod}(a,b)$
$\phi_k(r) = \max\{0,\min(kr,1),\min(r,k)\}$	$s_k(a,b) = \frac{1}{\Delta x}\,\max\{\text{minmod}(ka,b),\text{minmod}(a,kb)\}$ for $a,b \geqq 0$
$\phi_{CO}(r) = \max\{0,\min(r,k)\}$	$s_{CO}(a,b) = \frac{1}{\Delta x}\,\text{minmod}(ka,b)$
$\phi_{VL} = \frac{\lvert r\rvert+r}{1+\lvert r\rvert}$	$s_{VL}(a,b) = \frac{1}{\Delta x}\begin{cases} \frac{2ab}{a+b} & \text{for } ab > 0 \\ 0 & \text{otherwise} \end{cases}$
$\phi_M(r) = \text{minmod}(\frac{1+r}{2}, 2\,\text{minmod}(r,1))$	$s_M(a,b) = \frac{1}{\Delta x}\,\text{minmod}(\frac{a+b}{2}, 2\,\text{minmod}(a,b))$
$\phi_{VA}(r) = \frac{(r+\bar{c}^2)(1+r)}{1+r^2+2\bar{c}^2}$, $\bar{c}^2 = O(\Delta x)$	$s_{VA}(a,b) = \frac{1}{\Delta x}\frac{(ab+c^2)(a+b)}{a^2+b^2+2c^2}$, $c^2 = O(\Delta x^3)$
$1 \leqq k \leqq 2$, $r = b/a$	$a := u_{i+1} - u_i$, $b := u_i - u_{i-1}$

3. EULER EQUATIONS

The Euler equations may be written in conservation form

$$\mathbf{u}_t + \mathbf{f}(\mathbf{u})_x = 0 \tag{3.1}$$

where

$$\mathbf{u} = \begin{pmatrix} \rho \\ m \\ e \end{pmatrix} \quad , \quad \mathbf{f}(\mathbf{u}) = \begin{pmatrix} m \\ \rho v^2 + p \\ v(e+p) \end{pmatrix} \quad . \tag{3.2}$$

Here ρ is the density, v the velocity, p the pressure, e the total energy per unit volume and $m = \rho v$ is the momentum. We assume the gas is polytropic in which case the equation of state is

$$p = (\gamma-1)(e - \frac{1}{2}\rho v^2) \tag{3.3}$$

where γ is the ratio of sepcific heats.

The system (3.1) is hyperbolic because the Jacobian matrix $A = d\mathbf{f}/d\mathbf{u}$ has real eigenvalues

$$a_1(\mathbf{u}) = v - c \quad , \quad a_2(\mathbf{u}) = v \quad , \quad a_3(\mathbf{u}) = v + c \tag{3.4}$$

and a complete set of right eigenvectors

$$\mathbf{r}_1(\mathbf{u}) = \begin{pmatrix} 1 \\ v-c \\ H-vc \end{pmatrix} \quad , \quad \mathbf{r}_2(\mathbf{u}) = \begin{pmatrix} 1 \\ v \\ \frac{1}{2}v^2 \end{pmatrix} \quad , \quad \mathbf{r}_3(\mathbf{u}) = \begin{pmatrix} 1 \\ v+c \\ H+vc \end{pmatrix} \quad . \tag{3.5}$$

Here $c = \sqrt{\gamma p/\rho}$ is the sound velocity and $H = (e+p)/\rho$ is the enthalpy. The characteristic fields corresponding to the eigenvalues a_1, a_3 are genuinely nonlinear, the field corresponding to a_2 is linearly degenerate.

3.1 Upwind Schemes

In recent years several upwind schemes have been developed for solving the Euler equations. There are two groups of upwind schemes - the Godunov-type, also called flux-difference splitting schemes, and the flux-vector slpitting schemes. We will give here a short description of these methods, a detailed review can be found in the paper by Harten, Lax and van Leer [13].

The first upwind scheme was proposed by Godunov [9]. By the method of Godunov the discretized problem is interpreted as a sequence of Riemann problems. At every time level the numerical approximation is considered to be piecewise constant in every grid zone. To get an approximation at the next time level this problem is solved exactly. At each interface of the grid zones the break-up of the discontinuities is calculated by solving a local Riemann problem. For example, at the right boundary of the i-th grid-zone we have to solve the Cauchy problem with the piecewise constant initial values:

$$\mathbf{u}(x,0) = \begin{cases} \mathbf{u}_i^n & \text{for} \quad x \leqq x_{i+1/2} \\ \mathbf{u}_{i+1}^n & \text{for} \quad x > x_{i+1/2} \; . \end{cases} \tag{3.6}$$

The solution of this Riemann problem is a similarity solution depending on $\mathbf{u}_i^n$, $\mathbf{u}_{i+1}^n$ and on the ratio $\xi_{i+1/2}^n = (x - x_{i+1/2})/(t - t_n)$. We denote this solution by $\mathbf{w}(\xi^n_{i+1/2};\mathbf{u}^n_i,\mathbf{u}^n_{i+1})$.

The neighboring Riemann problems do not mutually influence the local solutions at $\xi^n_{i+1/2}=0$ for all values of i if the CFL-condition is satisfied:

$$a_{max}\frac{\Delta t}{\Delta x}<1 \tag{3.7}$$

where a_{max} is the largest signal velocity. Integrating the Euler equations over the rectangle $(x_{i-1/2},x_{i+1/2})\times(t_n,t_{n+1})$ we obtain in this case

$$\mathbf{u}^{n+1}_i=\mathbf{u}^n_i-\lambda[\mathbf{f}(\mathbf{w}(0;\mathbf{u}^n_i,\mathbf{u}^n_{i+1}))-\mathbf{f}(\mathbf{w}(0;\mathbf{u}^n_{i-1},\mathbf{u}^n_i))] \tag{3.8}$$

which corresponds to a scheme in conservation form where the numerical flux is the flux of the associated Riemann problems:

$$\mathbf{h}^n_{i+1/2}=\mathbf{f}(\mathbf{w}(0;\mathbf{u}^n_i,\mathbf{u}^n_{i+1})) \quad . \tag{3.9}$$

The exact solution of the Riemann problem for the Euler equations consists of four constant states separated by elementary waves. This solution can be calculated by a fixed point iteration introduced by Godunov [9]. Although very fast numerical iteration schemes have been developed for this procedure (see [6],[10]) its use at every interface still consumes much computer-time.

In recent years many efforts have been taken to construct faster upwind schemes. A possibility of reducing the computational effort required by Godunov's method is to replace the exact solution of the Riemann problem by a "simpler" approximate one. Numerical schemes based on such approximative Riemann solvers are called <u>Godunov-type</u> or <u>flux-difference splitting schemes</u> ([13]). We will describe next such a Godunov-type scheme, the <u>method of Roe</u> [28],[29].

Roe replaced the solution of the Riemann problem for the Euler equations by the exact solution of a Riemann problem for the linearized equation

$$\mathbf{w}_t+A_{lr}\mathbf{w}_x=0 \quad , \quad \mathbf{w}(x,0)=\begin{cases}\mathbf{u}_l & \text{for} \quad x<0 \quad ,\\ \mathbf{u}_r & \text{for} \quad x>0 \quad .\end{cases} \tag{3.10}$$

Here $A_{lr}=A_{lr}(\mathbf{u}_l,\mathbf{u}_r)$ is a matrix which is consistent with the Jacobian matrix A(u) of the Euler equations in the sense that A_{lr} has real eigenvalues, a complete set of eigenvectors, and $A_{lr}(\mathbf{u},\mathbf{u})=A(\mathbf{u})$. To assume the conservation property of the associated difference scheme, the matrix A_{lr} is required to have the "mean value" property in addition:

$$\mathbf{f}(\mathbf{u}_l)-\mathbf{f}(\mathbf{u}_r)=A_{lr}(\mathbf{u}_l-\mathbf{u}_r)\ . \tag{3.11}$$

For the Euler equations this matrix is $A_{lr}(\mathbf{u}_l,\mathbf{u}_r)=A(\overline{\mathbf{u}})$ where $\overline{\mathbf{u}}$ is defined by

$$\begin{aligned}\overline{v}&=(v_r\sqrt{\rho_r}+v_l\sqrt{\rho_l})/(\sqrt{\rho_r}+\sqrt{\rho_l})\\ \overline{H}&=(H_r\sqrt{\rho_r}+H_l\sqrt{\rho_l})/(\sqrt{\rho_r}+\sqrt{\rho_l})\\ \overline{c}&=(\gamma-1)(\overline{H}-\tfrac{1}{2}\overline{v}^2)\quad .\end{aligned} \tag{3.12}$$

The corresponding eigenvalues $\overline{a}_k=a_k(\overline{\mathbf{u}})$ and eigenvectors $\overline{\mathbf{r}}_k=\mathbf{r}_k(\overline{\mathbf{u}})$ are calculated by means of (3.4),(3.5).

The solution of the Cauchy problem (3.10) consists of four constant states seperated by three characteristic lines and can be easily computed (see [28],[13]). The numerical flux of the difference method corresponding with this approximate Riemann solver may be expressed as

$$\mathbf{h}_R(\mathbf{u}_l,\mathbf{u}_r) = \frac{1}{2}(\mathbf{f}(\mathbf{u}_l)+\mathbf{f}(\mathbf{u}_r)) - \frac{1}{2}\sum_{k=1}^{3} |\bar{a}_k|\alpha_k\bar{\mathbf{r}}_k \tag{3.13}$$

where the values α_k are the coefficients of the representation of $\mathbf{u}_r - \mathbf{u}_l$ in terms of the eigenvectors $\bar{r}_k$

$$\mathbf{u}_r - \mathbf{u}_l = \sum_{k=1}^{3} \alpha_k\bar{r}_k \quad . \tag{3.14}$$

Solving this linear system we obtain

$$\alpha_1 = \frac{1}{2}(C_1 - C_2),\ \alpha_2 = \rho_r - \rho_l - C_1\ ,\ \alpha_3 = \frac{1}{2}(C_1 + C_2) \tag{3.15}$$

where

$$C_1 = (\gamma-1)\{e_r - e_l + \frac{1}{2}\bar{v}^2(\rho_r - \rho_l) - \bar{v}(m_r - m_l)\}/\bar{c}^2$$

$$C_2 = \{m_r - m_l - \bar{v}(\rho_r - \rho_l)\} \quad .$$

By the method of Roe shock waves, contact discontinuities and rarefaction waves are approximated by linear discontinuities. This may cause trouble near sonic points ($a_k = 0$) where the method of Roe may admit the approximation of entropy violating "rarefaction shocks". But Roe's method can be modified by an "entropy-fix" to guarantee physically relevant solutions. The $|\bar{a}_k|$ term in (3.13) is replaced by

$$b_k = \begin{cases} |a_k| & \text{for} \quad |a_k| \geqq \delta_k \\ \delta_k & \text{for} \quad a_k < \delta_k \end{cases} \tag{3.16}$$

where δ_k is a positive number. This choice yields an encrease of entropy near sonic points and eliminates entropy violating shocks. For a proper choice see [12],[18].

Another upwind scheme was proposed by Osher and Solomon [26]. By this method the Riemann problem is approximately solved using rarefaction, compression and contact waves. This leads to a more complicated algorithm and we do not consider it here. Unlike in the Roe scheme the numerical flux of the Osher scheme is continuously differentiable which may be important for the construction of implicit schemes.

Another group of upwind schemes are <u>flux-vector splitting schemes</u>. These methods are based on splitting of the physical flux $\mathbf{f}(\mathbf{u})$ into a flux to the right $\mathbf{f}^+(\mathbf{u})$ and to the left $\mathbf{f}^-(\mathbf{u})$:

$$\mathbf{f}(\mathbf{u}) = \mathbf{f}^+(\mathbf{u}) + \mathbf{f}^-(\mathbf{u}) \quad . \tag{3.17}$$

The numerical flux of a flux-vector splitting scheme is then the sum of out- and incoming flux at each interface, e.g., at $x_{i+1/2}$

$$\mathbf{h}(\mathbf{u}_i^n,\mathbf{u}_{i+1}^n) = \mathbf{f}^+(\mathbf{u}_i^n) + \mathbf{f}^-(\mathbf{u}_{i+1}^n) \quad . \tag{3.18}$$

This corresponds to the approximation of $\mathbf{f}^+(\mathbf{u})_x, \mathbf{f}^-(\mathbf{u})_x$ by left-and right-hand differences, respectively. For the Euler equations flux-vector splitting methods were developed by Steger and Warming [32], and van Leer [19].

We will briefly outline this approach. Details can be found in the references above.

The method of Steger and Warming [32] is based on the property of the flux being a homogeneous function of degree one. So the flux can be expressed in terms of the Jacobian matrix $A(\mathbf{u})$:

$$\mathbf{f}(\mathbf{u}) = A(\mathbf{u})\mathbf{u} . \tag{3.19}$$

The matrix A can be diagonalized by a similarity transformation

$$R^{-1} A R = \Lambda = \mathrm{diag}(a_1, a_2, a_3) \tag{3.20}$$

where $R = (\mathbf{r}_1, \mathbf{r}_2, \mathbf{r}_3)$ is the matrix of the eigenvectors. Next we decompose every eigenvalue into a non-negative and a non-positive component a_k^+, a_k^-, e.g., via

$$a_k = a_k^+ + a_k^- = \max(0, a_k) + \min(0, a_k) \ , \ k = 1,2,3 \ . \tag{3.21}$$

So we can split the diagonal matrix into $\Lambda = \Lambda^+ + \Lambda^-$ and, using (3.20), the Jacobian matrix into

$$A = A^+ + A^- = R\Lambda^+R^{-1} + R\Lambda^-R^{-1} \tag{3.22}$$

where the matrices with the superscripts "+" and "-" possess non-negative and non-positive eigenvalues, respectively. Using the homogeneity property (3.19) we get the fluxes

$$\mathbf{f}^+(\mathbf{u}) = A^+\mathbf{u} \quad , \quad \mathbf{f}^-(\mathbf{u}) = A^-\mathbf{u} \tag{3.23}$$

corresponding to non-negative and non-positive eigenvalues, respectively. We note that the splitting of the eigenvalues in (3.21) are not unique. For another choice see [32].

The components of the general flux-vector $\tilde{f}(\mathbf{u}) = R\tilde{\Lambda}R^{-1}\mathbf{u}$ for any eigenvalues $\tilde{a}_1, \tilde{a}_2, \tilde{a}_3$ and corresponding diagonal matrix $\tilde{\Lambda}$ are

$$\begin{aligned} \tilde{f}_1(\mathbf{u}) &= \frac{\rho}{2\gamma}\,(2(\gamma-1)\tilde{a}_2 + \tilde{a}_1 + \tilde{a}_3) \quad , \\ \tilde{f}_2(\mathbf{u}) &= \tilde{f}_1 v + (\tilde{a}_3 - \tilde{a}_1)\,\frac{\rho c}{2\gamma} \quad , \\ \tilde{f}_3(\mathbf{u}) &= \tilde{f}_2 v + (\tilde{a}_1 + \tilde{a}_3)\,\frac{\rho c^2}{2\gamma(\gamma-1)} \quad . \end{aligned} \tag{3.24}$$

The flux-vector splitting method proposed by van Leer is not based on the homogeneity of the flux. As in the Steger and Warming scheme we have in supersonic $(M > 1)$ and sonic $(M < -1)$ regions $f^+ = f, f^- = 0$ and $f^+ = 0$, $f^- = f$, respectively. In terms of ρ, c and the Mach number $M = v/c$ the van Leer splitting is in the case $|M| < 1$

$$\begin{aligned} f_1^+(\mathbf{u}) &= \frac{\rho c}{4}\,(M+1)^2 \quad , & f_1^-(\mathbf{u}) &= -\frac{\rho c}{4}\,(1-M)^2 \quad , \\ f_2^+(\mathbf{u}) &= \frac{c}{\gamma}\, f_1^+((\gamma-1)M+2) \quad , & f_2^-(\mathbf{u}) &= -\frac{c}{\gamma}\, f_1^-(2-(\gamma-1)M) \quad , \\ f_3^+(\mathbf{u}) &= \frac{\gamma^2}{2(\gamma^2-1)}\,\frac{(f_2^+)^2}{f_1^+} \quad , & f_3^-(\mathbf{u}) &= \frac{\gamma^2}{2(\gamma-1)}\,\frac{(f_2^-)^2}{f_1^-} \quad . \end{aligned} \tag{3.25}$$

The split fluxes of van Leer are continuously differentiable whereas those of Steger-Warming are Lipschitz continuous but not differentiable at zeros of the eigenvalues.

3.2 Two-step Schemes

As in the scalar case, the upwind schemes can be modified to get higher resolution. Van Leer converted with his MUSCL scheme Godunov's method to second-order accuracy by replacing the piecewise constant by piecewise linear variables; then the Euler equations are solved with this piecewise constant initial values formulated as a Lagrangian step followed by a remap back to the Eulerian grid. The "piecewise parabolic method" of Woodward and Colella [35] in this context uses piecewise parabolic initial values to get a higher resolution. In the Eulerian MUSCL scheme of Colella [5] the Eulerian calculation is done directly (see also [6]). The two-step approach leads to a simpler algorithm and as building block every upwind scheme may be used.

The two-step algorithm for hyperbolic systems of conservation laws leads to the same procedure as in the scalar case. Only the scalar variables are replaced by vectors. But there are different ways to calculate the slopes in the grid zones. The simplest possibility is to apply the scalar theory to each of the Euler equations by calculating the slopes in terms of the conservative variables ρ,m,e. Another possible choice is in terms of the primitive variables ρ,u,p or any combination them. Van Albada et al. [1] have obtained the best results for a numerical example by using variables associated with the characteristic ones. We will adopt here a similar way by extending the scalar schemes to Euler equations using Roe's [29] and Huang's [15] formal generalizations to systems (see also [12]). These are based on a local linearization of the nonlinear system, which defines a local system of characteristic fields. The scalar schemes are then applied scalarly to each of the characteristic equations. This guarantees the favorable properties of the scalar schemes to be valid also for systems with constant coefficients.

For every grid zone an average value $\bar{\mathbf{u}}_i$, e.g., $\bar{\mathbf{u}}_i=(\mathbf{u}_{i+1}+2\mathbf{u}_i+\mathbf{u}_{i-1})/4$ is defined. Then we expand the right- and left-hand difference quotients in terms of the right eigenvectors $(\bar{\mathbf{r}}_k)_i$ of the Jacobian matrix $A(\bar{\mathbf{u}}_i)$

$$\begin{aligned} \frac{1}{\Delta x}(\mathbf{u}_{i+1}-\mathbf{u}_i) &= \sum_{k=1}^{3} (\alpha_k)_i(\bar{\mathbf{r}}_k)_i \quad , \\ \frac{1}{\Delta x}(\mathbf{u}_{i}-\mathbf{u}_{i-1}) &= \sum_{k=1}^{3} (\beta_k)_i(\bar{\mathbf{r}}_k)_i \quad . \end{aligned} \tag{3.26}$$

Here $(\alpha_k)_i,(\beta_k)_i$ denotes the components in the k-th characteristic field, calculated as in (3.15). A vector of slopes $\mathbf{s}$ is obtained when we apply to every characteristic field slopes s_k given by the scalar theory

$$\mathbf{s} = \sum_{k=1}^{3} (s_k)_i(\bar{\mathbf{r}}_k)_i \tag{3.27}$$

where

$$(s_k)_i = s_k((\alpha_k)_i,(\beta_k)_i) \quad . \tag{3.28}$$

The scalar slopes s_k applied need not be the same for each characteristic field.

A simplified version of this is obtained if the local linearization is performed at the interfaces of the grid zones. For this an average value at every interface is defined, e.g., $\bar{\mathbf{u}}_{i+1/2} = (\mathbf{u}_{i+1}+\mathbf{u}_i)/2$ or Roe's average value (3.12). The difference quotients may now be written as

$$\frac{1}{\Delta x}(\mathbf{u}_{i+1} - \mathbf{u}_i) = \sum_{k=1}^{3} (\alpha_k)_{i+1/2}(\mathbf{r}_k)_{i+1/2} \quad . \tag{3.29}$$

With the notation analogous to that above the vector of slopes is given analogously to (3.27) with

$$(s_k)_i = s_k((\alpha_k)_{i+1/2}, (\alpha_k)_{i-1/2}) \quad . \tag{3.30}$$

This procedure consumes less computational effort, but does not maintain symmetry properties of the solution.

4. NUMERICAL RESULTS

In this section some calculations will be presented and results indicated which have been obtained with the two-step schemes described previously. In all calculations we used a fixed uniform grid in space with step size $\Delta x = 0.01$, while Δt was calculated in each time step according to

$$\sigma \frac{\Delta t}{\Delta x} \max (|u| + c) = 1 \quad . \tag{4.1}$$

To minimize the computational effort for practical calculation σ should be as large as possible. We neglected the theoretical bounds for TVD-ness and choose $\sigma = 0.9$. All calculations were done on a Siemens 7881 computer using a four byte floating point arithmetic.

The first test problem we considered was the shock tube problem which was used by Sod [31] in 1978 comparing various finite difference methods. The initial condition is given by

$$(\rho,v,p)^T = \begin{cases} (1.0,0.0,1.0)^T & \text{for} \quad x < 0 \\ (0.125,0.0,0.1)^T & \text{for} \quad x \quad 0 \end{cases} \quad . \tag{4.2}$$

The exact solution consists in a shock wave propagating to the right and a rarefaction wave propagating to the left, separated by a contact discontinuity. In the figures we plotted at time $t = 0.2$ the density and velocity of the analytical solution by solid connecting lines, the values of the approximations by dots.

Figure 4.1 shows the typical results of an upwind scheme spreading the shock wave and especially the contact discontinuity over many grid points. Much better results are achieved by two-step schemes. Figure 4.2 represents the results where the slopes are calculated by s_1 in terms of the primitive variables ρ,v,p. More compressive slopes such as s_2 yield under- and overshoots. The best results are achieved by two-step schemes based on the method of Godunov; similar results we get by the schemes based on flux-vector splitting schemes; the schemes based on Roe's method still damped strongly the contact discontinuity. Slopes calculated in terms of the conservative variables ρ,m,e lead to slightly worse results. Using characteristic variables by extending the scalar theory to systems according to the methods of section 3.2, the two-step schemes are more robust and we obtained the best resolution using the compressive slope s_2 on each characteristic field. Only a small undershoot occurs at the contact discontinuity (see Figure 4.3). But the differences of the results between schemes based on different slopes and different upwind schemes are very small. So it turns out that Sod's

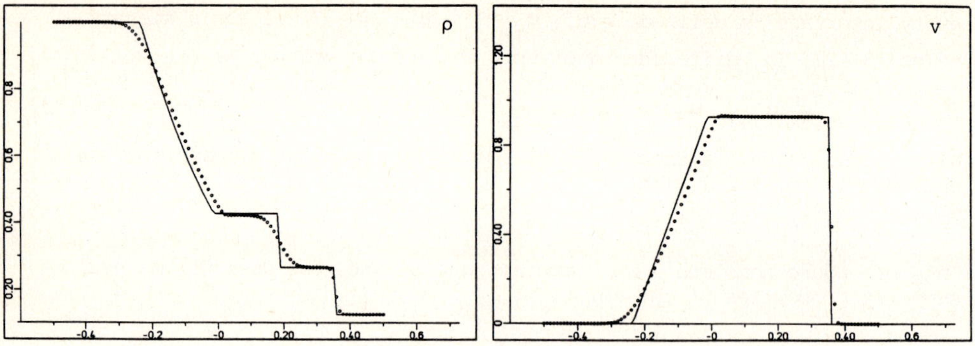

Fig.4.1 Method of Roe with entropy fix

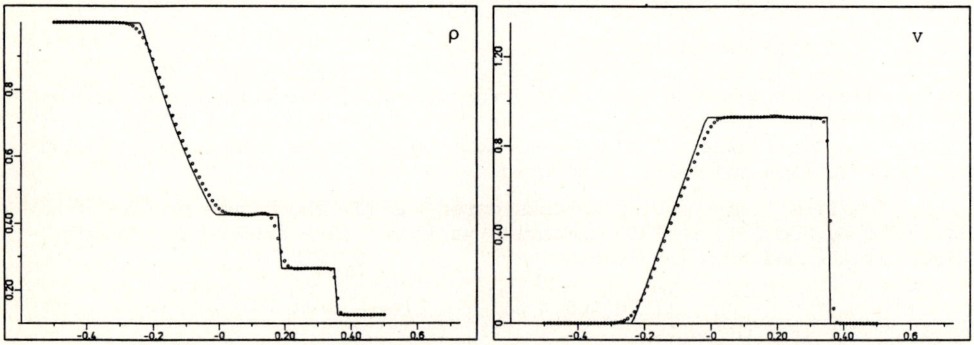

Fig.4.2 Flux-vector splitting of Steger and Warming,
slopes: s_1 calculated in terms of ρ,u,p

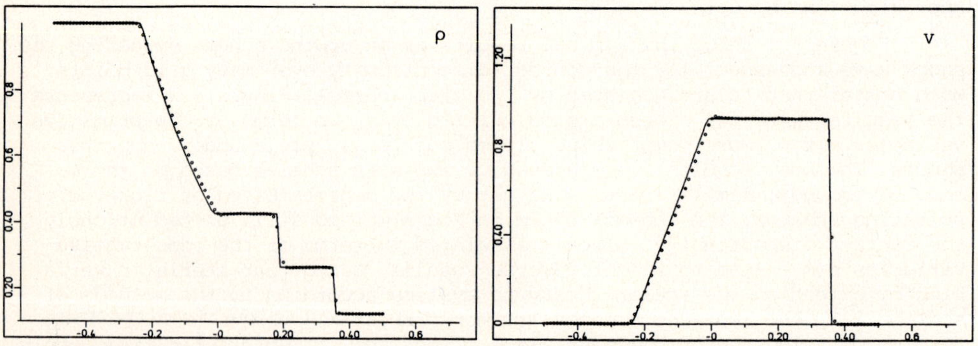

Fig.4.3 Flux-vector splitting of van Leer, slopes: s_2 calculated
in terms of characteristic variables

Riemann problem is no longer a convenient test problem to compare different high resolution schemes.

A more difficult test problem is the interaction of a shock wave with a contact discontinuity. Such a problem appears when the initial condition is given by

$$(\rho, v, p)^T = \begin{cases} (\rho_1,\ v_1,\ 10.0)^T & \text{for } x < -0.3 \\ (1.0,\ -1.0,\ 0.1)^T & \text{for } -0.3 < x < 0.3 \\ (10.0,\ -1.0,\ 0.1)^T & \text{for } x > 0.3 \end{cases} \tag{4.3}$$

with $v_1 = 601/106$ and $\rho_1 = \sqrt{9801/1202} = 1$. The exact solution of the local Riemann problem on the right consists in a contact discontinuity propagating to the left, while the solution on the left consists in a shock wave propagating to the right. These waves interact producing two shock waves and a contact discontinuity (see Figure 4.4).

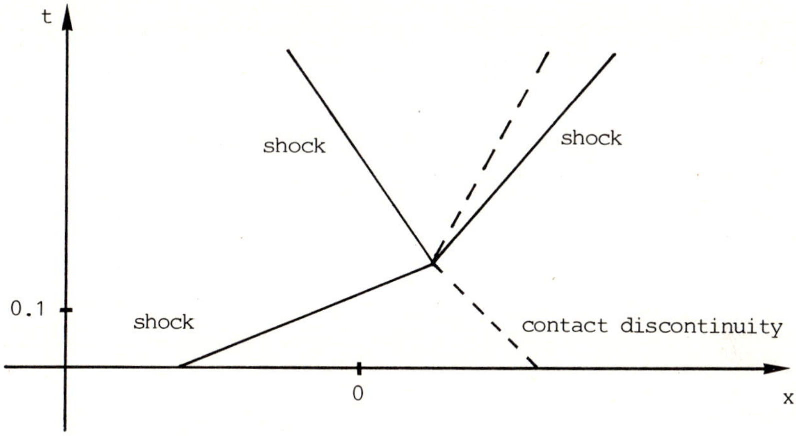

Fig.4.4 Test problem 2

This problem is not as complicated as the blast wave problem of Woodward and Colella [35], but the analytical solution can be easily computed by solving a Riemann problem (see Figure 4.4). The numerical schemes have to approximate this Riemann problem with approximated initial data. The solution contains a much stronger shock wave than in Sod's problem. We used the same discretization parameters as above; the figures were plotted at time $t = 0.4$.

Figures 4.5, 4.6 represent the results of two-step schemes calculating the slopes by s_1 in terms of the variables ρ, u, p. It is observed that van Leer's flux-vector splitting gives somewhat better results than that of Steger and Warming (see also [2]. This is evident especially at the shock wave to the left. Better results are achieved with the "non TVD slope" s_{VA} (Figure 4.7). More compressive slopes, such as s_2, s_M, may cause instabilities. This can be avoided by corrections as proposed in [7],[20], which reduce the compression of the slopes at critical points. Due to the rounding of initial data (4.3) all schemes produced a small wiggle by solving the left Riemann problem.

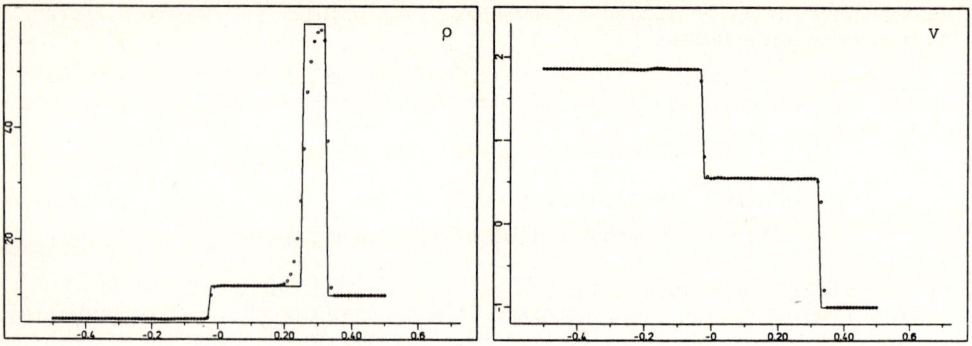

Fig.4.5 Flux-vector splitting of van Leer, slopes: s_1 calculated in terms of ρ, u, p

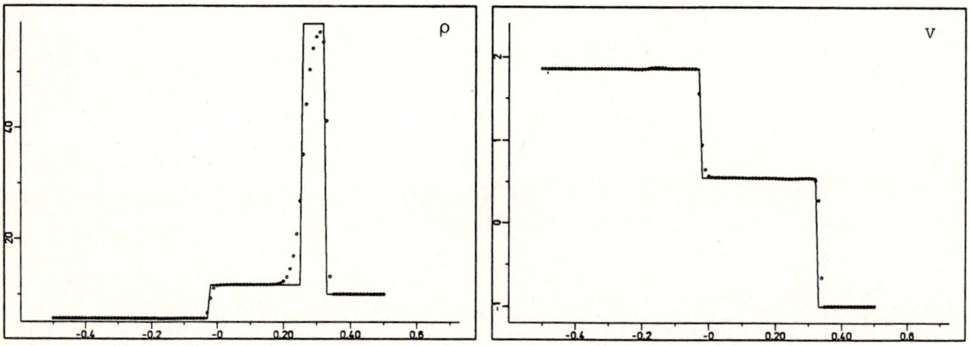

Fig.4.6 Flux-vector splitting of Steger and Warming, slopes: s_1 calculated in terms of ρ, u, p

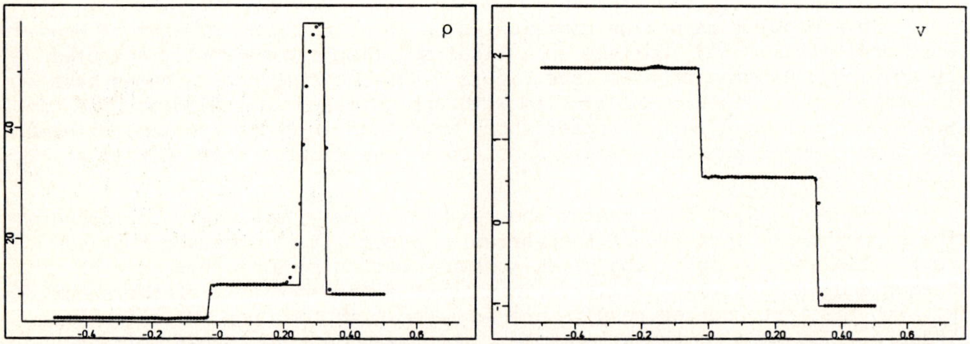

Fig.4.7 Flux-vector splitting of van Leer, slopes: s_{VA} calculated in terms of ρ, u, p

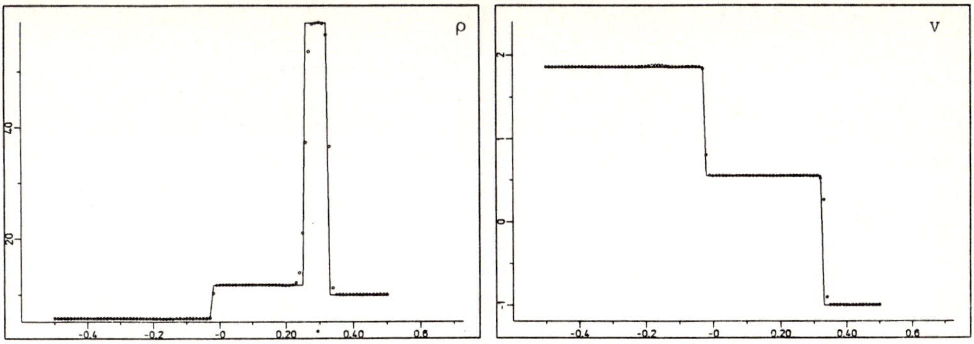

Fig.4.8 Method of Roe, slopes: $s_{1.5}$, s_2, $s_{1.5}$ calculated in terms of characteristic variables (3.29),(3.30)

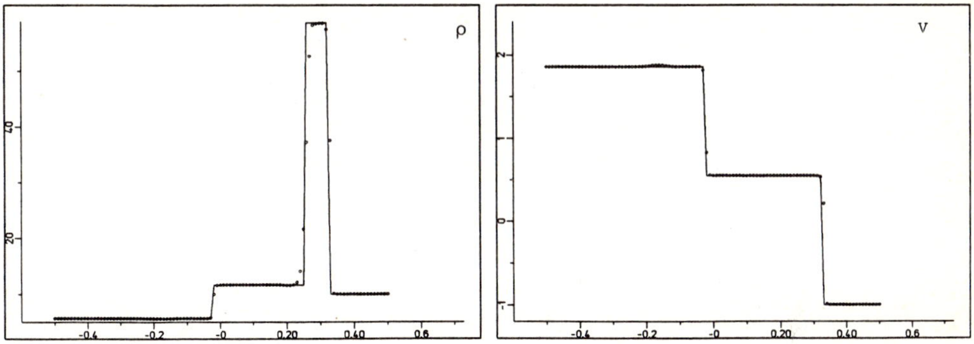

Fig.4.9 Method of Roe, slopes: $S_{1.5}$, s_2, $s_{1.5}$ calculated in terms of characteristic variables (3.26)-(3.28)

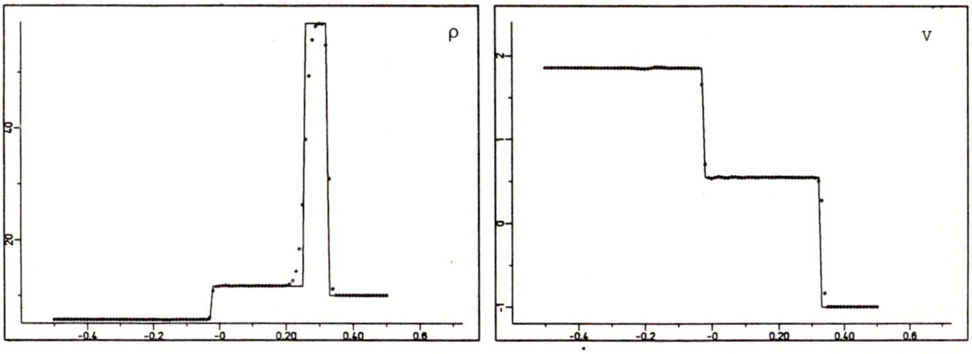

Fig.4.10 Flux-vector splitting of van Leer, slopes: (2.40) calculated in terms of characteristic variables (3.29),(3.30)

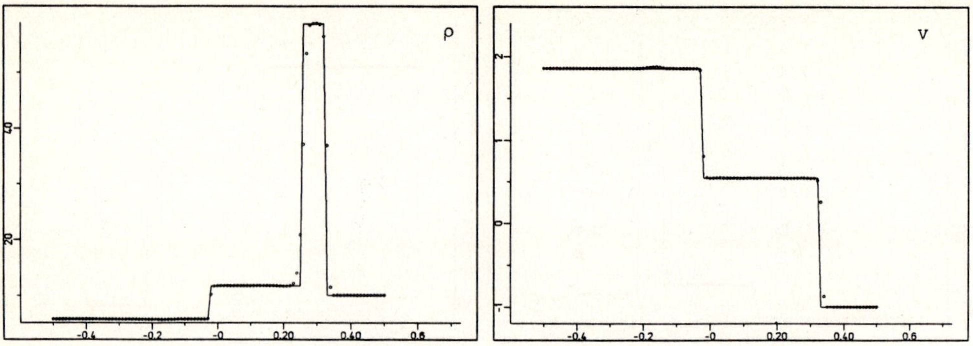

Fig.4.11 Method of Roe, slopes: (2.40),s_2,(2.40) calculated in terms of characteristic variables (3.29),(3.30)

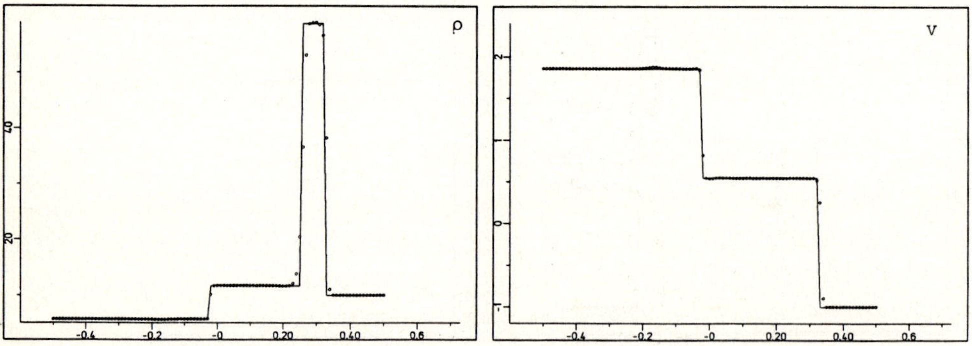

Fig.4.12 Method of Godunov, slopes: (2.40),s_2,(2.40) calculated in terms of characteristic variables (3.29),(3.30)

More robust algorithms and the possibility of using more compressive transition are obtained by making use of the field by field decompositions in Section 3.2. On the linearly degenerate field the most compressive slope given by the linear theory can be used; on the strictly nonlinear fields we obtained the best results without under- and overshoots by s_k with $k = 1.5$ (see Figure 4.8). Similar results are achieved by the slopes s_{VA}, s_{VL}. The approximations of schemes based on different upwind schemes are quite similar. The scheme of Steger and Warming produced a slight undershoot at the contact discontinuity which vanished for smaller values of σ. Both generalizations to systems of Section 3.2 with Huang's or Roe's mean value lead for this problem to almost identical results (see Figure 4.8,4.9). For symmetric problems we observed that the simplified version (3.29),(3.30) does not maintain the symmetry. Figure 4.10 indicates the results of van Leer's flux-vector splitting where the slope of Harten's and Osher's UNO scheme is employed on each characteristic field. In Figures 4.11, 4.12 this slope is replaced by the superbee s_2 on the linearly degenerate characteristic field. It is obvious from these figures that the resolution of shock waves by the two-step scheme based on the "UNO-slope" (2.40) is as good as those based on the slopes above, while at the contact discontinuity s_2 is more compressive.

But differences between similar compressive slopes are also barely visible at this test problem. In a parallel work [22] we studied extensively the numerical dissipation of various limiter and switching functions. The high resolution schemes are applied on a linear two-dimensional advection equation and "long time" calculations are made. The one-dimensional schemes are extended to two dimensions by operator splitting. At this problem the differences between various schemes become very illustrative. In agreement with the statement above it is observed that the slope (2.40) is not as compressive as the superbee. But the main advantage of the slope (2.40) is that the form of local extrema is much better maintained, because no clipping occurs. The results in [22] show in addition that (2.40) is more compressive than $\phi_{1.5}, \phi_{VA}, \phi_L$.

5. CONCLUSIONS

The two-step algorithm is a simple approach to convert every low order explicit upwind-scheme for the Euler equations to higher order accuracy. The approximation of test problems with schemes based on the upwind schemes of Godunov, Roe, Steger and Warming, and van Leer leads to the following conclusions.

For two-step schemes where the linear profiles of the dependent variables are calculated in terms of the primitive variables ρ, u, p, the best results were obtained by schemes based on Godunov's and van Leer's method. Slopes calculated in terms of the conservative variables led to slightly worse results. In agreement with the results of van Albada et al. [1] for van Leer's flux vector splitting, more robust and also more compressive schemes were obtained by calculating the slopes in terms of characteristic variables. Here it was achieved by using Roe's and Huang's generalization to systems. Different techniques using this concept which require different computational effort led for our problems to almost identical results. However, the simpler method does not maintain symmetry properties of the solution.

Numerical results of these schemes based on different upwind schemes are quite similar; but,schemes based on the flux-vector splitting of Steger and Warming did not seem to be as robust as the others (see also [2]). So the method of Roe and the flux vector splitting of van Leer are good alternatives to Godunov's method which needs more computational effort. Several slopes led to similar resolution of shock waves and contact discontinuities; more visible differences can only be seen at long time calculations.

ACKNOWLEDGEMENT

I would like to thank Professor E. Martensen for his encouragement of this work, Dr. D. Hänel and W. Schröder for helpful discussions, and L. Schmidt for his support in writing the computer codes.

REFERENCES

[1] van Albada, G.D., van Leer, B., Roberts, W.W.,Jr.: A comparative study of computational methods in cosmic gas dynamics, Astr.Astrophys. 108 (1982), 76-84.

[2] Andersen, W.K., Thomas, J.L., van Leer, B.: A comparison of finite volume flux vector splittings for Euler equations, AIAA Paper No. 85-0122 (1985).

[3] Boris, J.P., Book, D.L.: Flux corrected transport I. Shasta, A fluid transport algorithm that works, J. Comp. Phys. 11 (1973), 38-69.

[4] Chakravarthy, S.R., Osher, S.: A new class of high accuracy TVD schemes for hyperbolic conservation laws, AIAA Paper 85-0363 (1985).

[5] Colella, P.: A direct Eulerian MUSCL scheme for gas dynamics, SIAM J. Sci. Stat. Comp. 6 (1985), 104-117.

[6] Colella, P., Glaz, H.M.: Efficient solution algorithms for the Riemann problem for real gases, J. Comp. Phys. 59 (1985), 264-289.

[7] Colella, P., Woodward, P.R.: The piecewise parabolic method (PPM) for gas-dynamical simulation, J. Comp. Phys. 54 (1984), 174-201.

[8] Courant, R. Isaacson, E., Rees, M.: On the solution of nonlinear hyperbolic differential equations, Comm. Pure Appl. Math. 5 (1952), 243-255.

[9] Godunov, S.K.: Finite-difference method for numerical computation of discontinuous solutions of the equations of fluid dynamics. Mat. Sbornik 47 (1959), 271-306.

[10] Halter, E.: A fast solver for Riemann problems, Math.Meth. in the Appl. Sci. 7 (1985), 101-107.

[11] Harten, A.: On a class of high resolution total-variation-stable finite-difference schemes, SIAM J. Numer. Anal. 21 (1984), 1-23.

[12] Harten, A.: High resolution schemes for hyperbolic conservation laws, J. Comp. Phys. 49 (1983), 357-393.

[13] Harten, A., Lax, P.D., van Leer, B.: On upstream differencing and Godunov-type schemes for hyperbolic conservation laws, SIAM Rev. 25 (1983), 35-62.

[14] Harten, A., Osher, S.: Uniformly high-order accurate non-oscillatory schemes, I, MRC Technical Summary Report # 2823, May 1985.

[15] Huang, L.C.: Pseudo-unsteady difference schemes for discontinuous solutions of steady-state, one-dimensional fluid dynamics problems, J. Comp. Phys. 42 (1981), 195-211.

[16] Lax, P.D.: Hyperbolic Systems of Conservation Laws and the Mathematical Theory of Shock Waves, SIAM, Philadelphia (1972).

[17] Lax P.D., Wendroff, B.: Systems of Conservation Laws, Comm. Pure Appl. Math. 13 (1960), 217-237.

[18] van Leer, B.: On the relation between the upwind-differencing schemes of Godunov, Engquist-Osher and Roe, SIAM J. Sci. Stat. Comput. 5 (1984), 1-21.

[19] van Leer, B.: Flux-vector splitting for the Euler equations, Lecture Notes in Physics, Ed. Krause, Vol. 170 (1982), 507-512.

[20] van Leer, B.: Towards the ultimate conservative difference scheme, V. A second-order sequel to Godunov's method, J. Comp. Phys. 32 (1979), 101-136.

[21] van Leer, B.: Towards the ultimate conservative difference scheme, II. Monotonicity and conservation combined in a second-order scheme. J. Comp. Phys. 14 (1974), 361-370.

[22] Munz, C.-D.: On the numerical dissipation of high resolution schemes for hyperbolic conservation laws, to appear.

[23] Osher, S.: Convergence of generalized MUSCL schemes, SIAM J. Numer. Anal. 22 (1985), 947-961.

[24] Osher, S.: Riemannsolvers, the entropy condition, and difference approximations, SIAM J. Numer. Anal. 21 (1984), 217-235.

[25] Osher, S., Chakravarthy, S.: High resolution schemes and the entropy conditions, SIAM J. Numer. Anal. 21 (1984), 955-984.

[26] Osher, S., Solomon, F.: Upwind difference schemes for hyperbolic systems of conservation laws, Math. Comp. 38 (1982), 339-374.

[27] Richtmyer, R.D., Morten, K.W.: Difference Methods for Initial-Value Problems, Interscience, New York (1967).

[28] Roe, P.L.: The use of the Riemann problem in finite-difference schemes, Lecture Notes in Physics 141, Springer-Verlag, New York (1981), 354-359.

[29] Roe, P.L.: Approximate Riemann solvers, parameter vectors, and difference schemes, J. Comp. Phys. 43 (1981), 357-372.

[30] Roe, P.L., Baines, M.J.: Algorithm for advection and shock problems, Proceedings of the 4th GAMM Conference on Numerical Methods in Fluid Mechanics (1982), H. Viviand Ed., Vieweg (1982).

[31] Sod, G.A.: A survey of several finite difference methods for systems of nonlinear hyperbolic conservation laws, J. Comp. Phys. 27 (1978), 1-31.

[32] Steger, J.L., Warming, R.F.: Flux vector splitting of the inviscid gasdynamics equations with application to finite difference methods, J. Comp. Phys. 40 (1981), 263-293.

[33] Sweby, P.K.: High resolution schemes using flux limiter for hyperbolic conservation laws, SIAM J. Numer. Anal. 21 (1984), 995-1011.

[34] Tadmor, E.: Numerical viscosity and the entropy condition for conservative difference schemes, Math. Comp. 43 (1984), 369-381.

[35] Woodward, P., Colella, P.: The numerical simulation of two-dimensional fluid flow with strong shocks, J. Comp. Phys. 54 (1984), 115-173.

A new type of higher-order boundary integral approximation for potential flow problems in three dimensions

Z.P. Nowak

Institute of Applied Mechanics
and Aircraft Technology
Warsaw Technical University
00-665 Warsaw, Nowowiejska 24

1. Introduction

Higher-order panel methods have been used for solving the potential flow poblems for more than a decade. It is known that the results obtained with the aid of these methods are relatively insensitive to shapes and arrangement of surface elements. Approximately or strictly continuous singularity distributions are applied to construct the discrete schemes of higher-order accuracy, resulting in more stable calculations for supersonic flows [2] or flows with the leading edge vortex separation [4]. The final argument for the application of the higher-order panel methods is given in [3], where it is shown that the lower-order methods, using constant or even linear doublet distributions, provide incorrect data for the boundary layer calculations near the wing trailing edges. At least quadratic doublet distributions are needed to obtain the correct results for the lateral component of the local velocity in these areas.

The higher-order methods are more complicated and difficult to program, due mainly to the presence of the surface curvature corrections in the quadrature formulas. In particular, the kernels of the integral operators must be replaced by truncated series of analytically integrable terms. Such a "curvature expansion", when combined with other necessary approximations, requires a considerable human effort and a long computing time. The difficulties are greatly increased in supersonic flow cases, for which the domains of integration may be bounded by the curvilinear intersections of panels and the Mach cones. Therefore, in the PAN AIR method [9], currently used for calculating such flows, the curved surface of a body is replaced with a polyhedron and the curvature corrections are simply ignored. This, however, results in a loss of accuracy.

In the present paper a new type of the higher-order panel method is described, which takes account of the surface curvature by adding simple corrections to the source distribution. As in the PAN AIR method, the surface of a body is approximated with a closed polyhedron, composed of the plane panels, and the doublet distribution is sought in the form of a continuous piecewise quadratic function. An approximate integro-differential formulation of the flow problem, derived from Green's boundary formula, is used as a basis for the discretization procedure. The collocation points are located at the panel vertices and in the middle of the panel edges. This simplifies the procedure for the quadratic approximation of the doublet distribution over the panel areas and ensures the continuity of the doublet strength over the whole surface, including the abutments of various parts, such as the wing and the body of an aircraft. The method is almost as simple as the first-order methods, but its truncation error is $O(h^3)$, where h denotes the characteristic panel size.

In the present paper we confine our attention to non-lifting incompressible potential flows. The accuracy orders are examined for the problem of flow around a sphere. The generalizations to lifting linearized compressible potential flows will be presented elsewhere. The present approach seems to be especially attractive in connection with linearized supersonic flows, for which the other alternatives of increasing the order of accuracy are more difficult to implement.

2. The differential and integral formulations of the problem

The perturbation velocity potential φ for the incompressible flow in the region R_e outside an impermeable surface S is the solution of the Neumann boundary value problem

$$\Delta\varphi(p) = 0 \qquad , p \in R_e \quad , \tag{1a}$$

$$\frac{\partial\varphi}{\partial n}(p) = f(p) \qquad , p \in S \quad , \tag{1b}$$

$$\varphi(p) \to 0 \qquad \text{for } |p| \to \infty \; , \tag{1c}$$

where $f(p) = -V_\infty \cdot n_p$, V_∞ is the velocity of the undisturbed flow far from S, and n_p is the exterior normal at $p \in S$.

The recent panel methods (e.g.[10], [5], [9], [8]) favour Green's boundary formula

$$\varphi(p) = -\frac{1}{2\pi}\int_S \frac{f(q)}{|p-q|}\,dS(q) + \frac{1}{2\pi}\int_S \frac{\partial}{\partial n_q}\left(\frac{1}{|p-q|}\right)\varphi(q)\,dS(q) \qquad , p \in S, \tag{2}$$

as the integral counterpart of the problem (1a-c). Equation (2) can be written in the compact form

$$\varphi(p) = F(p) + A\varphi(p) \qquad , p \in S. \tag{3}$$

3. Previous work

In the following account of the previous work attention will be confined to these papers and reports, available to the present author, which contain sufficiently detailed presentations of the discretization procedures.

In [5], [9] and [8] the formula (2) is used as a basis for the discretization procedure. The continuous problem (2) is approximated with a system of linear algebraic equations

$$\varphi_n = F_n + A_n \varphi_n ,$$

where φ_n , F_n are the sequences of the approximate values for $\varphi(p_i)$, $F(p_i)$, $i=1,2,\ldots,n$, at the collocation points $p_i \in S$, and A_n is a linear n-dimensional operator. The discretization error, defined as the difference of the discrete and the exact solutions at the collocation points, depends on the accuracy of approximation of F with F_n and A with A_n.

To obtain sufficiently accurate approximations for F and A in the case of a thin and edged body, such as the wing of an aircraft, analytic formulae should be used for quadratures. In fact, in the neighbourhood of the trailing edge, where the kernels of F and A vary steeply over the opposite panel areas, numerical quadratures introduce unacceptable errors, unless a large number of nodal values are present in the quadrature formulae. An early higher-order panel method [11], based on the use of the Gauss quadrature formulae, requires lengthy calculations, as it can be seen from the efficiency comparisons given in [12], pp. 28-29. Since relatively simple analytic formulae for the integrals F and $A\varphi$ are available only for the plane polygonal domains, the surface S must be approximated with a system of panels of such a shape. In [5] and [8] the panels P_i , $i=1,2,\ldots,n$, are chosen to be tangent to S at the collocation points p_i (Fig. 1).

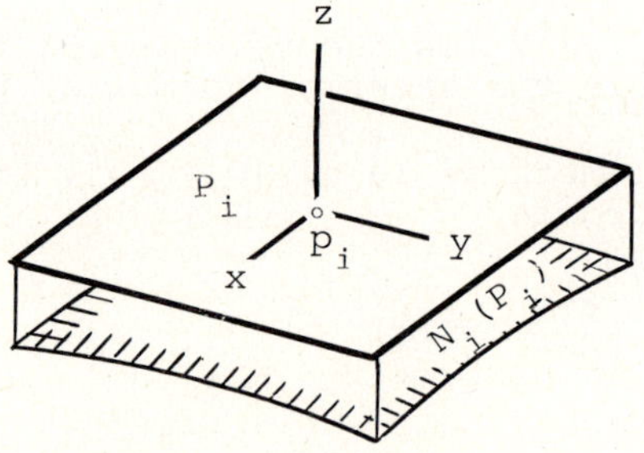

Fig. 1 An element of the approximating panel system

Their vertices are determined by projecting the nodes of a curvilinear grid, defined on S. Usually, the direction of the surface normal at p_i (z-axis in Fig. 1) is used for the projection. Let $N_i(q) \in S$ denote the image of a point $q \in P_i$, obtained by the projection from P_i to S. The image of the panel P_i on S will be denoted by $N_i(P_i)$.

The contributions of the image areas $N_i(P_i)$ to F(p) and $A\varphi(p)$ are

$$\int_{N_i(P_i)} a'(p,q) f(q) dS(q) \quad \text{and} \quad \int_{N_i(P_i)} a''(p,q) \varphi(q) dS(q), \tag{4}$$

respectively, where

$$a'(p,q) = -\frac{1}{2\pi} \frac{1}{|p-q|} \quad \text{and} \quad a''(p,q) = \frac{1}{2\pi} \frac{\partial}{\partial n_q} \left(\frac{1}{|p-q|}\right).$$

The integrals (4) can be written in the form

$$\int_{P_i} a'(p, N_i(q)) f(N_i(q)) J_i(q) \, dP_i(q)$$

and (5)

$$\int_{P_i} a''(p, N_i(q)) \varphi(N_i(q)) J_i(q) \, dP_i(q),$$

where $J_i(q) = dS(q)/dP_i(q)$ under the transformation $N_i(q)$. Let us note that for surfaces with the continuous curvature we have $J_i(q) = 1 + O(h^2)$, where h is the characteristic panel size.

In [5] and [8] the term $O(h^2)$ in the expression for $J_i(q)$ has been retained, which means that the methods are intended to reduce the truncation error to the level $O(h^3)$. Let us note, however, that F(p) and $A\varphi(p)$ are not identical with the sums of the integrals (4). In fact, the images $N_i(P_i)$ may leave gaps or may overlap on S. It is easy to see that the width of these gaps and overlaps is generally $O(h^3)$. Hence, their area is $O(h^4)$ for each panel, which is equal to $O(h^2) \cdot |P_i|$, where $|P_i|$ is the panel area. Consequently, the effect of the gaps and overlaps is such as the effect of an additional truncation error with a density $O(h^2)$, distributed over P_i. This means that the desired reduction of the truncation error to $O(h^3)$ has not been achieved. To reduce the truncation error to $O(h^3)$, the directions of the local surface normals, rather than the panel normals, should be used for the projection. More precisely, the panel constructions and the quadratures in (5) should be performed using the projection N from $q \in P_i$ to $N(q) \in S$, such that $q - N(q)$ lies along the normal to S at $N(q)$. It can be shown that the width of gaps and overlaps is then diminished to $O(h^4)$, resulting in an acceptable density $O(h^3)$ of the additional error.

The factors $f(N(q))$, $\varphi(N(q))$ and $J_i(q)$ in the corrected integrals (5) can be approximated with quadratic functions. As in

[5], the approximation for φ (N(q)) on a panel P_i can be obtained by the method of weighted least squares, using the values of φ_n at the nodes of the 9-point star with the central point $p_i \in P_i$. In [5] and [9] the source strength f(N(q)) is approximated with linear functions over the panel areas. However, a simple analysis, given in [8], shows that a quadratic approximation for f(N(q)) is necessary to reduce the error in F_n (and hence φ_n) to $O(h^3)$.

Relatively simple analytic formulae ([5], pp. 122) are available for the integrals of the form

$$\int_P \frac{(x_q-x_p)^m (y_q-y_p)^n}{|p-q|^k} \, dx_q \, dy_q, \quad m,n=0,1,2,\ldots;k=1,3,5,\ldots, \tag{6}$$

where x_p , y_p are the coordinates of a fixed point $p=(x_p,y_p,z_p)$ in the Cartesian coordinate system connected with a panel P (Fig. 1), and $q=(x_q,y_q,0) \in P$. For the quadratures, the kernels a'(p,N(q)) and a"(p,N(q)) of the corrected integrals (5) must be approximated with combinations of quotients of the same form as the integrand in (6). Let us consider the simpler case of the kernel a'(p,N(q)). We have

$$|p-N(q)|^2 = (p-N(q)) \cdot (p-N(q)) = |p-q|^2 (1+\varepsilon(q)),$$

where

$$\varepsilon(q) = [2(p-q)+q-N(q)] \cdot (q-N(q)) / |p-q|^2 .$$

Consequently,

$$a'(p,N(q)) = \frac{1}{|p-q|} (1+\varepsilon)^{-1/2} = \frac{1}{|p-q|} (1-\frac{\varepsilon}{2} + \frac{3}{8}\varepsilon^2 - \ldots), \tag{7}$$

which gives a'(p,N(q)) as the sum of the terms

$$\frac{\omega_k(x_p, y_p; x_q, y_q)}{|p-q|^k}, \quad k=1,3,5,\ldots,$$

where $\omega_k(x_p, y_p; x_q, y_q)$ are regular functions, which can be approximated with polynomials. A similar expansion is obtained also for the kernel a"(p,N(q)). We can now multiply all the polynomial approximations, present in the integrands in (5), and after complicated algebraic calculations represent the integrals (5) as combinations of the integrals (6). The analytic formulae are then used for quadratures.

A detailed analysis for the surfaces of the class C^4 shows that reducing the truncation error to $O(h^3)$ requires the calculation of 30 integrals (6) for each panel, with $m + n \leq 3$ and $k=1,3,5$. When the first integral with $m=n=0$ and $k=1$ has been obtained from a formula involving four logarithms and four arctangents for a quadrilateral panel, the further ones follow by recursion. Various critical situations, connected with the divisions by small numbers, arise when a collocation point p lies too close to the plane of P (even when it is far from P), or when it lies too close to an edge or vertex of P. In such cases additional series expansions are needed. The calculations of the influence coefficient matrix, using various strategies for various critical situations, are lengthy even when the analytic integration is confined to the panels in a neighbourhood of each collocation point.

The above method cannot be used for determining linearized supersonic flows, since it introduces small discontinuities in the geometry of the panel system and in the approximation for $\varphi(N(p))$. Such discontinuities are conducted along the characteristics in the flowfield, creating a noise which distorts the results of the calculations. Furthermore, additional difficulties appear, when domains of integration are bounded by the curvilinear intersections of panels and the Mach cones. In the PAN AIR method [9], designed mainly for the calculations of supersonic flows, the surface S is replaced with a closed polyhedron, composed of the plane panels, and the doublet strength is approximated with a continuous piecewise quadratic function. Accordingly, $J_i(q)=1$ in (5) and no expansions, such as (7), are needed for $a'(p,N(q))$ and $a''(p,N(q))$. Such a procedure introduces a truncation error $O(h^2)$.

4. The present method

4.1. The approximate integro-differential formulation of the problem

We have seen that reducing the truncation error to $O(h^3)$ is difficult and costly, even for incompressible flow problems, if the exact Green formula (2) is used for derivation of the discrete system.

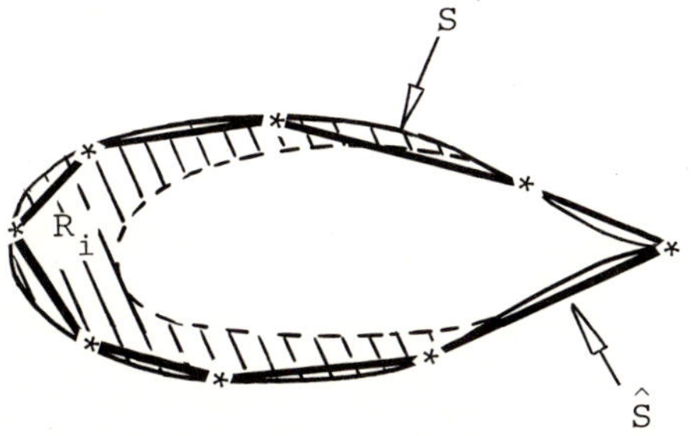

Fig. 2 A cross-section of S and $\hat{S}$

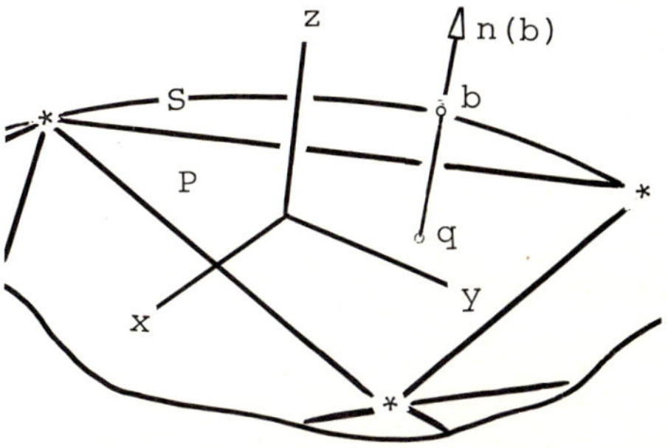

Fig. 3 The correspondance between b and q

Proceeding now to the description of our method, let us suppose that the function φ, which is defined in $R_e \cup S$, has been continued into a part R_i of the interior region (hatched area in Fig. 2) in such a way that it belongs to the class C^3 in $R_e \cup S \cup R_i$ (the smoothness assumption could be slightly relaxed). Near the concave parts of S, such as the neighbourhood of the cusp in Fig. 2, no continuation will be needed.

We approximate the surface S with a closed polyhedron $\hat{S}$ with vertices on S. It will be convenient to use polyhedra composed of the triangular panels. For a point $q \in \hat{S}$ let $b \in S$ be such that b-q lies along the normal to S at b (Fig. 3). In the notation of the previous Section $b=N(q)$. Let $\delta(q)$ denote the signed distance between b and q:

$$\delta(q) = (b-q)\cdot n(b) = O(h^2) ,$$

and let $n_x(q)$, $n_y(q)$, $n_z(q)$ be the components of the unit normal $n(b)$ in the Cartesian coordinate system connected with the panel P (Fig. 3). Using

$$\frac{\partial\varphi}{\partial n}(b) = \frac{\partial\varphi}{\partial n}(q) + \delta(q)\frac{\partial^2\varphi}{\partial n^2}(q) + O(\delta(q)^2) \tag{8}$$

and

$$\frac{\partial\varphi}{\partial n}(q) = n_x(q)\frac{\partial\varphi}{\partial x}(q) + n_y(q)\frac{\partial\varphi}{\partial y}(q) + n_z(q)\frac{\partial\varphi}{\partial z}(q) ,$$

in combination with (1b), we obtain

$$\frac{\partial\varphi}{\partial z}(q) = \frac{f(n(q))}{n_z(q)} - \frac{n_x(q)}{n_z(q)}\frac{\partial\varphi}{\partial x}(q) - \frac{n_y(q)}{n_z(q)}\frac{\partial\varphi}{\partial y}(q) \tag{9}$$

$$- \frac{\delta(q)}{n_z(q)}\frac{\partial^2\varphi}{\partial n^2}(q) + O(h^4).$$

By the continuity of $\Delta\varphi$ and $\partial(\Delta\varphi)/\partial n$ we have

$$\Delta\varphi(q) = \Delta\varphi(b) + O(\delta(q)) = O(h^2)$$

and, consequently,

$$\frac{\partial^2\varphi}{\partial z^2}(q) = - \Delta_{\hat{S}}\varphi(q) + O(h^2),$$

where $\Delta_{\hat{S}} = \frac{\partial^2}{\partial x^2} + \frac{\partial^2}{\partial y^2}$. Since

$$\frac{\partial^2\varphi}{\partial n^2}(q) = \frac{\partial^2\varphi}{\partial z^2}(q) + O(h),$$

we obtain

$$\frac{\partial^2\varphi}{\partial n^2}(q) = -\Delta_{\hat{S}}\varphi(q) + O(h). \tag{10}$$

Now, (9) takes the form

$$\frac{\partial\varphi}{\partial z}(q) = \frac{f(N(q))}{n_z(q)} - \frac{n_x(q)}{n_z(q)}\frac{\partial\varphi}{\partial x}(q) - \frac{n_y(q)}{n_z(q)}\frac{\partial\varphi}{\partial y}(q)$$
$$+ \frac{\delta(q)}{n_z(q)}\Delta_{\hat{S}}\varphi(q) + O(h^3). \tag{11}$$

A general representation formula ([1], p. 257) for the functions of the class C^2 in the region $\hat{R}_e$ outside $\hat{S}$ can be led to the form:

$$\varphi(p) = -\frac{1}{4\pi}\int_{\hat{S}}\frac{1}{|p-q|}\frac{\partial\varphi}{\partial z}(q)\, d\hat{S}(q)$$
$$+ \frac{1}{4\pi}\int_{\hat{S}}\frac{\partial}{\partial z}\left(\frac{1}{|p-q|}\right)(\varphi(q)-\varphi(p))\, d\hat{S}(q) \tag{12}$$
$$- \frac{1}{4\pi}\int_{\hat{R}_e}\frac{\Delta\varphi(q)}{|p-q|}\, d\hat{R}_e(q) \quad , \; p\in\hat{R}_e ,$$

where z is normal to $\hat{S}$ at q. The above formula is valid also for $p\in\hat{S}$, since the doublet integral is continuous in $\hat{R}_e\cup\hat{S}$. A similar regularization of the doublet integral is used in [7]. The volume integral in (12) is $O(h^4)$, since $\Delta\varphi=O(h^2)$ in the parts of R_i of the width $O(h^2)$, lying between $\hat{S}$ and S, and $\Delta\varphi=0$ outside. Substituting (11) into (12) we obtain the following approximate integro-differential formulation of the problem (1a-c):

$$\varphi(p) = -\frac{1}{4\pi}\int_{\hat{S}}\frac{1}{|p-q|}\left[\frac{f(n(q))}{n_z(q)} - \frac{n_x(q)}{n_z(q)}\frac{\partial\varphi}{\partial x}(q)\right.$$

(contd)

$$- \frac{n_y(q)}{n_z(q)} \frac{\partial\varphi}{\partial y}(q) + \frac{\delta(q)}{n_z(q)} \Delta_{\hat{S}}\varphi(q)] \, d\hat{S}(q)$$

$$+ \frac{1}{4\pi} \int_{\hat{S}} \frac{\partial}{\partial z}\left(\frac{1}{|p-q|}\right) (\varphi(q)-\varphi(p)) \, d\hat{S}(q) \tag{13}$$

$$+ O(h^3), \quad p \in \hat{S} .$$

If the surfaces S and $\hat{S}$ are identical, and p is not a vertex, then (12) can be led to the form (2). For curved surfaces, the source strength has been corrected in (13) to include the surface curvature effect.

Let us note that (12) and (13) are valid also when p is a vertex of $\hat{S}$.

4.2. The discretization method

The truncation error will be reduced to $O(h^3)$ by discretizing (13) instead of (2).

As usual, the surface S is defined as an image of a rectangle R on an auxiliary (ξ,η)-plane, under a transformation $(\xi,\eta)\rightarrow(X, Y,Z)$, or as a union of such images. The nodes of a rectangular grid on R are transformed into a set of grid points on S (circles and stars in Fig. 4).

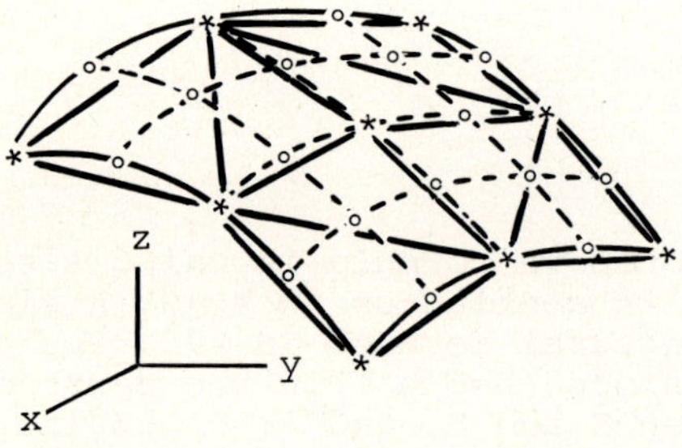

Fig. 4 The construction of $\hat{S}$

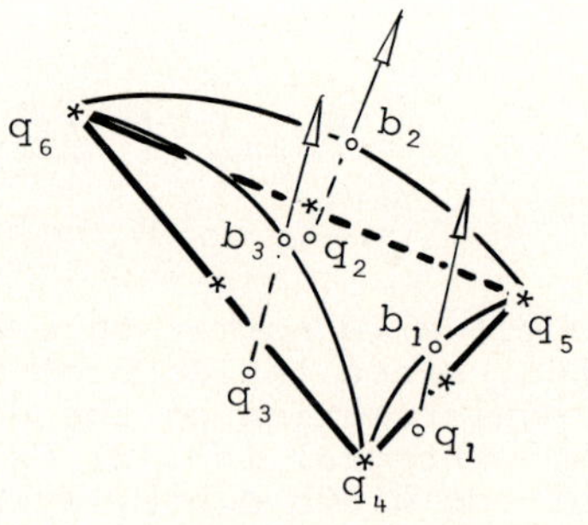

Fig. 5 The characteristic points on S and $\hat{S}$

The points denoted by stars in Fig. 4 serve as the vertices of $\hat{S}$, and also as the collocation points for the discretization procedure. The points denoted by circles will be used for improving the approximation of the integrated functions. As shown in Fig. 5, the collocation points are introduced also in the middle of the panel edges.

The solution φ is approximated over each panel with a quadratic function, using values of φ_n at six collocation points on the panel boundaries. Similar approximations will also be needed for the quotients

$$\frac{f(n(q))}{n_z(q)}\ ,\quad \frac{n_x(q)}{n_z(q)}\ ,\quad \frac{n_y(q)}{n_z(q)}\ ,\quad \frac{\delta(q)}{n_z(q)}\ ,\qquad (14)$$

appearing in the expression for the source strength in (13). As in Fig. 5, let b_1, b_2, b_3 denote the three grid points, lying on S close to the edges of a panel, and let q_1, q_2, q_3 be their projections on the panel plane. The quadratic approximations for the functions (14) are obtained with the aid of their values at q_1, q_2, q_3 and at the vertices q_4, q_5, q_6.

All the quadratic approximations are substituted into condition (13), which is enforced, after neglecting the term $O(h^3)$, at the collocation points. The functions $\partial\varphi/\partial x$, $\partial\varphi/\partial y$ and $\Delta_{\hat{S}}\varphi$ are obtained by differentiating the quadratic expression for φ. The third order polynomial for the source strength in (13) is truncated by omitting the third order terms. Consequently, only the potentials with quadratic singularity distributions over the panels areas are calculated. Since the panels are flat, no additional kernel expansions are needed.

The required potentials are combinations of 12 integrals (6) with $m + n \leq 2$ and $k=1,3$. The divisions 0/0 occur in the recursion formulas for the component terms of the doublet integral in (13), when $k=3$ and p coincides with a vertex or lies on an edge of a panel P. Fortunately, the panels surrounding p make no contribution to the doublet integral at p, due to the special form of the density function: $\varphi(q)-\varphi(p)$. Since the remaining panels do not contain p, no divisions 0/0 will ever occur.

The usual practice (e.g. [5], [9], [8]) is to put the collocation points in the middle of the panel areas. Additional interpolations are then required to obtain the approximate values of φ at nodal points on the panel boundaries. For example, in the PAN AIR method 21 values of φ_n may influence the nodal values. Additional "outer spline" matrices must then be calculated for each panel, containing 9 x 21 elements. As admitted in [9], this part of the PAN AIR program is one of the most complicated. No additional interpolations are needed for the present method: the values of φ_n are available at six nodal points.

The truncation error for the present method is composed of the following parts:

1. The error introduced by the quadratic approximation for φ and for the functions (14), and by replacing $\partial\varphi/\partial x$, $\partial\varphi/\partial y$, $\Delta_{\hat{S}}\varphi$ with the corresponding differentiated expressions.

2. The error introduced by truncating the third order polynomial for the source strength in (13).
3. The error introduced by dropping the last term in (13).

The contributions of all these parts are $O(h^3)$.

The velocity $V = V_\infty + \nabla\varphi$ may be obtained by differentiating the piecewise quadratic functions for φ. A correction term, similar to that present in (8), may be added to improve the results for V on S. Since the quadratic approximations are differentiated, the error in V is $O(h^2)$.

5. Results of calculations

Results of calculations will be presented for the incompressible potential flow around the sphere $X^2 + Y^2 + Z^2 = 1$, with the velocity $V_\infty = 1$, directed along the X-axis.

The sphere is represented as the image of the rectangle $-\pi/2 \leq\xi\leq\pi/2$, $0\leq\eta\leq 2\pi$, under the transformation

$$(X,Y,Z) = (\cos\xi\ \cos\eta,\ \sin\xi,\ \cos\xi\ \sin\eta\ ).$$

In view of the symmetries of the body and the flowfield, the set of collocation points may be confined to 1/8 of the sphere, corresponding to the rectangle $0\leq\xi,\eta\leq\pi/2$. This part of the sphere will be denoted by S. Exemplary grids, containing m=5 grid lines in the ξ and η directions are shown in Figs. 6a,b, together with the corresponding system of 6 panels.

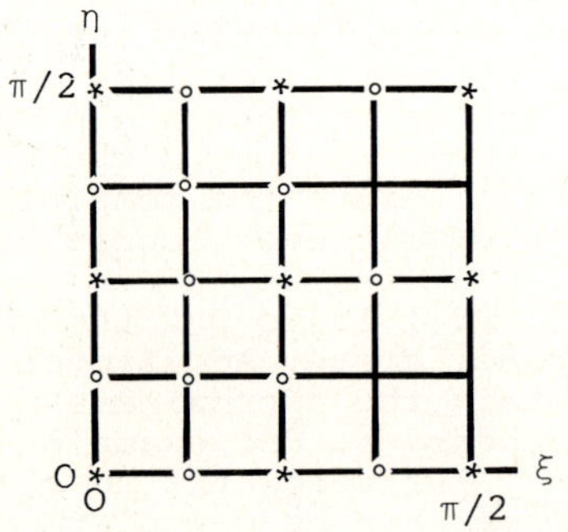

Fig. 6a The parameter plane

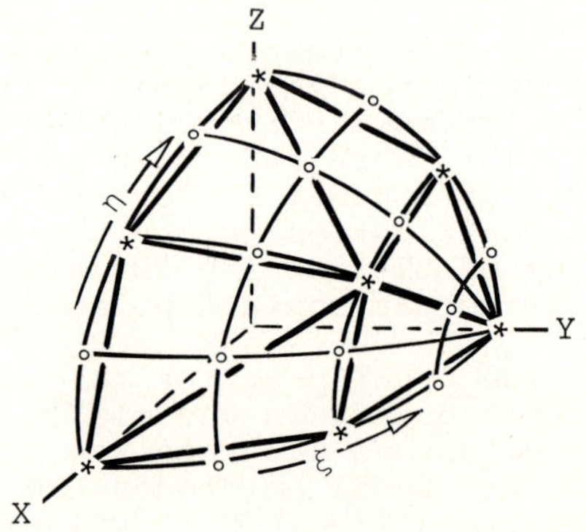

Fig. 6b The physical space

The calculations were made for m=3,5,9 and 13 grid lines in the ξ and η directions. Since ξ and η are the arc-length coordinates, measured as shown in Fig. 6b, the characteristic panel size can be defined as

$$h = \pi/(m-1).$$

Table I presents the general data for the systems of panels and collocation points.

Table I

Case no. i	Number of grid lines m_i	Number of panels	Number of colloc. points	Characteristic panel size h_i
1	3	1	6	$\pi/2$
2	5	6	19	$\pi/4$
3	9	28	69	$\pi/8$
4	13	66	151	$\pi/12$

Let i denote the case number, as in the first column of Table I. For the case i, let the discrete solution for φ, obtained at the vertices of the approximating polyhedron, be interpolated over S with a sufficiently smooth function $\tilde{\varphi}_i$. As a measure of the discretization error we shall use the mean square norm

$$\|\tilde{\varphi}_i-\varphi\|_2 = \left(\int_S |\tilde{\varphi}_i(q)-\varphi(q)|^2 \, dS(q)/|S|\right)^{1/2}, \tag{15}$$

where φ is the given exact solution ([6], p. 287) and $|S|=\pi/2$ denotes the area of S. The integral in (15) was calculated approximately, using the vertex values of $\tilde{\varphi}_i$. The velocity was calculated at the points obtained by the radial projection of the panel centroids on S. Assuming that the resulting discrete distribution has been interpolated over S with a vector-valued function $\tilde{V}_i$, we can define the analogous error norm for the velocity distribution

$$\|\tilde{V}_i-V\|_2 = \left[\int_S |\tilde{V}_i(q)-V(q)|^2 \, dS(q)/|S|\right]^{1/2}. \tag{16}$$

The integral in (16) was calculated approximately by the midpoint rule, using the given values of $\tilde{V}_i$ at the projection points.

The values of $\|\tilde{\varphi}_i-\varphi\|_2$ and $\|\tilde{V}_i-V\|_2$, i=1,2,3,4, are given in Table II, which shows the effect of introducing the source strength corrections, present in (13). To obtain the comparative set of results ("no corrections" columns), the source strength in (13) was set equal to f(N(q)). It can be seen that omitting the curvature corrections introduces significant errors into the results for φ and V.

Table II

Case no.	no corrections		corrections added	
i	$\|\tilde{\varphi}_i-\varphi\|_2$	$\|\tilde{V}_i-V\|_2$	$\|\tilde{\varphi}_i-\varphi\|_2$	$\|\tilde{V}_i-V\|_2$
1	0.1568	0.5193	$0.5274_{10}-1$	0.1948
2	$0.4881_{10}-1$	0.1133	$0.6976_{10}-2$	$0.6594_{10}-1$
3	$0.1398_{10}-1$	$0.2854_{10}-1$	$0.4887_{10}-3$	$0.1145_{10}-1$
4	$0.6498_{10}-2$	$0.1285_{10}-1$	$0.9958_{10}-4$	$0.4228_{10}-2$

In order to estimate the accuracy orders we introduced the quotients

$$q_i(\varphi) = \log(\|\tilde{\varphi}_i-\varphi\|_2/\|\tilde{\varphi}_{i-1}-\varphi\|_2)/\log(h_i/h_{i-1}) \text{ , } i=2,3,\ldots,$$

and the analogous quotients $q_i(V)$. If $h_1>h_2>h_3>\ldots$ and $q_i(\varphi)\geqq\alpha$ for $i>i_o$, then

$$\|\tilde{\varphi} -\varphi\|_2 \leqq C\ h_i^\alpha$$

with some constant C, i.e., $\|\tilde{\varphi}_i-\varphi\|_2 = O(h_i^\alpha)$.

The results for $q_i(\varphi)$ and $q_i(V)$ are given in Table III.

Table III

	no corrections		corrections added	
i	$q_i(\varphi)$	$q_i(V)$	$q_i(\varphi)$	$q_i(V)$
2	1.6842	2.1965	2.9184	1.5630
3	1.8033	1.9888	3.8354	2.5252
4	1.8900	1.9683	3.9234	2.4578

Without curvature corrections the norms $\|\tilde{\varphi}_i-\varphi\|_2$ and $\|\tilde{V}_i-V\|_2$ seem to be of the order $O(h_i^2)$. If the corrections are introduced then $\|\tilde{\varphi}_i-\varphi\|_2$ becomes $O(h_i^4)$, which is less than the truncation error $O(h_i^3)$. Is seems that the main part of the truncation error, which is a combination of the third order monomials changing signs over the panel areas, has been cancelled as a result of the quadratures. The corresponding results for $q_i(V)$ suggest that $\|V_i-V\|_2 = O(h_i^2)$, as predicted in Section 4 for the present method of calculating the velocity by differentiating the quadratic approximations of the potential. In order to obtain better accuracy, comparable with the accuracy of the results for φ, a differentiated form of Green's formula should be used for the velocity calculations.

Acknowledgement. Thanks are due to Professor Wolfgang Hackbusch for his kind help and encouragement during the phase of this research.

References

[1] Courant, R., Hilbert, D., Methods of Mathematical Physics, II, Interscience Publishers, New York 1962.

[2] Ehlers, F.E., Epton, M.A., Johnson, F.T., Magnus, A.E., Rubbert, P.E., An improved higher order panel method for linearized supersonic flows, AIAA Paper 78-15 (1978).

[3] Hirschel, E.H., Fornasier, L., Flowfield and vorticity distribution near wing trailing edges, AIAA Paper 84-241 (1984).

[4] Johnson, F.T., Tinoco, E.M., Lu,P., Epton, M.A., Recent advances in the solution of three-dimensional flows over wings with leading edge vortex separation, AIAA Paper 79-0282 (1979).

[5] Johnson, F.T., A general panel method for the analysis and design of arbitrary configurations in incompressible flow, NASA CR 3079 (1980).

[6] Karamcheti, K., Principles of Ideal Fluid Aerodynamics, John Wiley and Sons, Inc., New York 1966.

[7] Kleinman, R.E., Wendland, W.L. On Neumann's method for the exterior Neumann problem for the Helmholtz equation, Journal of Mathematical Analysis and Applications, 57 (1977), pp. 170-202.

[8] Lötstedt, P., A three dimensional higher order panel method for subsonic flow-problems - Descriptions and applications, SAAB-SCANIA Report L-O-1 R 100 (1984).

[9] Magnus, A.E., Epton, M.E., PAN AIR - A computer program for predicting subsonic or supersonic linear potential flows about arbitrary configurations using a higher order panel method, NASA CR 3251 (1980).

[10] Morino, L., Kuo, C.C., Subsonic potential aerodynamics for complex configurations: A general theory, AIAA Journal 12 (1974), pp. 191-197.

[11] Roberts, A., Rundle, K., Computation of incompressible flow about bodies and thick wings using the spline-mode system, BAC (CAD) Report Aero Ma 19 (1972).

[12] Sytsma, H.S., Hewitt, B.L., Rubbert, P.E., A comparison of panel methods for subsonic flow computation, AGARDograph 241 (1979).

A STUDY OF TURBULENT MOMENTUM AND HEAT TRANSPORT IN A BOUNDARY LAYER USING LARGE EDDY SIMULATION TECHNIQUE

L.Schmitt, K.Richter, R.Friedrich

Lehrstuhl für Strömungsmechanik
Technische Universität München
Arcisstraße 21, 8000 München 2
Federal Republic of Germany

SUMMARY

This paper presents results of a large eddy simulation of high Reynolds number turbulent boundary layer flow with zero pressure gradient and heat transport. The numerical model and the specification of initial and boundary conditions are discussed. Using statistical evaluation methods and 'visualization' methods for the instantaneous flow quantities we demonstrate the efficiency of our approach.

INTRODUCTION

Still today turbulent flows of practical importance lie far beyond the scope of direct numerical simulation (DNS). The main difficulty is the large number of degrees of freedom (being of the order of $Re^{9/4}$), as measured by the number of grid points or grid volumes. In order to reduce this number and to bring it within the range of existing computers large eddy simulation (LES) technique resolves only the large (grid) scales (GS) by DNS and accounts for the small (subgrid) scales (SGS) either semiempirically or with the help of more or less sophisticated mathematical models.

LES is thus intermediate between DNS and statistical turbulence theory where quantities are defined as averages over the whole range of scales and are modeled in order to close the differential equations. Differing from statistical modeling, LES turns into a DNS when the resolution gets fine enough to capture even the smallest scales.

LES allows not only to calculate statistical averages from the resolved fluctuations but also to specify the contribution from each scale (power spectra) and makes it possible to predict one-point correlation functions which cannot be measured nowadays like e.g. pressure-velocity and pressure-strain correlations. It produces 'snapshots' of the whole flow field from which any desired information can be drawn and this is what no experiment will ever be able to provide.

Among the two ways of generating transport equations for the GS quantities Schumann's volume averaging approach [16] seemed us more transparent than the nonlinear filtering of Leonard [11] especially with respect to the approximations necessary to close these equations. Finally, Schumann's pioneering idea of modeling the SGS terms was a valuable guide for our work. For more details regarding the different methods and their applications we refer to the review articles of Schumann, Grötzbach & Kleiser [17], Ferziger [5], Rogallo & Moin [13] and Grötzbach [6].

Since developing flows like wall boundary layers contain the inherent difficulty of nonhomogeneity of flow variables in the main flow direction, they have till now not been a paying object for LES technique. We know of only one recently published paper considering boundary layer flow. In a DNS Spalart & Leonard [18] investigate equilibrium boundary layers at a Reynolds number one order of magnitude lower than ours. But, using self-similar coordinates the authors avoid nonperiodic streamwise boundary conditions. In contrast to this work our aim is to study explicit inflow and outflow boundary conditions for which we develop a suitable concept.

DESCRIPTION OF THE NUMERICAL MODEL

Our approach to simulate turbulence is based on the integral conservation equations for a fluid with constant material properties. They read

$$\iint_A (\underline{n} \circ \underline{v})\, dA = 0 \quad , \tag{1a}$$

$$\iiint_V \rho \frac{\partial \underline{v}}{\partial t}\, dV = -\iint_A \rho (\underline{n} \circ \underline{v}\underline{v})\, dA - \iint_A (p\underline{n})\, dA - \iint_A (\underline{n} \circ \underline{\underline{\tau}})\, dA + \iiint_V \underline{F}\, dV \quad , \tag{1b}$$

$$\iiint_V \rho c_p \frac{\partial T}{\partial t}\, dV = -\iint_A \rho c_p (\underline{n} \circ \underline{v} T)\, dA - \iint_A (\underline{n} \circ \underline{q})\, dA + \iiint_V \dot{Q}\, dV \quad . \tag{1c}$$

The first equation expresses continuity for a volume V fixed in space with surface A through which the fluid flows and the second describes the rate of change of momentum for this volume. The last equation is the enthalpy equation in terms of the fluid temperature. The shear stress $\underline{\underline{\tau}}$ follows Newton's law of viscosity and the heat flux $\underline{q}$ Fourier's law of heat conduction.

These equations provide a natural decomposition into resolvable GS and unresolvable SGS turbulence quantities if we apply them to small volumes and define averages over such volumes and their surfaces of the form

$$^{\Delta\varphi}\overline{\phi} = \frac{1}{\Delta\varphi} \int_{\Delta\varphi} \phi \, d\varphi \tag{2a}$$

as 'resolvable' quantities and departures from these values, denoted by

$$\phi' = \phi - {}^{\Delta\varphi}\overline{\phi} \quad , \tag{2b}$$

as 'unresolvable' quantities.

For numerical evaluation of equations (1) and (2) we use a Cartesian mesh system (see [14], for a curvilinear orthogonal grid) with grid volumes ΔV and surfaces $\Delta A_\beta^\pm$. The notation ΔA_β^+, ΔA_β^- distinguishes between surfaces with normal vectors pointing into positive or negative coordinate directions. In terms of Cartesian coordinates the properly nondimensionalized transport equations take the general form:

$$\frac{\partial\, {}^{\Delta V}\overline{\phi}}{\partial t} = \frac{1}{\Delta V} \sum_{\beta=1}^{3} \left(\Delta A_\beta^+ \, {}^{\Delta A_\beta^+}\overline{FLUX_\phi} - \Delta A_\beta^- \, {}^{\Delta A_\beta^-}\overline{FLUX_\phi} \right) + {}^{\Delta V}\overline{SOURCE_\phi} \tag{3}$$

where

Table 1 Definition of the flux and source terms.

ϕ	$FLUX_\phi$	$SOURCE_\phi$	Eq. no.
1	v_β	0	3a
v_α (α = 1,2,3)	$-v_\beta v_\alpha - p\,\delta_{\beta\alpha} - \tau_{\beta\alpha}$	F_α	3b
T	$-v_\beta T - q_\beta$	$\dot{Q}$	3c

and

$$\tau_{\alpha\beta} = -\frac{1}{Re} D_{\alpha\beta} , \tag{4a}$$

$$D_{\alpha\beta} = \frac{\partial v_\alpha}{\partial x_\beta} + \frac{\partial v_\beta}{\partial x_\alpha} , \tag{4b}$$

$$q_\alpha = -\frac{1}{Pe} \frac{\partial T}{\partial x_\alpha} . \tag{5}$$

Equations (3) together with (4) and (5) form an unclosed set of equations which demands for specification of closure relations. First, we have to model the SGS contributions ${}^{\Delta A\alpha}\overline{v'_\alpha v'_\beta}$ and ${}^{\Delta A\alpha}\overline{v'_\alpha T'}$ to turbulent convection. These fluxes arise from (3b) and (3c) by introducing (2b) and are important if the grid does not resolve the detailed structure of the flow field. This is especially the case in high Reynolds number turbulent flows. Next, convection and diffusion terms make it necessary to assume relations between GS quantities at different surfaces, since their number is larger than the number of equations.

For the SGS fluxes we adopt eddy viscosity concepts in a way suggested by Schumann [16] and Grötzbach [7]:

$${}^{\Delta A\alpha}\overline{v'_\alpha v'_\beta} = - {}^{\alpha\beta}\mu_{iso} \left({}^{\Delta A\alpha}\overline{D_{\alpha\beta}} - \langle {}^{\Delta A\alpha}\overline{D_{\alpha\beta}} \rangle \right) - {}^{\alpha\beta}\mu_{inh} \langle {}^{\Delta A\alpha}\overline{D_{\alpha\beta}} \rangle , \tag{6a}$$

$${}^{\Delta A\alpha}\overline{v'_\alpha T'} = - {}^{\alpha}a_{iso} \, {}^{\Delta A\alpha}\overline{\partial (T - \langle T \rangle)/\partial x_\alpha} - {}^{\alpha}a_{inh} \, {}^{\Delta A\alpha}\overline{\partial \langle T \rangle / \partial x_\alpha} . \tag{6b}$$

$\langle \; \rangle$ indicates a proper ensemble mean value. These models account for possible bad resolution, flow inhomogeneities and grid volume anisotropies. The 'isotropic' eddy diffusivities ${}^{\alpha\beta}\mu_{iso}$ and ${}^{\alpha}a_{iso}$ are mainly responsible for the correct drainage of turbulence energy, respectively intensity of temperature fluctuations, from the large scales and have to be determined such, that the dissipation in the simulated large-scale motion is of the same magnitude as the viscous dissipation in reality. They can be calculated without any further empirical assumption from the theory of locally isotropic turbulence and the validity of the Kolmogorov spectrum. The second parts in (6) account for the effect of unresolved small scales that do not only take a passive and dissipative role but actively contribute to turbulence production. This part of the model is of importance essentially in the near wall region. The 'inhomogeneous' diffusivities are derived from common mixing length models and are formulated such that for very coarse grid volumes the whole SGS model (6) turns into a statistical model.

The resolved GS variables are the grid surface averaged velocity components and the grid volume averaged pressure, temperature and SGS kinetic energy, as illustrated below.

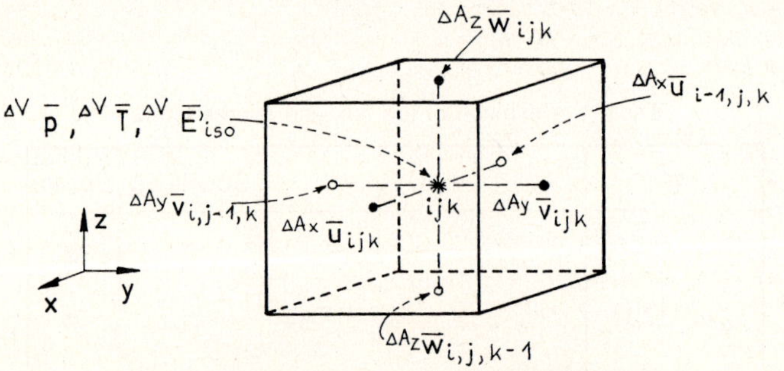

Fig.1 Location of variables in the staggered grid.

They are the dependent variables in our numerical scheme. Unknown GS quantities (e.g., momentum and heat fluxes) are related to these variables by proper interpolation. This leads to second-order accurate energy-conserving 'finite-difference' equations in time and space on a staggered grid. They are integrated in time for given initial and boundary conditions (see below) using explicit time steps. Each integration cycle starts with an Euler step, is followed by a couple of (say 50) leapfrog steps and ends with an averaging step in order to control spurious $2\Delta t$-oscillations.
Continuity is enforced at each time level using a split-step method. First, a tentative velocity field $\tilde{v}_\alpha$ is calculated by neglecting the pressure term. This field is not yet divergence-free. Then, the pressure field is determined from a three-dimensional Poisson-like equation which is obtained from introducing the correction relation

$$v_\alpha = \tilde{v}_\alpha - \tau \cdot \delta_\alpha p \qquad (7)$$

into the continuity equation (3a). The resulting pressure equation is solved by direct methods (Fast Fourier Transform, Cyclic Reduction, Tridiagonal Matrix Algorithm). The last step is to calculate the divergence-free velocity field from (7).

The whole theory is implemented in a FORTRAN 77 code. An important feature of this code is that it is not restricted to periodic and wall boundary conditions (possible for a channel flow) but that it allows for various combinations of periodic, velocity, pressure and wall boundary conditions.
All of our computations are performed on a CDC CYBER 175 (and CYBER 875). Since the core memory (130000 words) of both computers is in general too small to store all the required data, the fields are segmented in blocks and kept either in the core memory or in an external direct access file. An efficient dynamic management controls the transfer of data, minimizing the input/output requirements. Typical CPU-times to calculate all flow variables (three velocity components, pressure, temperature, SGS energy) per time step range between 2 seconds for 32x16x16 and 30 seconds for 64x32x32 mesh volumes.

The simulations provide three-dimensional time-dependent vector and scalar fields with quasi-random behavior. In order to analyse these fields we have developed extensive evaluation and representation tools.
For most of the flow variables a large number of different statistical data can be calculated as there are ensemble mean values, second and higher order moments like skewness and flatness, but also moments of velocity derivatives or vorticity components, cross-correlations and correlation coefficients, two-point correlations, power spectra, and finally terms appearing in statistical turbulence equations (e.g., pressure strain terms).
Among the more advanced statistical evaluation methods we apply, are the quadrant analysis technique, the Karhunen-Loeve expansion (see [15]) and the histogram of the vorticity inclination angle ([15]).
'Visualization' of the flow structure is possible by means of vector, perspective and contour plots of instantaneous velocity, pressure, temperature, SGS energy or vorticity in different cross sections of the flow field. Instantaneous two-dimensional and three-dimensional streamlines and vorticity lines ([15]) can be determined and displayed. We are able to demonstrate the time dependency of the flow in motion pictures of fluid markers (simulated hydrogen bubbles). Here, we have provided means to show path lines, time lines or streak lines.
Only these detailed tools of data analysis enable us to check the simulation results against experimental findings, to determine quantities which cannot be measured and to investigate the structure of the simulated turbulence fields. But more than that these tools are also a presupposition for the study of variations in the SGS model, for the influences of the boundary conditions or for modifications of the numerical method.

APPLICATION OF THE NUMERICAL MODEL TO BOUNDARY LAYER FLOW

In this section we discuss the simulation of turbulent flat plate boundary layer flow with zero pressure gradient. Heat transport is included, but temperature is treated as a passive scalar and does therefore not affect the momentum transport.
It would be ideal to calculate a long stretch of a boundary layer extending tens of boundary layer thicknesses in mean flow direction and as far as possible in y-direction and to let turbulence develop naturally. Especially because of the limited computer resources this is not feasible. We therefore restrict the simulations to a relatively small integration domain and try to adopt boundary conditions such that physically meaningful results are obtained.
Figure 2 shows the studied flow domain, which is fixed in space (with the inflow edge at x_{or}) and extends 8 units in mean flow direction, x, 4 in the spanwise direction, y, and 2 in the direction normal to the wall, z.

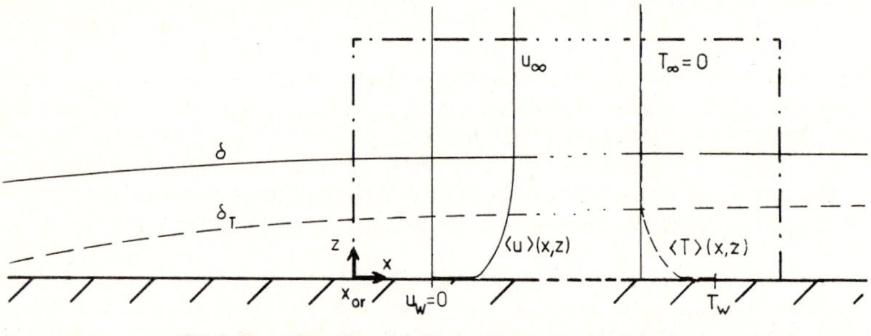

Fig.2 Boundary layer with computational domain.

The thermal boundary layer grows inside the momentum layer and possesses the thickness δ_T = 0.66 at x_{or}. The free-stream velocity u_∞ and the wall temperature T_W are 27.74 and 20.29. All quantities are made dimensionless by means of the momentum boundary layer thickness δ_{ref}, the friction velocity $u_{\tau ref} = (\langle \tau_w \rangle_{ref} / \varrho)^{1/2}$ and the heat flux temperature $T_{ref} = \langle q_w \rangle_{ref} / (\varrho c_p u_{\tau ref})$; where δ_{ref}, the time mean wall shear stress $\langle \tau_w \rangle_{ref}$ and the time mean wall heat flux $\langle q_w \rangle_{ref}$ are taken at the inflow edge. In terms of these quantities the Reynolds number is 3240. This value is very close to the one of Klebanoff [9]. The Prandtl number taken was 0.7 .

Because of the elliptic nature of the basic equations we have to prescribe boundary conditions at all surfaces of the computational domain. These boundary conditions are necessary for the determination of the flux terms appearing in the transport equation (3) and can in general be realized only under certain assumptions.
The flux terms a t t h e w a l l are illustrated below.

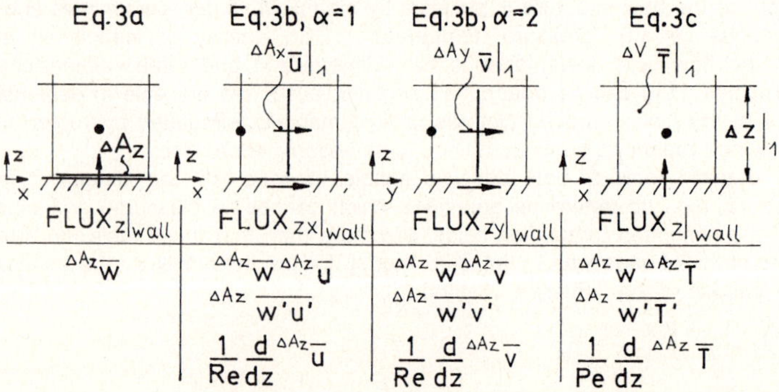

Fig.3a Flux terms at the wall.

Consistent with the no-slip condition is

$$w\Big|_{wall} = 0.0 \tag{8}$$

which also means ${}^{\Delta A_z}\overline{w} = 0$, $w' = 0$ and no approximations in the convection terms are needed (all terms are zero). In the diffusion terms assumptions are necessary. Here too, we follow along the line of Schumann [16] and Grötzbach [7]. For the y-component of the wall shear stress we write

$$^{\Delta Az}\overline{\tau_{zy}}\Big|_{wall} = -\frac{1}{Re}\,\frac{^{\Delta Ay}\overline{v}\big|_1 - v_w}{\Delta z\big|_1 / 2} \quad . \tag{9}$$

This relation is acceptable since there is no mean momentum transport in y-direction. For ${}^{\Delta Az}\overline{\tau_{zx}}$ it is more appropriate to formulate

$$^{\Delta Az}\overline{\tau_{zx}}\Big|_{wall} = -c_\tau\left(^{\Delta Ax}\overline{u}\big|_1 - u_w\right) \quad , \tag{10a}$$

$$c_\tau = \frac{|\langle \tau_w \rangle|}{|\langle ^{\Delta Ax}\overline{u}|_1 - u_w \rangle|} \quad . \tag{10b}$$

A similar relation holds for the wall heat flux

$$^{\Delta Az}\overline{q}\Big|_{wall} = -c_q\left(^{\Delta V}\overline{T}\big|_1 - T_w\right) \quad , \tag{11a}$$

$$c_q = \frac{|\langle q_w \rangle|}{|\langle ^{\Delta V}\overline{T}|_1 - T_w \rangle|} \quad . \tag{11b}$$

c_τ and c_q are specified using logarithmic profiles for $\langle u \rangle$ and $\langle T \rangle$ averaged over the wall adjacent mesh cell:

$$|\langle ^{\Delta Ax}\overline{u}|_1 - u_w \rangle| = \frac{\sqrt{|\langle \tau_w \rangle|}}{\varkappa}\left[\ln\left(E \cdot Re \cdot \sqrt{|\langle \tau_w \rangle|} \cdot \Delta z|_1\right) - 1\right] \quad , \tag{12a}$$

$$|\langle ^{\Delta V}\overline{T}|_1 - T_w \rangle| = \frac{|\langle q_w \rangle|}{\sqrt{|\langle \tau_w \rangle|}}\left\{\frac{1}{\varkappa_H}\left[\ln\left(Re \cdot \sqrt{|\langle \tau_w \rangle|} \cdot \Delta z|_1\right) - 1\right] + B_T(h^+, Pr)\right\} \tag{12b}$$

(for smooth wall, $Pr = 0.7$: $\varkappa = 0.4$, $\varkappa_H = 0.465$, $E = 9.025$, $B_T = 3.799$).
$\langle \tau_w \rangle$, $\langle q_w \rangle$ are either calculated from the simulated statistical profiles or prescribed empirically. It should be noted that our wall adjacent mesh volume covers the viscous and buffer layer and extends well into the logarithmic regime. The maximum of turbulence is produced in it. Though this model does not take care of the details, the above boundary conditions account for the proper global behavior.
In order to determine the flux terms at free edges of the domain we explicitly prescribe values for the flow quantities (see fig.3b and formulae (13)–(16)).

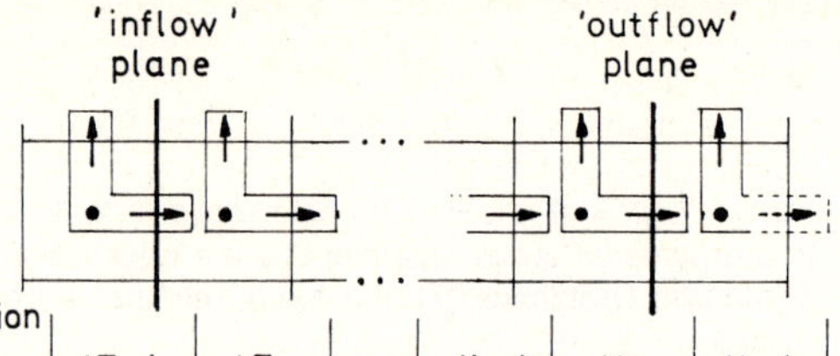

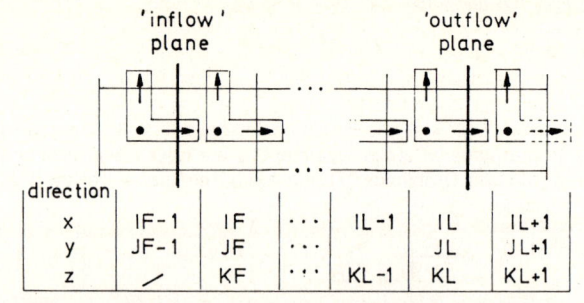

direction						
x	IF-1	IF	...	IL-1	IL	IL+1
y	JF-1	JF	...		JL	JL+1
z	/	KF	...	KL-1	KL	KL+1

Fig.3b Specification of grid indices for 'inflow/outflow' planes.

Crucial for boundary layer simulations are boundary conditions at the inflow plane. What we would need there are the velocity vector and the temperature as a function of time. It is clear that no experiment can provide us these data. Therefore our way of specifying these quantities is as follows: Fluctuations ${}^{\Delta V}\overline{\phi}'' = {}^{\Delta V}\overline{\phi} - {}^{\Delta V}\overline{\langle\phi\rangle}$ simulated in a plane (IP = const) far enough downstream so that streamwise two-point correlations have died out, are transferred to the inflow plane, weighted such that their rms values coincide with experiment ([9,3]) and are superimposed with the statistical mean values, viz:

$$
\begin{aligned}
{}^{\Delta V}\overline{\phi}(IF-1,J,K)\Big|^{n+1} = {} & {}^{\Delta V}\overline{\langle\phi\rangle}(IF-1,K)\Big|_{Exp.} \\
& + \left({}^{\Delta V}\overline{\phi}(IP,J,K)\Big|^{n} - \langle{}^{\Delta V}\overline{\phi}\rangle(IP,K)\right)\frac{{}^{\Delta V}\overline{\phi}_{rms}(IF-1,K)\big|_{Exp.}}{{}^{\Delta V}\overline{\phi}_{rms}(IP,K)} \; . \qquad (13)
\end{aligned}
$$

In the outflow plane quantities are extrapolated according to

$$
{}^{\Delta A x}\overline{u}(IL,J,K)\Big|^{n+1} = {}^{\Delta A x}\overline{u}(IL-1,J,K)\Big|^{n} \; , \qquad (14a)
$$

$$
{}^{\Delta V}\overline{\phi}(IL+1,J,K)\Big|^{n} = {}^{\Delta V}\overline{\phi}(IL,J,K)\Big|^{n} \; . \qquad (14b)
$$

Because of staggering u needs a special treatment. In order to fulfil continuity it must be prescribed for the new time level n + 1. ϕ stands for v, w, T, E_{SGS}. Other extrapolation formulae are tried and tested.

At the free stream edge u, v, T and E_{SGS} are specified (Dirichlet boundary conditions) and w follows from a relation similar to (14a) allowing for vertical outflow:

$$
\left.
\begin{aligned}
& {}^{\Delta A x}\overline{u}(I,J,KL+1) = u_\infty \; , \quad {}^{\Delta A y}\overline{v}(I,J,KL+1) = 0 \; , \\
& {}^{\Delta V}\overline{T}(I,J,KL+1) = T_\infty \; , \quad {}^{\Delta V}\overline{E}_{SGS}(I,J,KL+1) = 0 \; ,
\end{aligned}
\right\} \qquad (15a)
$$

$$
{}^{\Delta A z}\overline{w}(I,J,KL)\Big|^{n+1} = {}^{\Delta A z}\overline{w}(I,J,KL-1)\Big|^{n} \; . \qquad (15b)
$$

Homogeneity of the turbulent flow in y-direction allows for periodic boundary conditions for all quantities:

$$
{}^{\Delta V}\overline{\phi}(I,JF-1,K) = {}^{\Delta V}\overline{\phi}(I,JL,K) \; , \quad {}^{\Delta V}\overline{\phi}(I,JL+1,K) = {}^{\Delta V}\overline{\phi}(I,JF,K) \; . \qquad (16)
$$

With the above boundary conditions for the velocities equation (7) provides proper boundary conditions for the pressure.

To start a simulation initial conditions for velocity and temperature have to be generated. Our procedure to do this consists in superimposing statistical mean values with random Gaussian fluctuations fulfilling the experimental rms values and zero velocity divergence. Using this procedure we have performed simulations in two different ways: 1) We started with a coarse grid of 32x16x16 mesh volumes calculating a few problem times and then switched over to the fine grid (64x32x32) and 2) We used the fine grid right from the beginning. It turned out that the statistical data were equivalent in both cases.

In order to obtain statistical averages we refer to the ergodic hypothesis and evaluate mean values by averaging in x – and y – directions (neglecting small downstream variations). Additional time averaging is used to improve the statistics.

The following figures are unsmoothed reproductions of computer results. Marked points represent calculated values and are interconnected only for better visibility.

Figures 4a – 4l show profiles of statistical quantities reflecting typical characteristics of a zero pressure gradient smooth wall boundary layer flow. The mean velocity profile (fig.4a) has a shape factor H (= displacement thickness/momentum thickness) of 1.27 typical for a Reynolds number of 3240. The temperature profile develops inside the momentum boundary layer due to a discontinuity in the wall temperature upstream of the inflow edge. The temperature difference between wall and free – stream is so small that our assumption of treating temperature as a passive scalar is justified (cf.[3]).

The turbulence kinetic energy and the rms velocity fluctuations (fig.4b,4c) compare fairly well with Klebanoff's [9] measurements. While entered experimental profiles correspond to the conditions at the inflow edge, calculated profiles reflect the average development of the simulated flow in x – direction and thus explain the differences between experiment and computation. Similar effects are found in other correlations of velocity and temperature fluctuations.

The resemblance of the T_{rms} – profile (fig.4d) to the u_{rms} – profile (fig.4c) supports the idea that the velocity field controls the temperature field leading to a high correlation between u – and T – fluctuations (see fig.4i).

Starting with initially uncorrelated velocity and temperature fluctuations physically reasonable Reynolds stress and heat flux profiles have built up as a consequence of the used boundary condition at the inflow edge (figures 4e,4f). The major part of these quantities results from the resolved fluctuations whereas the subgrid scales (represented by the inhomogeneous parts of the SGS models) contribute not more than 15 per cent. From figure 4f it can be concluded that molecular fluxes are not significant at the computed z – positions.

The rapid variations in the skewness and flatness factors of velocity and temperature fluctuations near the edge of the momentum boundary layer (figures 4g,4h) suggest a region of intermittent turbulence. As one expects from experiment (e.g., [9,2]) skewness of the v – component is zero and skewness of u – and w – components have opposite sign at the edge pointing to isolated regions of low speed upward moving fluid. Again the coincidence of maximum values in the skewness and flatness profiles of u – and T – fluctuations expresses the strong coupling between these two quantities.

Figure 4i demonstrates the well established correlations between resolvable u –, w – and T – fluctuations in the whole boundary layer. As it is expected in a flow situation where momentum is transported towards the wall and heat away from the wall, the correlation coefficients of $\langle u''w'' \rangle$ and $\langle w''T'' \rangle$ are negative and positive, respectively.

The diagonal elements of the pressure strain correlation tensor (fig.4j), govern the exchange of energy among the three velocity components [8]. The longitudinal part is negative as it should be and thus transfers energy from the streamwise component of turbulence velocity fluctuations to the cross – stream components.

These quantities like the rms values of the pressure fluctuations (fig.4k) belong to those which cannot be measured nowadays. At the wall the simulated value for p_{rms} of 2.9 is in fair agreement with experiment (see [20]).

Like in channel flow [12,15] rms vorticity fluctuations away from the wall are virtually identical (fig.4l). This is in contrast to the rms velocity fluctuations and may be explained by the fact that the small scales which are more likely to show isotropy contribute relatively more to vorticity than to velocity fluctuations [12].

Figures 5 show the numerical analogue of the hydrogen-bubble technique. At regular intervals (10 time steps) particles are generated along lines either parallel or normal to the wall and are followed over 100 time steps (corresponding to a problem time of 0.1).

The instantaneous three-dimensional view of 480 different particles in figure 5a gives an impression of the diffusion of the particles due to the 'random' turbulent motion. The sideward movement of particles is apparent from figure 5b. Large displacements near the wall are consistent with a maximum of the turbulence intensity in y-direction (see fig.4c).

Particles originating from a horizontal wire (figures 5c,5d) not only penetrate into the outer turbulent region but also approach the wall. Their motion suggests ejections of slowly moving fluid away from the wall and inrushes (sweeps) of rapidly moving fluid towards the wall, as becomes clear from time lines and streak lines in figures 5e-5i.

While the before discussed figures correspond to 'snapshots' of the flow taken from several sequences of a computer film, figure 5k illustrates the time development of one line of (48 different) particles. The corresponding path lines (fig.5j) reflect the Lagrangian aspect of this type of flow 'visualization'.

A more detailed analysis of ejection and sweep type events and their relation to the Reynolds shear stress is possible using the quadrant analysis technique. This statistical technique devides the contributions to $\langle -u''w'' \rangle$ into four classes depending on the sign of the velocity fluctuations, as shown in the following table (see e.g., [1,4]):

Table 2 Definition of quadrants.

Quadrant	Vector symbol	Sign of u''	Sign of w''	Sign of $u''w''$	Type of motion
Q1	↗	+	+	+	outward interaction
Q2	↖	−	+	−	ejection
Q3	↙	−	−	+	inward interaction
Q4	↘	+	−	−	sweep

Figures 6a,6b show for two vertical positions the contributions Q1,...,Q4 as functions of a threshold value H which is a multiple of the wall shear stress. It turns out, that in fact, ejections and sweeps dominate over the interactive motions and that both are associated with relatively high values of the fluctuating shear stress. A histogram would show that for only few of these events $|u''w''|$ is large pointing to a spotty occurence of this quantity. Close to the wall sweeps are the dominant events whereas in the outer region ejections prevail. This is especially prominent for higher values of the threshold level. These findings are in agreement with the skewness and flatness factors of velocity fluctuations (see above).

In figures 7 and 8 the instantaneous flow field is illustrated with the help of vector plots, streamlines and contour lines. We have chosen cross sections where extremely high values of the fluctuating shear stress, $u''w''$, appear (more than 20 times the Reynolds stress), associated in figure 7 with an ejection type event in the outer region ($z_e \approx 0.66$) and in figure 8 with a sweep near the wall ($z_s \approx 0.13$).

We found that ejections of high intensity are long living events which could be followed over nearly one problem time. Figures 7a-7c give the impression that the volume of fluid involved in this event covers a relatively large region in the (x,z)-plane but does not extend very much into the y-direction.

The instantaneous 'streamlines' in figure 7d correspond to the vector field in figure 7c and are obtained from numerical integration of $v\,dz - w\,dy = 0$. The figure shows regions with vortical motion and with converging and diverging streamlines because the vector field (v,w) is not divergence-free. The ejection event (near $y_e \approx 1.44$) is very distinct. In the upper part of the figure streamlines are practically vertical indicating the displacement effect of the slowly growing boundary layer.

Contour lines of temperature fluctuations are displayed in figure 7e. Comparing these with the vector plot of figure 7a we can verify their negative correlation with u" and their positive correlation with w".
Figure 7f shows a pattern of instantaneous streamlines in a vertical plane parallel to the main flow as viewed by an observer moving at a speed 10 per cent slower than the free-stream velocity. Similar looking flow patterns are produced by experimental techniques [19]. Note that the flow patterns vary depending upon the relative velocity of the observer (compare figures 7f and 8g) and that characteristic structures appear particularly in regions where the observer moves with the mean flow velocity.
The distributions of velocity vectors in figures 8a and 8d reveal the flow region ($x_s \approx 2.35$, $y_s \approx 1.19$, $z_s \approx 0.13$) where the sweep type event occurs. Its spanwise and vertical extension becomes apparent from contour lines of the fluctuating u-component in figure 8b. In contrast to a nearby happening ejection ($y_e \approx 1.5$) we observe here converging streamlines transporting fluid downward (fig.8e). These events lead to remarkable downward and upward deformations of contour lines of the total u-component (fig.8c). The sweep transporting cooler fluid towards the wall, produces a typical structure in the temperature fluctuations (fig.8f) and leads to temperature-velocity correlations of the same sign as the ejection.

CONCLUSIONS

The presented results show that our technique of simulating momentum and heat transport in a turbulent boundary layer at high Reynolds number predicts turbulence statistics in fair agreement with experiment.
With the help of specific evaluation methods we could draw information from the instantaneous flow data about the turbulence structure. We detected events similar to an eruption of slow moving fluid away from the wall and an inrush of rapidly moving fluid towards the wall which are in the literature sometimes called ejection and sweep. We made these findings though we treated the near wall region globally, i.e. we did not resolve the details of this region. It seems that such events can be simulated (at least) in qualitative agreement with experiment and that their occurence is independent of the details of the near wall turbulence structure.

On the whole the results are supported by our channel flow simulations [15]. However, there we could show (using Galilean transformation, e.g.) that small scales are influenced by the numerical treatment of the mean convection terms. This effect is also noted in [10]. Further simulations (not reported here) indicate a remarkable interdependence between numerical method, SGS model and resolved GS values especially in flow cases with strong mean convection.
Finally a crucial point in our boundary layer simulation is the specification of boundary conditions. Our concept for explicit inflow boundary conditions works reasonably well.

Though several questions remain open we feel that the present technique and the obtained experiences form a useful basis for further developments.

ACKNOWLEDGEMENT. The authors wish to thank the German Research Society (DFG) for supporting this work.

LITERATURE

[1] Alfredsson,P.H., Johansson,A.V. On the detection of turbulence generating events. J.Fluid Mech.139, 325 – 345, 1984.

[2] Andreopoulos,J., Durst,F., Zaric,Z., Jovanovic,J. Influence of Reynolds number on characteristics of turbulent wall boundary layers. Experiments in Fluids 2, 7 – 16, 1984.

[3] Antonia,R.A., Danh,H.Q., Prabhu,A. Response of a turbulent boundary layer to a step change in surface heat flux. J.Fluid Mech.80, 153 – 177, 1977.

[4] Brodkey,R.S., Wallace,J.M., Eckelmann,H. Some properties of truncated turbulence signals in bounded shear flows. J.Fluid Mech.63, 209 – 224, 1974.

[5] Ferziger,J.H. Higher – level simulations of turbulent flows. In: Essers,J.A. (ed.): Computational methods for turbulent, transonic, and viscous flows. Washington: Hemisphere Publ. Co., 93 – 182, 1983.

[6] Grötzbach,G. Direct numerical and large eddy simulation of turbulent channel flows. In: Cheremisinoff,N.P. (ed.): Encyclopedia of Fluid Mechanics, Vol.6. West Orange, New Jersey, USA: Gulf Publ., in press. Extended version in: Primärbericht 01.02.06P49A, Kernforschungszentrum Karlsruhe, 1984.

[7] Grötzbach,G., Schumann,U. Direct numerical simulation of turbulent velocity –, pressure –, and temperature – fields in channel flows. Proc. of the Symp. on Turbulent Shear Flows. Penn. State Univ., Apr.18 – 20, 1977.

[8] Hinze,J.O. Turbulence. New York: McGraw Hill, 1975.

[9] Klebanoff,P.S. Characteristics of turbulence in a boundary layer whith zero pressure gradient. NACA – TR – 1247, 1955.

[10] Laurence,D. Advective formulation of large eddy simulation for engineering type flows. In: Proc. of Euromech 199 on Direct and Large Eddy Simulation of Turbulence, TU München, Sept.30 – Oct.2, 1985. Braunschweig, Wiesbaden: Vieweg. Notes on Numerical Fluid Mechanics, to appear.

[11] Leonard,A. On the energy cascade in large eddy simulations of turbulent fluid flows. Adv. Geophys. 18A, 237 – 248, 1974.

[12] Moin,P., Kim,J. Numerical investigation of turbulent channel flow. J.Fluid Mech.118, 341 – 377, 1982.

[13] Rogallo,R.S., Moin,P. Numerical simulation of turbulent flows. Ann.Rev.Fluid Mech. 16, 99 – 137, 1984.

[14] Schmitt,L., Friedrich,R. Large eddy simulation of turbulent boundary layer flow. In: Pandolfi,M., Piva,R. (eds.): Proc. of the Fifth GAMM – Conf. on Numerical Methods in Fluid Mechanics. Braunschweig, Wiesbaden: Vieweg. Notes on Numerical Fluid Mechanics 6, 299 – 306, 1984.

[15] Schmitt,L., Richter,K., Friedrich,R. Large eddy simulation of turbulent boundary layer and channel flow at high Reynolds number. In: Proc. of Euromech 199 on Direct and Large Eddy Simulation of Turbulence, TU München, Sept.30 – Oct.2, 1985. Braunschweig, Wiesbaden: Vieweg. Notes on Numerical Fluid Mechanics, to appear.

[16] Schumann,U. Subgrid scale model for finite difference simulations of turbulent flows in plane channels and annuli. J.Comp. Phys.18, 376 – 404, 1975.

[17] Schumann,U., Grötzbach,G., Kleiser,L. Direct numerical simulation of turbulence. In: Kollmann,W. (ed.): Prediction methods for turbulent flows. Washington: Hemisphere Publ. Co., 123 – 258, 1980.

[18] Spalart,P.R., Leonard,A. Direct numerical simulation of equilibrium turbulent boundary layers. Proc. of the 5th Symp. on Turbulent Shear Flows. Ithaka, N.Y., Aug.7 – 9, 1985.

[19] Utami,T., Ueno,T. Visualization and picture processing of turbulent flow. Experiments in Fluids 2, 25 – 32, 1984.

[20] Willmarth,W.W. Pressure fluctuations beneath turbulent boundary layers. Ann.Rev.Fluid Mech.7, 13 – 38, 1975.

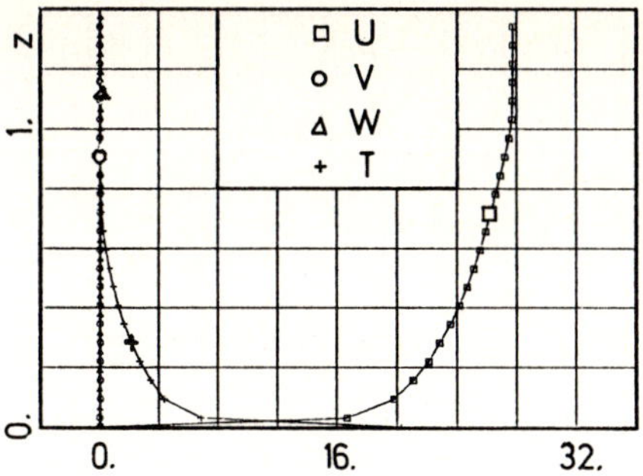

Fig.4a Mean velocity profiles, mean temperature profile.

$V_\alpha = \langle {}^{\Delta A\alpha}\overline{V_\alpha} \rangle, \; T \triangleq \langle {}^{\Delta V}\overline{T} \rangle$

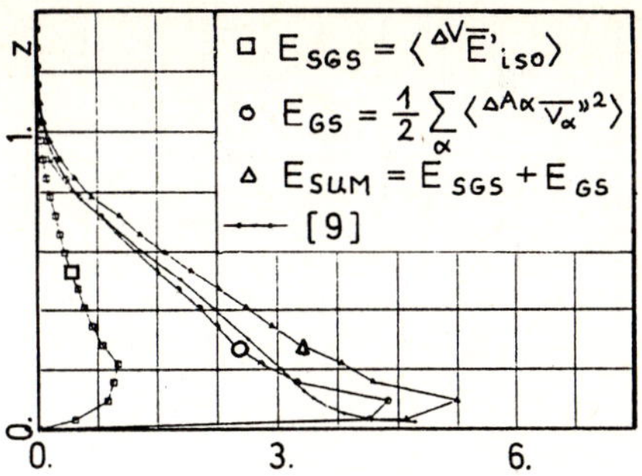

Fig.4b Contributions to turbulence kinetic energy.

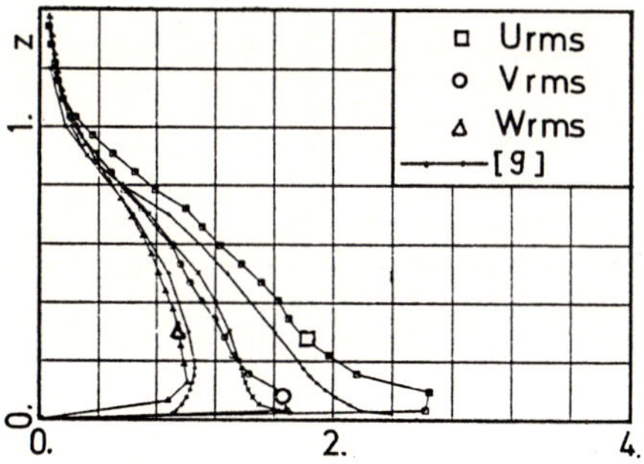

Fig.4c Resolvable rms velocity fluctuations.

$V_{\alpha\, rms} = \langle {}^{\Delta A\alpha}\overline{V_\alpha''^{\,2}} \rangle^{1/2}$

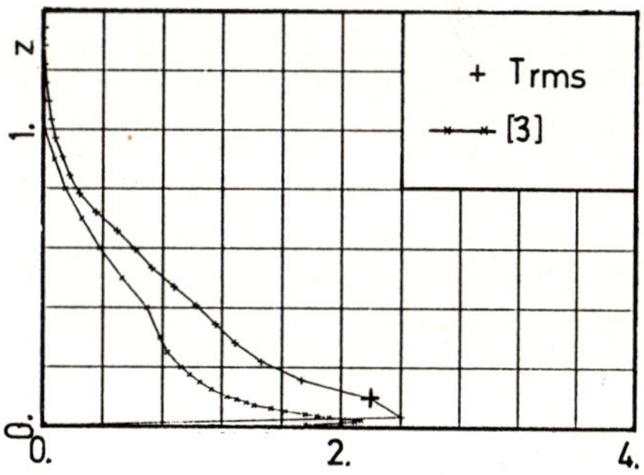

Fig.4d Resolvable rms temperature fluctuations.

$T_{rms} \triangleq \langle {}^{\Delta V}\overline{T''^{\,2}} \rangle^{1/2}$

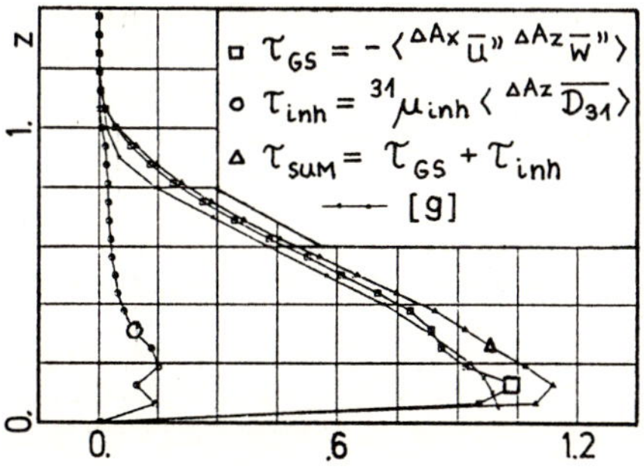

Fig.4e Contributions to turbulence shear stress.

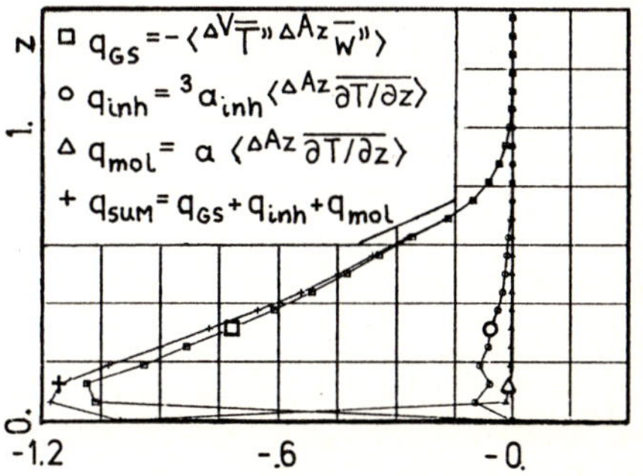

Fig.4f Contributions to turbulence heat flux.

Fig.4 Statistical mean values.

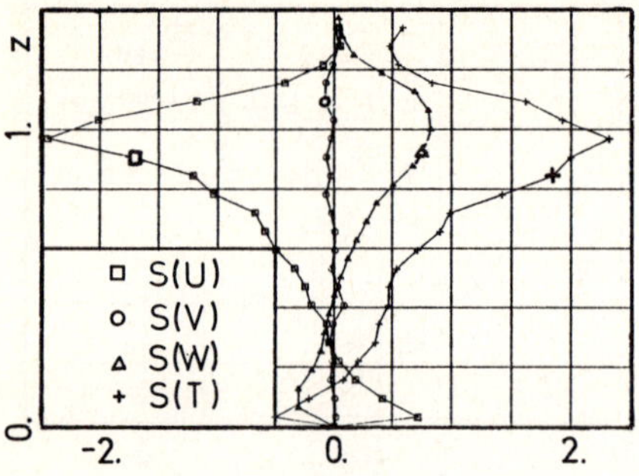

Fig.4g Resolvable skewness factors.

$$S(\phi) = \frac{\langle {}^{\Delta\varphi}\overline{\phi''^3} \rangle}{\phi^3_{rms}}$$

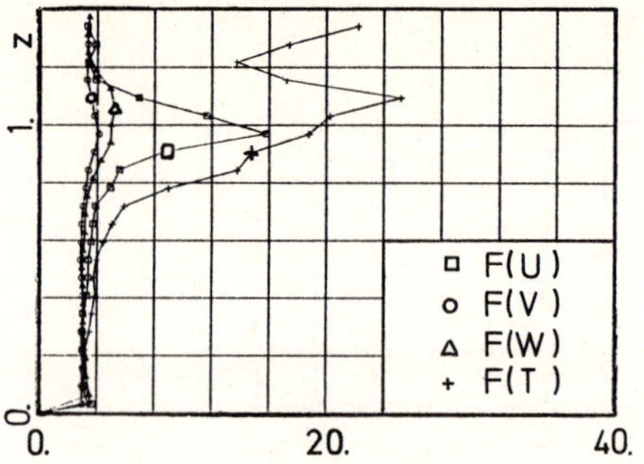

Fig.4h Resolvable flatness factors.

$$F(\phi) = \frac{\langle {}^{\varphi}\overline{\phi''^4} \rangle}{\phi^4_{rms}}$$

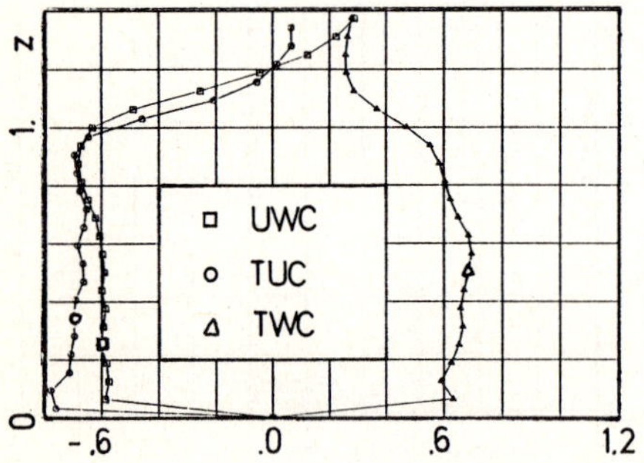

Fig.4i Resolvable correlation coefficients.

$$\phi_1 \phi_2 C = \frac{\langle {}^{\Delta\varphi_1}\overline{\phi_1''} \; {}^{\Delta\varphi_2}\overline{\phi_2''} \rangle}{\phi_{1\,rms} \; \phi_{2\,rms}}$$

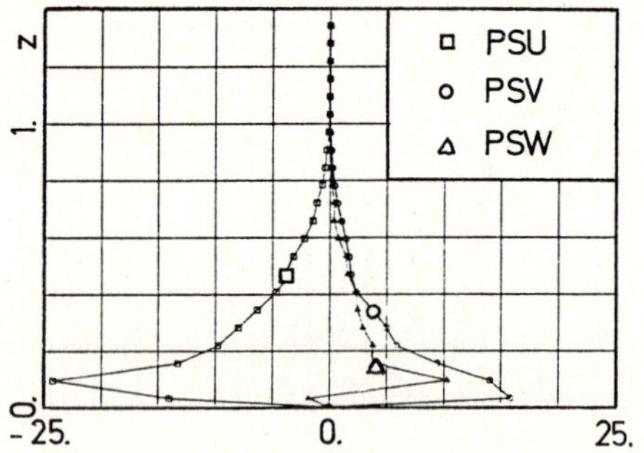

Fig.4j Resolvable portion of the trace of the pressure strain correlation tensor.

$$PSV_\alpha = 2 \langle {}^{\Delta V}\overline{p''} \; \delta_{(\alpha)} ({}^{\Delta A_\alpha} \overline{v''_{(\alpha)}}) \rangle$$

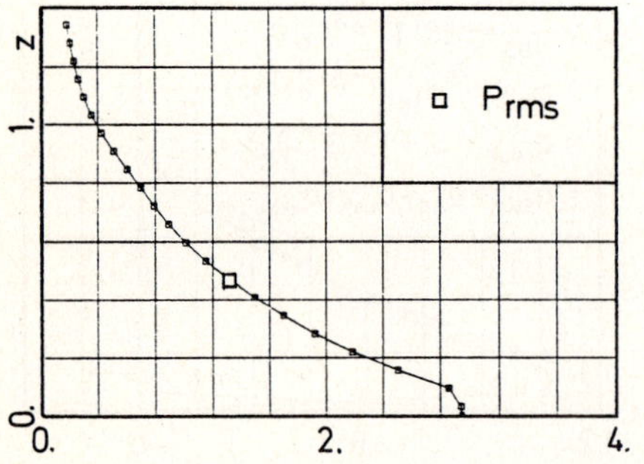

Fig.4k Resolvable rms pressure fluctuations.

$$P_{rms} = \langle {}^{\Delta V}\overline{p''^2} \rangle^{1/2}$$

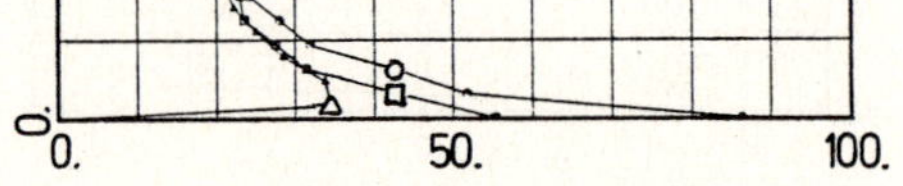

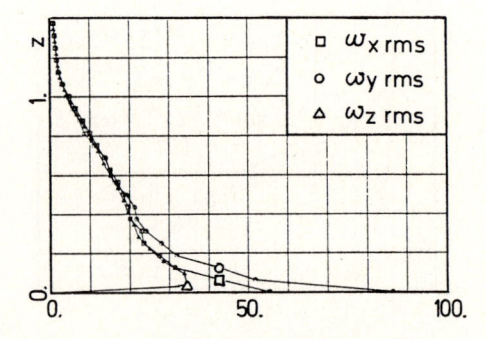

Fig.4l Resolvable rms vorticity fluctuations.

$$\omega_{\alpha\, rms} = \langle {}^{\Delta V}\overline{\omega''^2_{(\alpha)}} \rangle^{1/2}$$

Fig.4 Statistical mean values (continued).

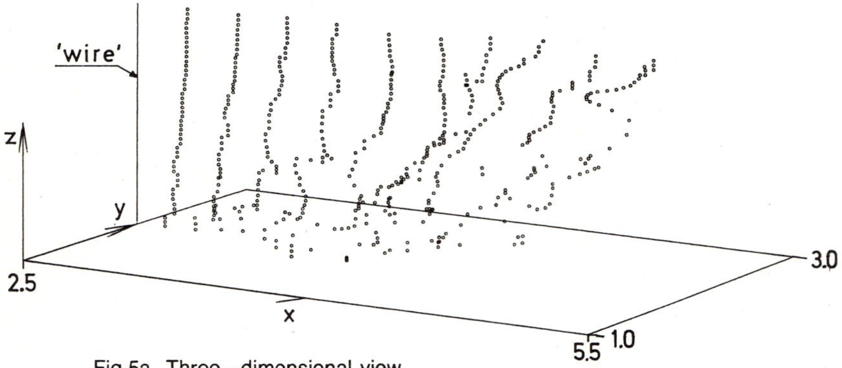

Fig.5a Three – dimensional view

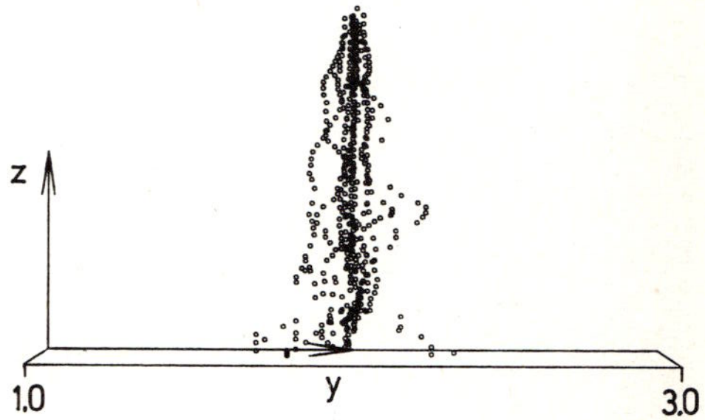

Fig.5b Front view

of particles generated from a 'z – wire' located at y = 2.0 and extending from the wall to z = 1.0 .

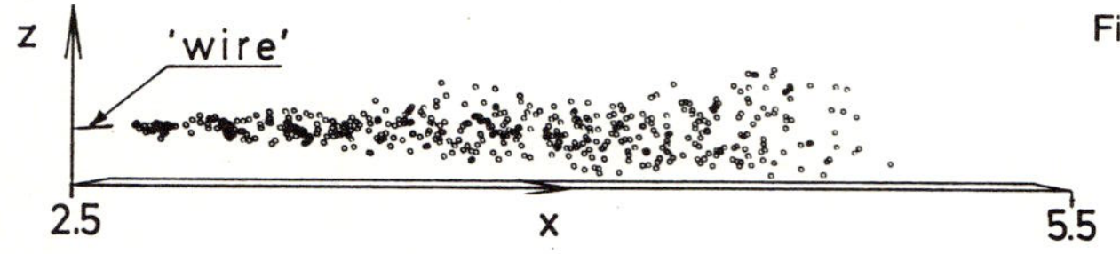

Fig.5c Side view

Fig.5d Front view

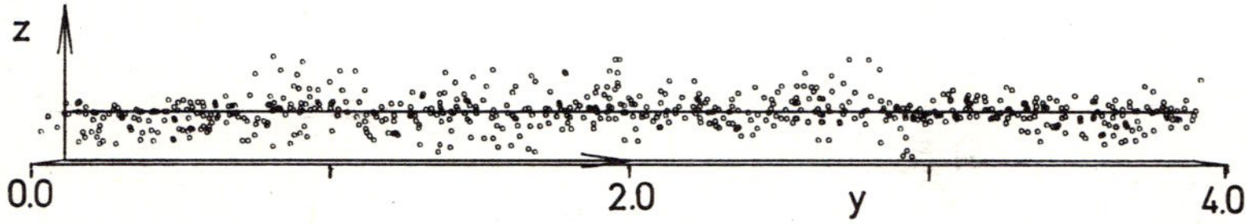

of particles generated from a 'y – wire' located at z = 0.15625 .

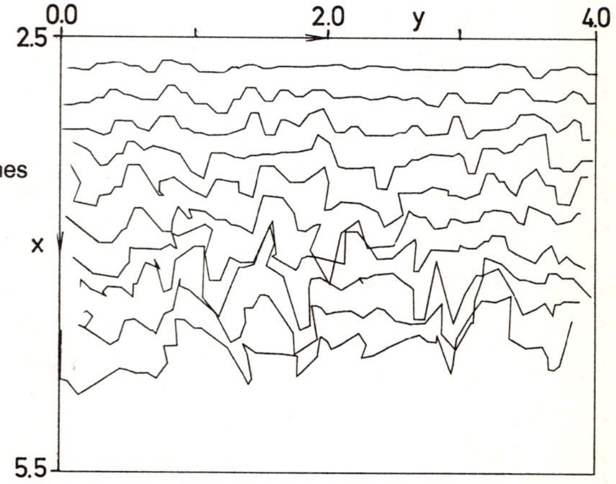

Fig.5e Top view of time lines

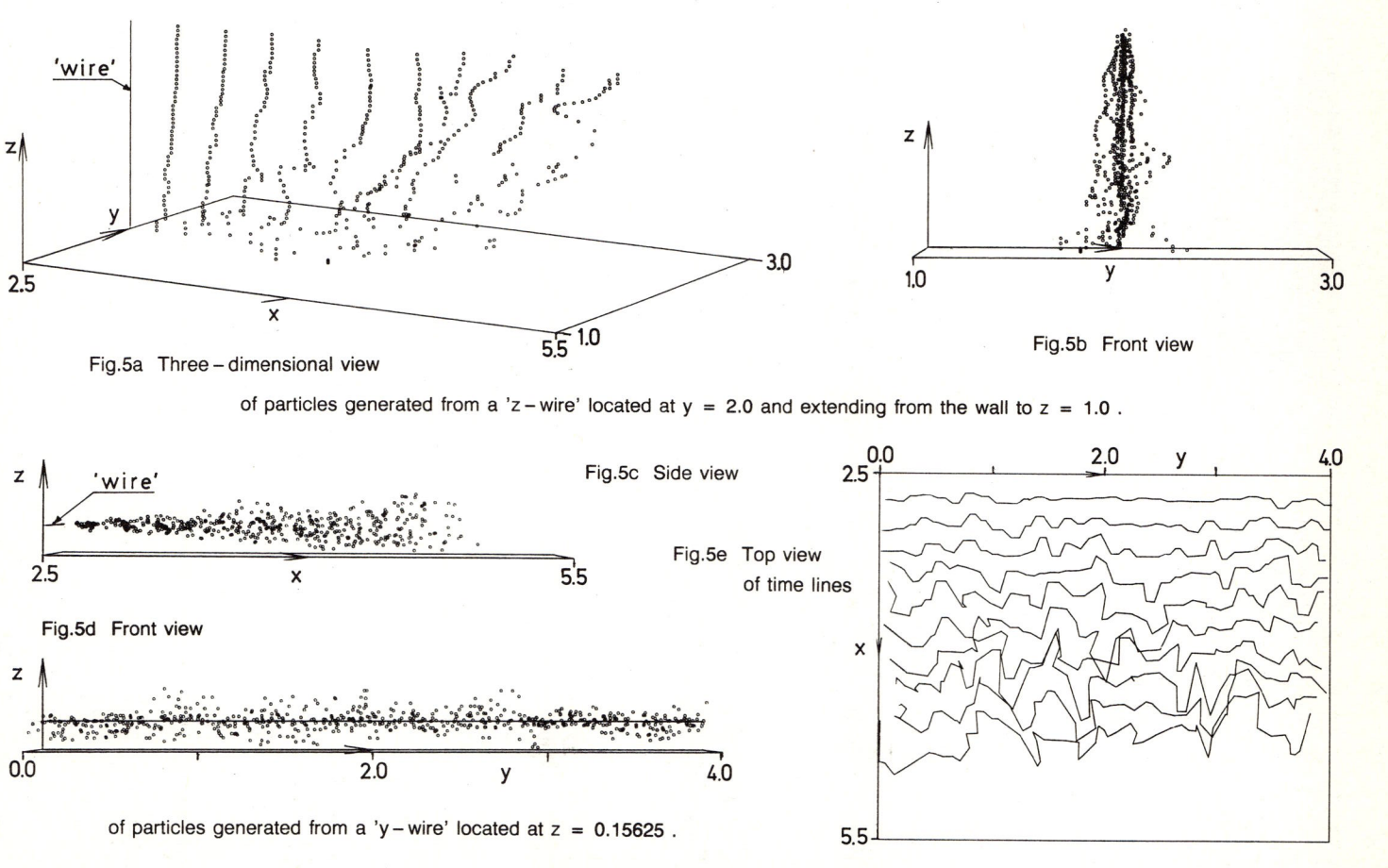

Fig.5 Marker particle simulation.

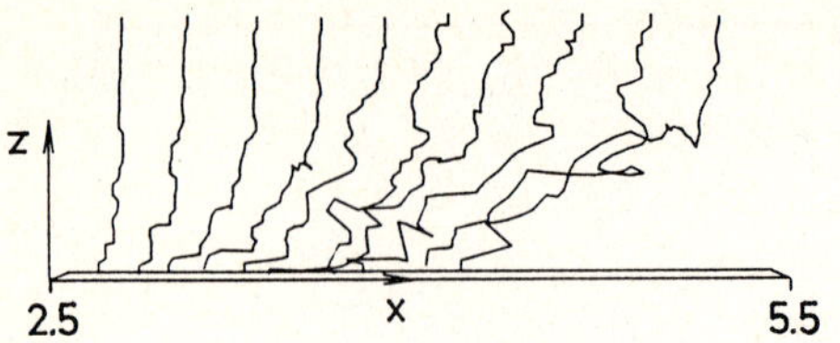

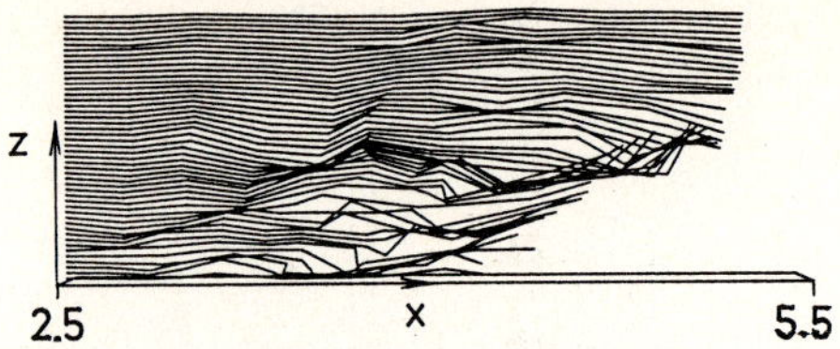

Fig.5f Time lines Fig.5g Streak lines
of particles generated from a 'z – wire' located at y = 1.0
and extending from the wall to z = 1.0 (side views).

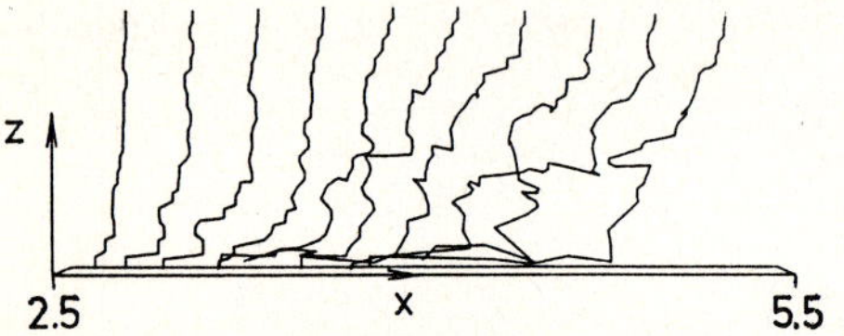

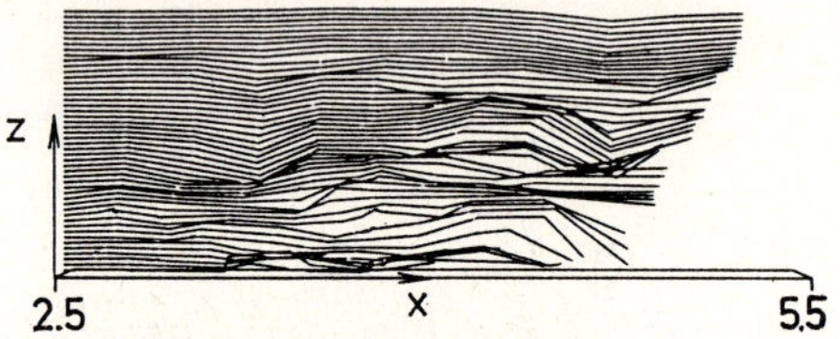

Fig.5h Time lines Fig.5i Streak lines
of particles generated from a 'z – wire' located at y = 3.0
and extending from the wall to z = 1.0 (side views).

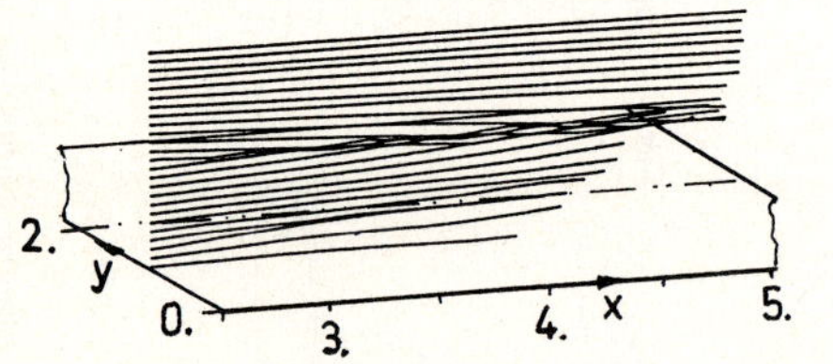

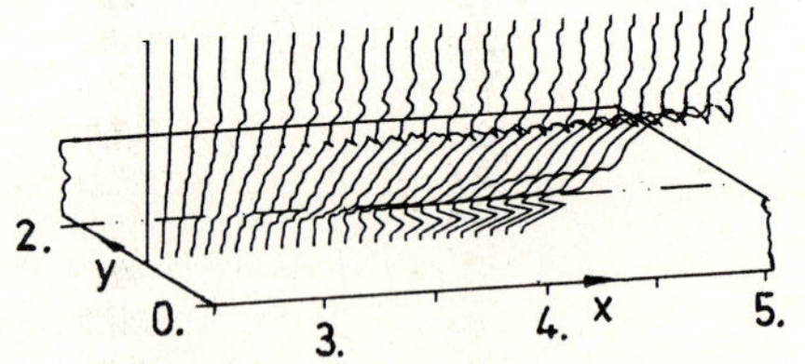

Fig.5j Path lines Fig.5k Time development of a line
of particles generated from a 'z – wire' located at y = 1.0
and extending from the wall to z = 1.0 (3-d views).

Fig.5 Marker particle simulation (continued).

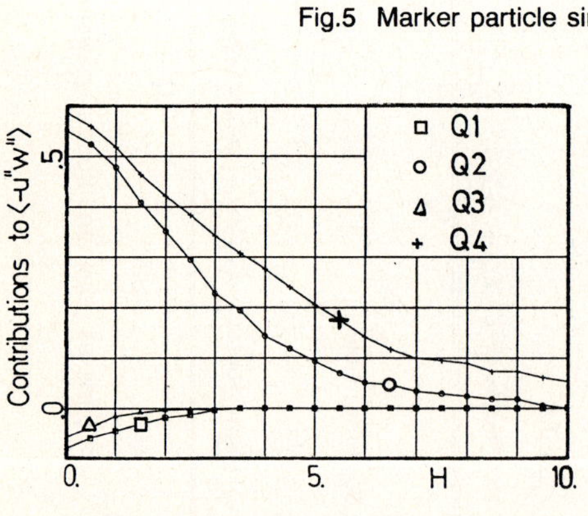

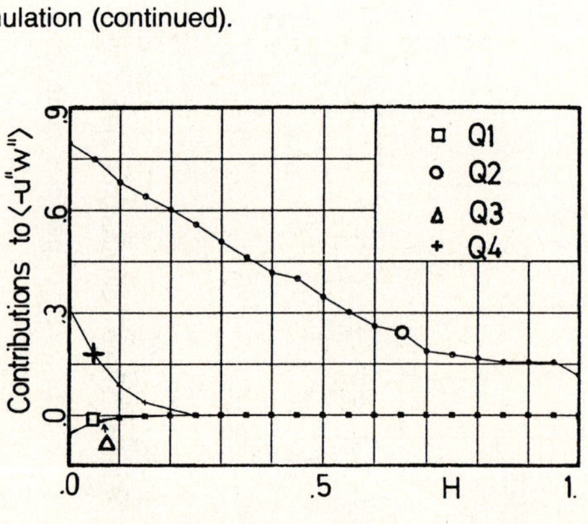

Fig.6a z = 0.0625 (< – uw> = 0.96). Fig.6b z = 1.0 (< – uw> = 0.04).

Fig.6 Quadrant analysis, fractional contributions to resolvable turbulence shear stress.

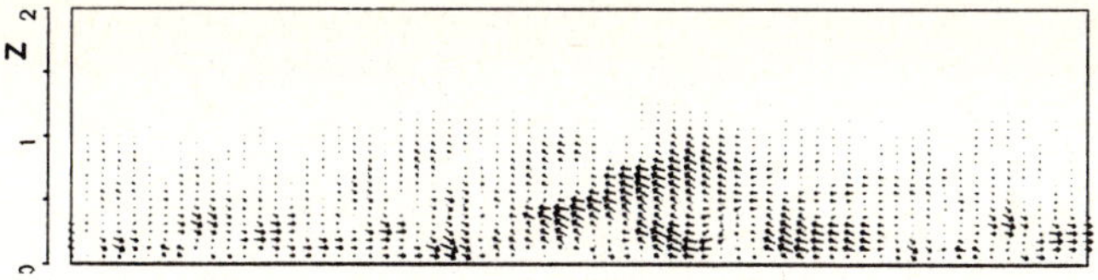

Fig.7a (u",w") – vectors in the (x,z) – plane at y_e.

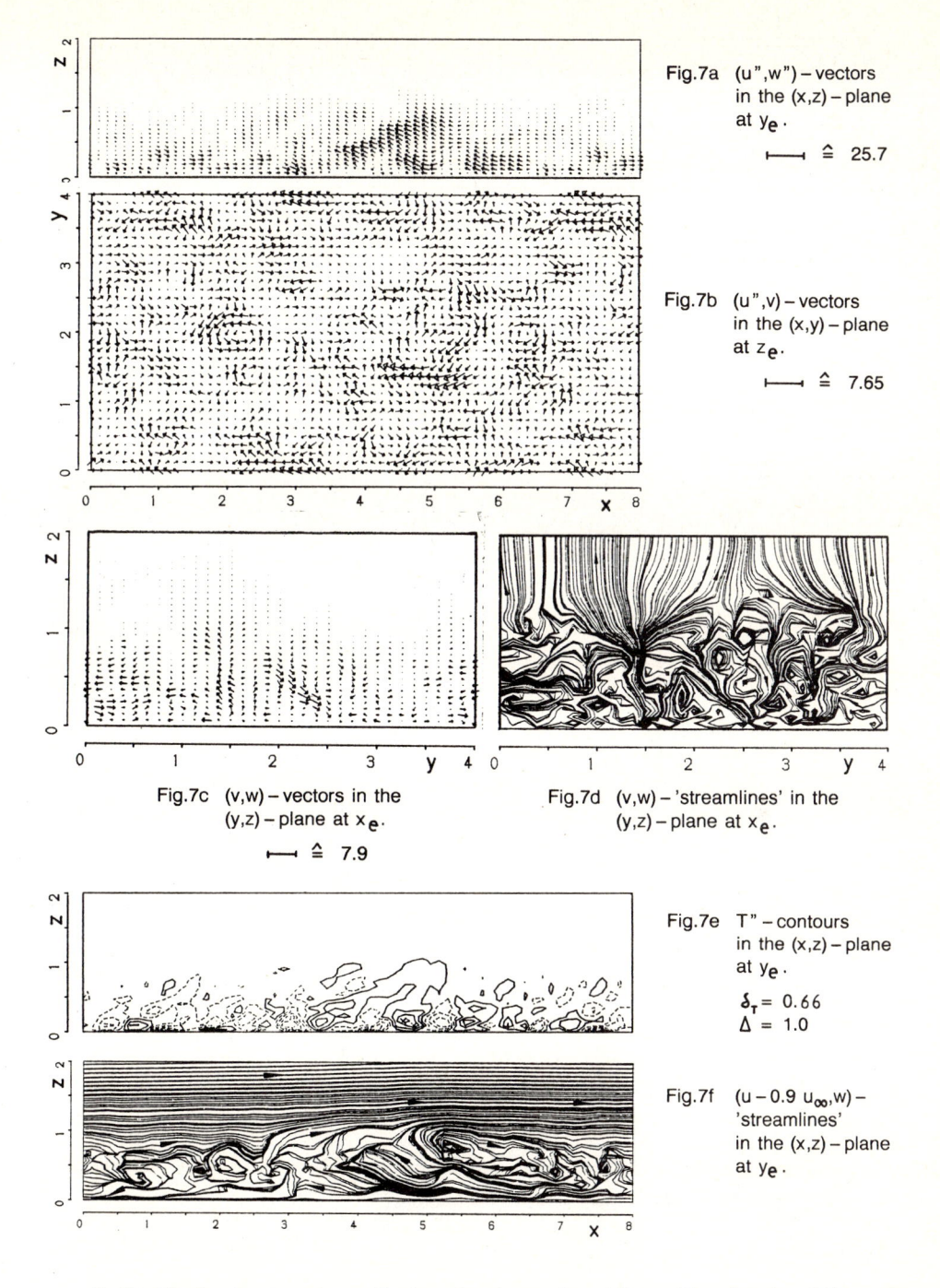

⊢—⊣ ≙ 25.7

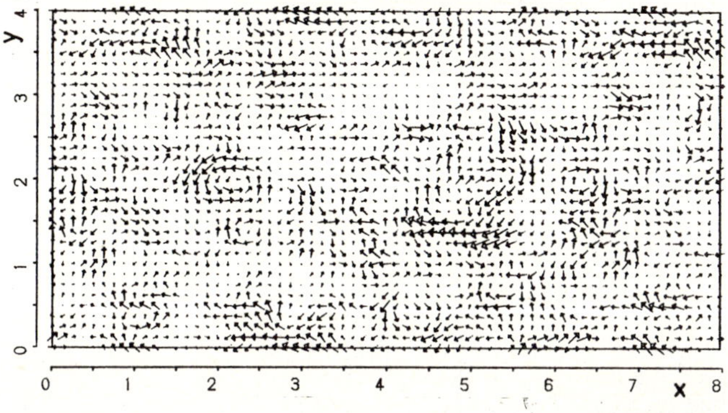

Fig.7b (u",v) – vectors in the (x,y) – plane at z_e.

⊢—⊣ ≙ 7.65

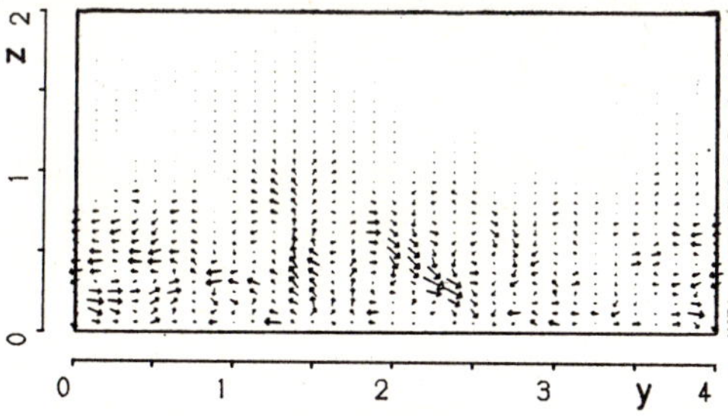

Fig.7c (v,w) – vectors in the (y,z) – plane at x_e.

⊢—⊣ ≙ 7.9

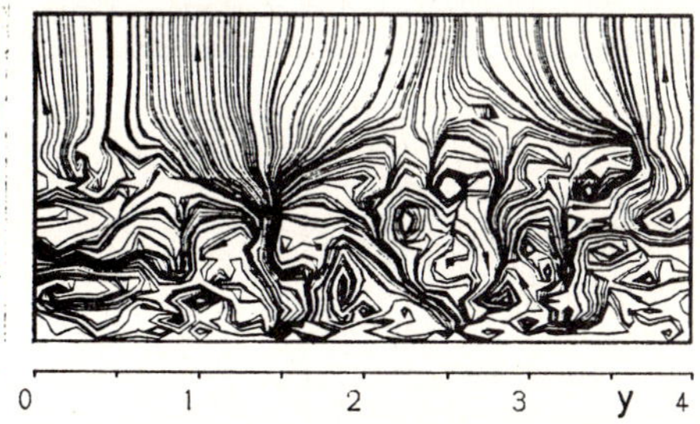

Fig.7d (v,w) – 'streamlines' in the (y,z) – plane at x_e.

Fig.7e T" – contours in the (x,z) – plane at y_e.

$\delta_T = 0.66$
$\Delta = 1.0$

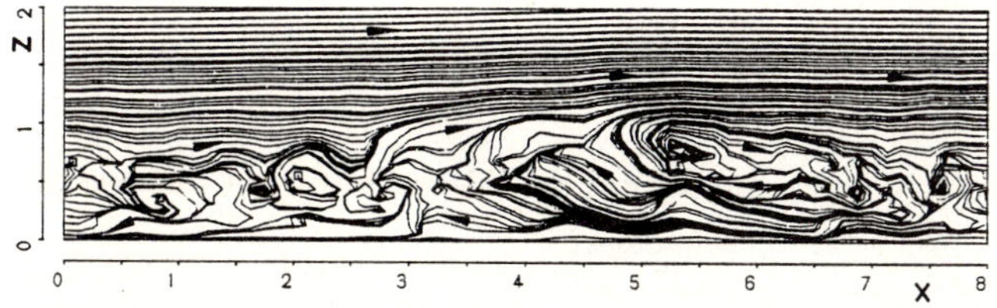

Fig.7f $(u - 0.9\, u_\infty, w)$ – 'streamlines' in the (x,z) – plane at y_e.

Fig.7 Ejection type event near the boundary layer edge at the problem time t = 9.7 (x_e = 4.5625, y_e = 1.4375, z_e = 0.65625, Δ : contourline increment, dashed: negative values).

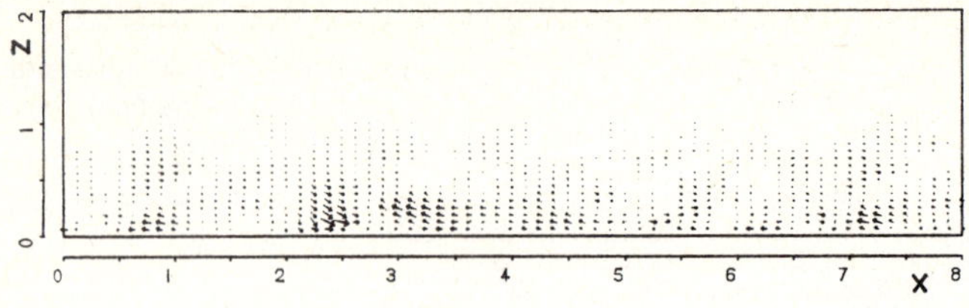

Fig.8a (u",w") – vectors in the (x,z) – plane at y_s.

⊢—⊣ ≙ 34.2

Fig.8b u" – contours in the (y,z) – plane at x_{su}.

Δ = 2.0

Fig.8c u – contours in the (y,z) – plane at x_{su}.

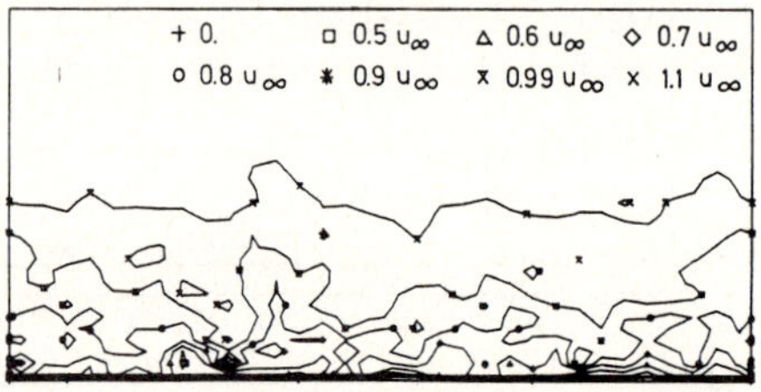

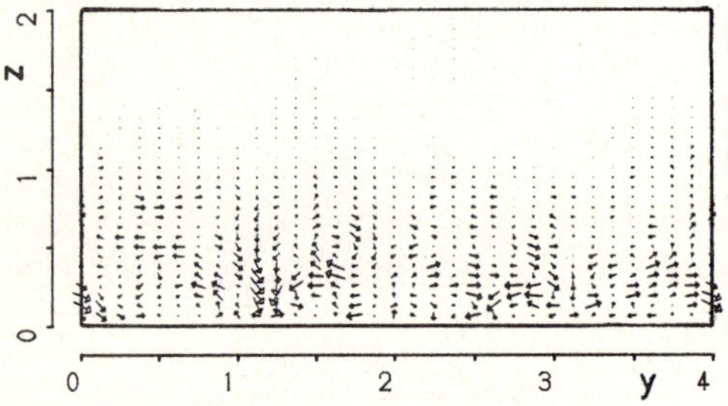

Fig.8d (v,w) – vectors in the (y,z) – plane at x_s.

⊢—⊣ ≙ 6.54

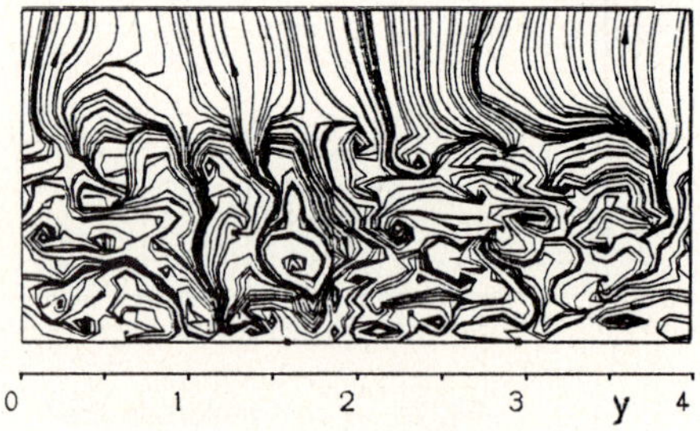

Fig.8e (v,w) – 'streamlines' in the (y,z) – plane at x_s.

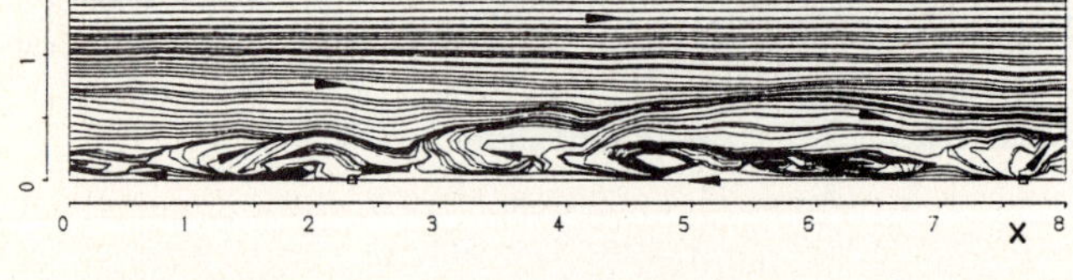

Fig.8f T" – contours in the (x,z) – plane at y_s.

Δ = 1.0

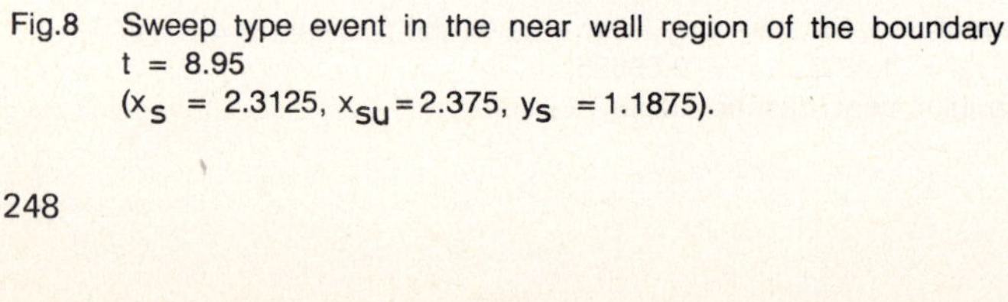

Fig.8g $(u - 0.7\, u_\infty, w)$ – 'streamlines' in the (x,z) – plane at y_s.

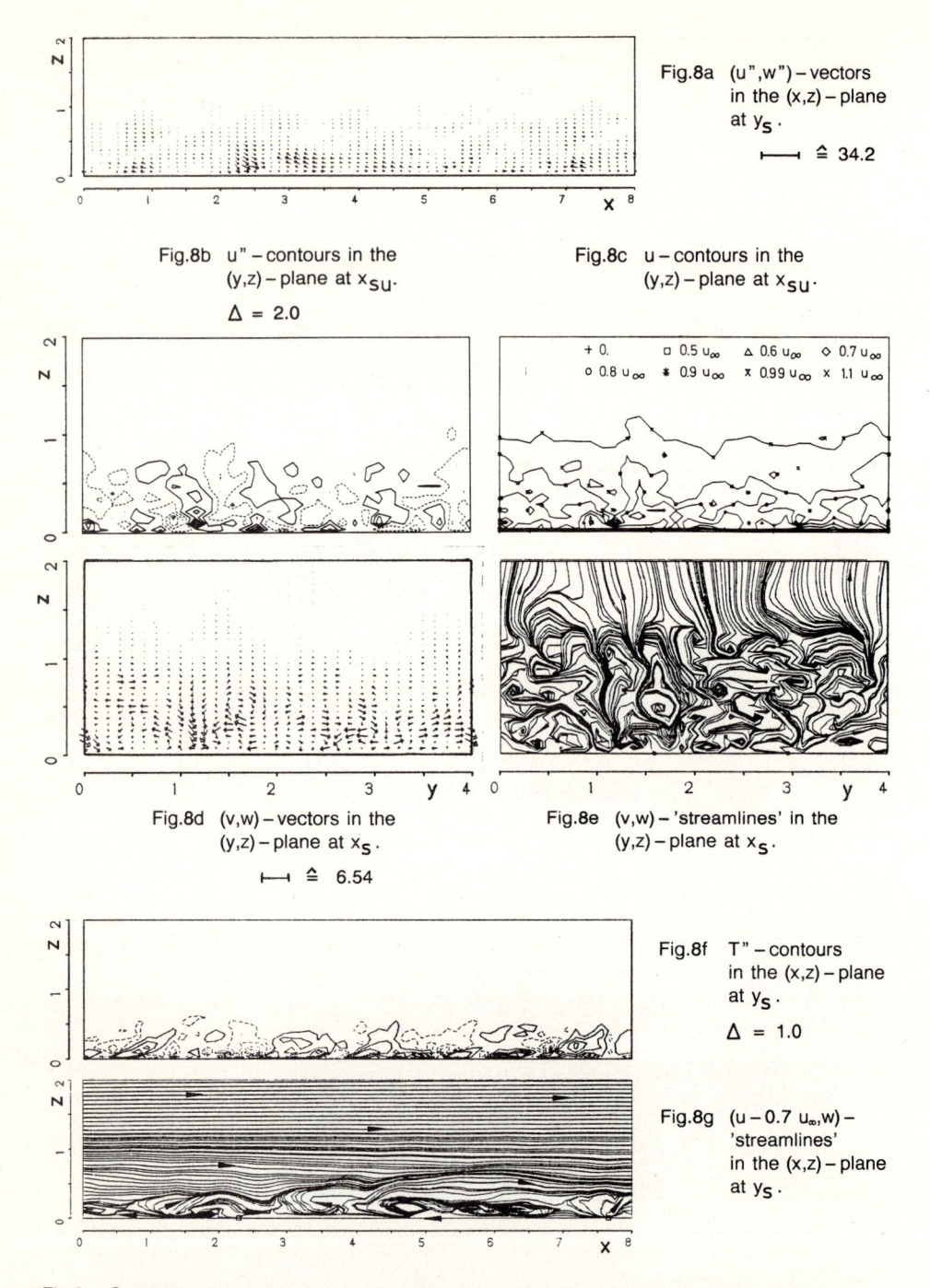

Fig.8 Sweep type event in the near wall region of the boundary layer at the problem time t = 8.95
(x_s = 2.3125, x_{su} = 2.375, y_s = 1.1875).

APPLICATION OF THE MULTIGRID METHOD TO THE SOLUTION OF PARABOLIC DIFFERENTIAL EQUATIONS

W. Schröder, D. Hänel

Aerodynamisches Institut, RWTH Aachen, Germany

SUMMARY

Two multigrid concepts for the solution of time dependent differential equations are presented. The heat conduction equation and the Navier-Stokes equations are solved with the methods proposed. A comparison of the rates of convergence of the multigrid methods with that of a single grid method shows the efficiency of the multigrid procedures.

INTRODUCTION

The multigrid method has been successfully applied to the solution of elliptic differential equations. The application to parabolic equations requires a reformulation in order to incorporate the time dependence of the solution. Two integration procedures were developed so far [1] . In the first, the direct multigrid method, the solution advances in time on the finest and on the coarser grids. The accuracy of the finest grid is maintained on all coarser grid levels [2,3] . The second method, the indirect multigrid method, employs the multigrid concept in an implicit scheme to accelerate the iterative matrix inversion process for two time levels [4,5] .

The purpose of this paper is to compare both methods by means of test calcuations for the linear and nonlinear heat conduction equation, and also for the Navier-Stokes equations. The aim of the investigation is to develop an efficient method of solution for a compressible viscous flow.

In the following section of this paper the methology of the multigrid method will be briefly reviewed. The next sections then show how the direct and indirect method can be used in the numerical integration of the equations mentioned above. In the last section the rate of convergence of a single grid scheme is compared with that of the direct and indirect multigrid method.

A BRIEF REVIEW OF THE MULTIGRID METHOD

In simple numerical integration procedures the differential equations are discretized and the resulting difference equations are solved on a single grid. In the multigrid method a sequence of coarse and fine grids is used for the solution of the difference equations. The advantage of using several grids is that a correction of the solution established on a fine mesh can be evaluated with little computational effort on the coarse grid [1] .

The local mode analysis shows, that the error belonging to one grid level contains predominately high frequent components. Since their wavelength is usually less than the fourfold step size of the fine grid these high frequency components are not visible on sufficiently coarse grids. Consequently, the error of the fine grid solution can only be represented on the coarse grid if the high-frequency fluctuations are smoothed before the transfer to the coarse grid. An effective way of damping high-frequency oscillations is to use relaxation methods, e.g. Gauß-Seidel or Jacobi procedures. They reduce the low-frequency components, only very slowly, such, that solutions based on relaxation techniques would be very inefficient. In the multigrid method the lower frequencies of the error of the fine grid are effectively damped by relaxation sweeps on the coarser grids. This is possible since the low-frequency components of the fine grids are high-frequency fluctuations on the coarser grids. As a consequence, the role of relaxation procedure is not reduction of the error but smoothing, and to determine an approxiate correction to the fine-grid solution.

Several algorithms were developed which employ the multigrid concept: In the "Correction-Cycle" only the correction to the fine-grid approximation is stored on the coarser grids which restricts its application to linear or linearized operators [1] . In contrast to the "Correction-Cycle" the "Full-Approximation-Storage" (FAS) mode of operation is suitable for general nonlinear problems. The latter algorithm is also applicable to composite grids which are the basis of local mesh refinement. In the FAS mode the approximate values are stored on the coarse grid. The approximation consists of the transfered approximation of the fine, and of the correction of the coarse grid. Compared with the "Correction-Cycle" the FAS mode needs slightly more computational work [1] .

Since for problems in fluid mechanics the solution of nonlinear systems is of interest the FAS mode was employed. This procedure will therefore be described in detail for a two-level iteration.

The discrete problem on a fine grid G_h may be written as

$$L_h \hat{\varphi}_h = \hat{f}_h . \tag{1}$$

The subscript h characterizes the fine grid. The quantity L_h designates any explicit or implicit operator, and $\hat{\varphi}_h$ is the solution of Eq.(1). After some relaxation sweeps with the initial distribution for $\hat{\varphi}_h$ on G_h one obtains a smoothed error $\hat{V}_H = \hat{\varphi}_h - \varphi_h$. Then equation (1) can be written as a defect equation

$$L_h(\varphi_h + \hat{V}_h) - L_h\varphi_h = r_h \tag{2}$$

$$r_h = \hat{f}_h - L_h\varphi_h .$$

Equations (1) and (2) are formally identical, but in contrast to the approximation φ_h the error $\hat{V}_h$ is smooth. Consequently, $\hat{V}_h$ can be approximated on a coarser grid.

On the coarser grid G_H equation (2) reads

$$L_H(I_h^H \varphi_h + \hat{V}_H) - L_H(I_h^H \varphi_h) = \mathbb{I}_h^H r_h . \tag{3}$$

Then the FAS coarse-grid equation can be formulated in the following form

$$L_H \hat{\varphi}_H = \hat{f}_H \tag{4}$$

with

$$\hat{\varphi}_H = I_h^H \varphi_h + \hat{V}_H$$

$$\hat{f}_H = \mathbb{I}_h^H r_h + L_H(I_h^H \varphi_h) .$$

I_h^H , II_h^H represent restriction operators which, in general, differ from each other. With a known approximate solution φ_H for equation (4) the approximate coarse-grid correction reads

$$V_H = \varphi_H - I_h^H \varphi_h . \tag{5}$$

The correction V_H is interpolated for the fine grid G_h, yielding a corrected appoximation φ_h^{new}

$$\varphi_h^{new} = \varphi_h + I_H^h (\varphi_H - I_h^H \varphi_h) . \tag{6}$$

Brandt [1] calls the interpolation according to equation (6) "FAS-Interpolation". In general, equation (6) cannot be identified with the shorter form

$$\varphi_h^{new} = I_H^h \varphi_H \tag{7}$$

since the product $I_H^h \cdot I_h^H$ is not the identity operator. There are cases, however, e.g. near interior discontinuities, where equation (7) is to be preferred to the FAS-Interpolation (Eq.(6)). The process, described in equations (1) to (7), has to be contained up to a certain prescribed accuracy.

The procedure presented for the two level iteration can be extended to K-grids, resulting in a multigrid iteration. The transfer from fine to coarse grids and back to the fine grids is called a cycle.

Equation (4) can be rewritten in the following form:

$$L_H \hat{\varphi}_H = f_H + \tau_h^H \tag{8}$$

where

$$\tau_h^H = L_H(I_h^H \varphi_h) - II_h^H (L_h \varphi_h)$$

$$f_H = II_h^H f_h .$$

Without the term τ_h^H, equation (8) represents the discrete problem directly formulated on the coarse grid G_H with a modified right hand side f_H. In this case (without τ_h^H) equation (8) would show a larger truncation error than equation (1) according to the coarser step sizes on G_H. This would prevent the multigrid efficiency come into play. Consequently, τ_h^H is a "fine to coarse defect correction" which maintains the truncation error of the fine grid G_h on the coarse grid level G_H. τ_h^H is also called relative local truncation error. Moreover, the correction term τ_h^H leads to the so-called τ extrapolation which resembles the Richardson extrapolation.

Instead of regarding the coarse grid as a subsidiary grid in order to determine a correction to the fine grid approximation, the fine grid can also be considered as a device to compute the defect correction τ_h^H to the coarse grid equations. Only a few relaxation sweeps are performed on the fine grid G_h in every cycle in order to renew τ_h^H while most of the solution process is carried out on the coarse grid G_H. Brandt

[1] calls this the dual point of view. It is the basis of the direct multigrid strategy for the solution of time-dependent problems.

ELEMENTS OF THE MULTIGRID-METHOD

In computations carried out so far by the authors the V-cycle was always used. It is illustrated below

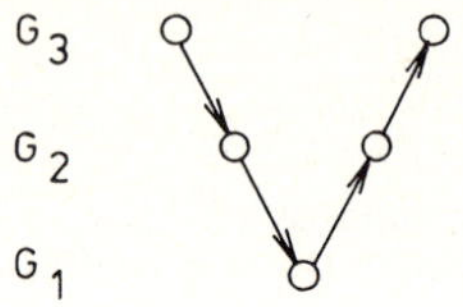

G_3 characterizes the finest and G_1 the coarsest grid. The W-Cycle is more useful for theoretical investigations.

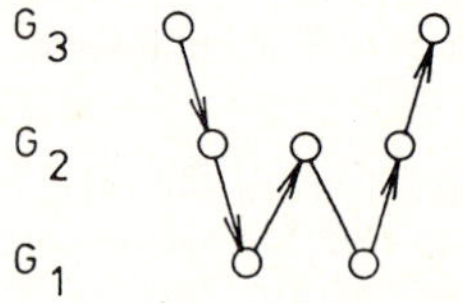

Standard coarsening that is to say two space steps on the fine grid form one space step on the coarse grid was applied between different grid levels. The number of relaxations on every mesh was fixed. As restriction operators I_h^H, II_h^H, either the simple point-to-point injection or the full weighting operator were used, given here for an uniform grid

$$II_h^H = \frac{1}{16}\begin{bmatrix} 1 & 2 & 1 \\ 2 & 4 & 2 \\ 1 & 2 & 1 \end{bmatrix}. \qquad (9)$$

This operator can be derived from the following requirement for the distributions of the function r(x,y) on the grids G_h and G_H:

$$\int_A r_h \, dx\, dy = \int_A (II_h^H r_h)\, dx\, dy. \qquad (10)$$

The interpolation of the corrections back to the finer grids was performed by a bilinear or bicubic FAS-Interpolation.

DIRECT MULTIGRID METHOD

The direct multigrid method for the solution of time dependent problems was employed in conjunction with explicit difference schemes. The operator L_h, e.g. Euler forward or explcit MacCormack-scheme [6] is determined by the method of solution. It also has to damp out the high-frequency components of the error. Since for explicit schemes the time step depends on the space step, larger time steps can be used on coarser grids. Consequently, the total amount of computational work is reduced in the direct multigrid method for two reasons: 1) According to the dual point of view, the main part of the iteration process is performed on the coarse grids, while only some "visits" on the fine grid are needed in order to update the relative local truncation error. 2) On the coarse grids the solution proceeds faster than on the fine grid level, since on the coarse grids the time step allowed due to a stability analysis is increased compared with the time step of the fine mesh.

The FAS direct multigrid strategy is given by the following equations. On the finest grid h = m the discrete problem reads

$$L_m \varphi_m = f_m . \tag{11}$$

On coarser grids $m \geq c \geq 1$ the corrected difference equations are written

$$L_{m-c}\, \varphi_{m-c} = f_{m-c} + \tau_m^{m-c} \tag{12}$$

with

$$f_{m-c} = \Pi_{m-c+1}^{m-c}\, f_{m-c+1} \tag{12a}$$

$$\tau_m^{m-c} = \tau_{m-c+1}^{m-c} + \Pi_{m-c+1}^{m-c}\, \tau_m^{m-c+1} \tag{12b}$$

and

$$\tau_{m-c+1}^{m-c} = L_{m-c}(\Pi_{m-c+1}^{m-c}\, \varphi_{m-c+1}) - \Pi_{m-c+1}^{m-c}(L_{m-c+1}\, \varphi_{m-c+1}) . \tag{12c}$$

Note that equation (12c) represents the fine-to-coarse-defect correction in time and space, the term τ_{m-c+1}^{m-c} contains the time step of the coarse grid G_{m-c} in its first term and the time step of the fine mesh G_{m-c+1} in the second term.

A complete V-Cycle of the direct multigrid method for time dependent problems consists of the following parts:

a) Perform a fixed number of time steps on the finest grid (Eq.(11)).
b) Transfer the variables and the right-hand side from the fine to the coarse grid and compute the fine-to-coarse-defect correction (Eq.(12a,b,c)).
c) Carry out a fixed number of time steps on the coarser grid $m \geq c \geq 1$ (Eq.(12)).
d) Continue with steps b) and c) until the coarsest grid is reached.
e) Perform a fixed number of time steps on the coarsest grid; this number of time steps should be very large in order to obtain a high rate of convergence.
f) Interpolate the variables back to the finer grid level (Eq.(6)).
g) Carry out a few time steps on the finer grid in order to reduce interpolation error.
h) Repeat steps f) and g) until the finest grid is reached.
i) Continue with step a).

INDIRECT MULTIGRID METHOD

In many problems with different characteristic length scales the time step restriction of the explicit difference schemes is prohibitive for convergence to a steady state within reasonable computational time. One way to reduce the computational effort is to improve the efficiency of the explicit difference schemes by the direct multigrid strategy as presented in the previous section. Another possibility of improving the rate of convergence is to resort to implicit formulation, so that the time step is only weakly or not at all restricted. One disadvantage of the implicit difference schemes, however, is that the inversion of the solution matrix for each time step is very costly if elimination methods are used. This holds especially for the solution of large systems of equations as e.g. the time dependent Navier-Stokes equations in two or three dimensions. It is at this point where the multigrid principle is inserted in the implicit difference schemes in order to perform the inversion to the next time level with little computational work. This leads to the indirect application of the multigrid procedure in a time dependent problem. Its key feature is that the elimination method is replaced by an iterative relaxation process which is accelerated by the multigrid method as presented in section 2. In contrast to the direct method the (implicit) time step in the indirect multigrid process is frozen during the multigrid procedure. This will be explained in the following:

Consider the discrete time dependent problem on the finest grid G_m

$$L_m \varphi_m = f_m \tag{13}$$

where L_m represents the operator of the method of solution e.g. backward Euler in time and central or upstream differences in space. In the indirect multigrid method the solution of equation (13) is obtained iteratively for every time step. Therefore Eq.(13) is approximated by

$$\tilde{L}_m \varphi_m = \tilde{f}_m . \tag{14}$$

The quantity $\tilde{f}_m$ is a modified right-hand side $\tilde{f}_m = f_m + (\varphi^n_{i,j} / \Delta t)_m$ - the subscripts i,j denote the grid point, the superscript h characterizes the old time level and Δt represents the implicit time step, and τ_m is any relaxation operator as e.g. Jacobi or Gauß-Seidel. To accelerate the iteration process between equations (14) and (13) the multigrid method is applied to Eq.(14). Then the corrected difference equations on coarser grids $m \geq c \geq 1$ read:

$$\tilde{L}_{m-c} \varphi_{m-c} = \tilde{f}_{m-c} + \tilde{\tau}^{m-c}_m \tag{15}$$

with

$$\tilde{f}_{m-c} = \text{II}_{m-c+1} \tilde{f}_{m-c+1} \tag{15a}$$

$$\tilde{\tau}^{m-c}_m = \tilde{\tau}_{m-c+1} + \text{II}^{m-c}_{m-c+1} \tilde{\tau}^{m-c+1}_m \tag{15b}$$

and

$$\tilde{\tau}^{m-c}_{m-c+1} = \tilde{L}_{m-c} (\text{II}^{m-c}_{m-c+1} \varphi_{m-c+1}) - \text{II}^{m-c}_{m-c+1} (\tilde{L}_{m-c+1} \varphi_{m-c+1}) . \tag{15c}$$

The interpolation of the corrections computed on the coarse grids back to the finer grids is performed by an FAS-Interpolation Eq.(6). Equation (13) and the iterated equation (14) do coincide if $/ \tilde{L}_m \varphi_m - \tilde{f}_m / \leq \varepsilon$, where ε is a very small number. If this condition is satisfied the computation can be continued with the next time step. If only the steady state solution is of interest ε can be chosen larger than for time accurate solutions provided the solution for $t \to \infty$ does not depend on the initial conditions. In general, for steady state computations only a few iterations on every grid are sufficient to approximate Eq.(13) by Eq.(14).

Note that in the indirect multigrid approach the operator of the method of solution

L_m and the relaxation operator $\tilde{L}_{m-c}$ $m \geq c \geq 0$ differ from each other. The multigrid process, which is applied in every time step, depends primarily on the smoothing properties for the high-frequency components of the operator $\tilde{L}_{m-c}$ $m \geq c \geq 0$. One therefore has to see to it that $\tilde{L}_{m-c}$ $m \geq c \geq 0$ has a very high smoothing rate in order to obtain an effective method. In contrast to the often used implicit approximate factored scheme of Beam and Warming [7] the indirect multigrid method can be formulated without any time step restriction.

The complete indirect multigrid method consists of the following steps:

a) Formulate the discrete problem on the finest grid for the time level $t^{n+1} = t^n + \Delta t$ (Eq.(13)).

Perform the multigrid process.

b) Approximate Eq.(13) by Eq.(14) and perform a number of relaxation sweeps on Eq.(14).
c) Inject the variables and the right-hand side from the fine to the coarse grid and compute the coarse grid correction (Eq.(15a,b,c)).
d) Perform a number of relaxation sweeps on the coarse grid equation (15).
e) Repeat c) and d) until the coarsest grid is reached.
f) Carry out some relaxation sweeps on the coarsest grid.
g) Interpolate the corrections back to the finer grid (Eq.(6)).
h) Relax the difference equation of the finer grid a few times.
i) Repeat f) and g) until the finest grid is reached.
j) If the condition $/\tilde{L}_m \varphi_m - \tilde{f}_m/ \leq \varepsilon$ is fulfilled continue with step a) otherwise repeat b) - j).

RESULTS

SOLUTION OF THE HEAT CONDUCTION EQUATIONS

In the first part of the investigation the efficiency of the direct and the indirect multigrid method was investigated and compared with an explicit single grid method for the linear and nonlinear heat conduction equation.

The explicit predictor-corrector scheme of MacCormack was used as L_m operator in the direct multigrid approach and for the single grid computations.

In the indirect multigrid method the backward Euler scheme was employed, and the pointwise Gauß-Seidel relaxation in lexicographical order as relaxation operator $\tilde{L}_m$.

In order to study the influence of constant or locally varying time steps on the direct multigrid method and of variable space steps on both multigrid procedures the computations were performed on a nonuniform grid. For the linear and for the nonlinear heat conduction equation both multigrid approaches resulted in an extreme improvement of the rate of convergence compared with a single grid method for all grids used.

The direct multigrid method reduced the computational work both for the minimum time step and for the local time step to less than 10 percent of a single grid computation as shown in Fig. 1 for the relative error as function of the work units.

It was found that the efficiency of the indirect multigrid method depends strongly on the work necessary for the inversion of the solution matrix. If the time steps of the explicit single grid method and of the indirect multigrid method are of the same order of magnitude the explicit single grid scheme has a better convergence behaviour than the multigrid procedure as is demonstrated in Fig. 2a for the history of the relative error of the solution.

The rate of convergence of the indirect multigrid method, however, can be improved if larger time steps are used. This is shown in Fig. 2b. For a time step which is 100 times larger than the time step of the explicit scheme the rate of convergence of the indirect multigrid method is comparable to the convergence history of the direct multigrid method.

Since the time step of the indirect method is unrestricted further improvement of the convergence can be achieved, if the time step is increased.

SOLUTION OF THE NAVIER-STOKES EQUATIONS

Both procedures the direct and the indirect multigrid approach were used for the solution of the compressible Navier-Stokes equations in two dimensions. The time-dependent compressible Couette flow was chosen as test case in order to be able to compare the numerical results with the exact solution. In Fig. 3 the convergence histories of the different methods used for a computation with a Reynolds number of $Re = 10$ and a Mach number $Ma = 0.5$ are presented. In this case and for calculations with even lower Reynolds numbers $Re \leq 10$ the indirect multigrid method had the best rate of convergence. If, however, the Reynolds number was increased to $Re \geq 1000$ the indirect multigrid approach as formulated above became ineffective since the time step is restricted at high Reynolds numbers for centrally discretized convective terms. For this reason upwind differencing through flux-vector splitting for the Euler terms was employed to obtain a diagonally dominant matrix in particular for very large time steps. The flux-splitting formulation of van Leer [8] was used. In order to achieve a second-order accurate discretization for the convective terms the flux-splitting was combined with the MUSCL approach of van Leer, and Woodward [9]. The time step was increased with decreasing residual. Thus, the used collective pointwise Gauß-Seidel relaxation became a Switched Evolution/Relaxation scheme [10]. Consequently, very large time steps could be employed for the nearly converged state.

In Fig. 4 the rates of convergence of the single grid method, of the direct and of the modified indirect multigrid method are presented for the computation of the compressible Couette flow with a Reynolds number $Re = 1000$ and a Mach number $Ma = 0.5$. There is no doubt that the indirect multigrid procedure in conjunction with the flux-splitting method of van Leer for the Euler terms has the best rate of convergence. Fig. 5 demonstrates the convergence histories of the single grid explicit MacCormack-scheme and the indirect multigrid method for the computation of the subsonic flow past a flat plate ($Re = 1000$, $Ma = 0.8$). For this more complex flow problem the reduction of the computational work by the indirect multigrid approach compared with the single grid method is even more dramatic than for the simpler Couette flow problem.

CONCLUSIONS

Two multigrid concepts for solving time-dependent problems were presented and tested for the solution of the two-dimensional linear and nonlinear heat conduction equation and for the two-dimensional compressible Navier-Stokes equations. For the heat conduction equation the direct and the indirect multigrid approach reduce the computational effect to less than 10 percent of an explicit single grid method. The computations performed for the Navier-Stokes equations showed that the efficiency of the indirect multigrid procedure depends strongly on the difference scheme used. The best rates of convergence were obtained with the indirect multigrid method in which flux-vector splitting of van Leer for the Euler terms in conjunction with a collective Switched/Evolution Relaxation (SER) scheme was used. Further investigations have to demonstrate whether the favourable convergence histories of the indirect multigrid concept still hold for more complex problems.

REFERENCES

[1] BRANDT, A.: "Guide to multigrid development", Lecture Notes in Mathematics, 960 (1981) pp. 220-312.

[2] NI, R. H.: "A multiple-grid scheme for solving the Euler equations", AIAA Paper 81-1025 (1981).

[3] JAMESON, A.: "Solution of the Euler equations for two dimensional transonic flow by a multigrid method", presented at the International Multigrid Conference, Copper Mountain, 1983.

[4] MULDER, W. A.: "Multigrid relaxation for the Euler equations", Lecture Notes in Physics, 218 (1985) pp. 417-421.

[5] SCHRÖDER, W., HÄNEL, D.: "A comparison of several MG-methods for the solution of the time-dependent Navier-Stokes equations", presented at the Second European Conference on Multigrid Methods, Köln-Bonn, October 1985.

[6] MACCORMACK, R. W.: "The effect of viscosity on hypervelocity impact cratering", AIAA Paper 69-354 (1969).

[7] BEAM, R. M., WARMING, R. F.: "An implicit factored scheme for the compressible Navier-Stokes equations", Proceedings of the AIAA 3rd. Computational Fluid Dynamics Conference, Albuquerque, New Mexiko 1977.

[8] VAN LEER, B.: "Flux-vector splitting for the Euler equations", Lecture Notes in Physics, 170 (1982) pp. 507-512.

[9] VAN LEER, B.: "Towards the ultimate conservative difference scheme V. A second-order sequel to Godunov's method", J. Comp. Phys., 32 (1979) pp. 101-136.

[10] VAN LEER, B., MULDER, W. A.: "Relaxation methods for hyperbolic equations", Proceedings of the INRIA Workshop on Numerical Methods for the Euler Equations for Compressible Fluids, Le Chesnay, France, Dec. 1983; to be published by SIAM.

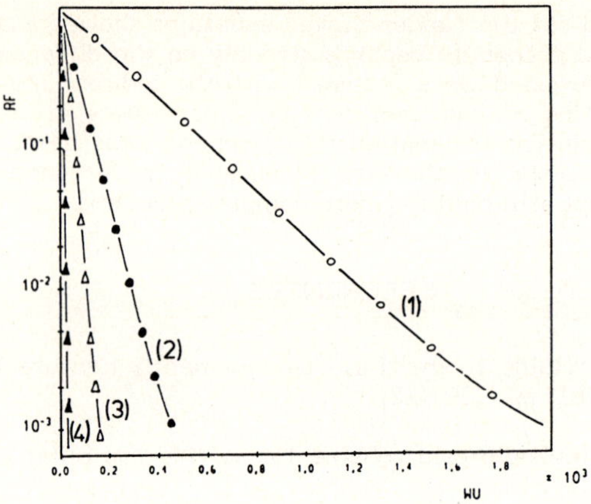

Fig. 1 Relative error as function of the work units (WU; 1 WU = 1 time step with the explicit MacCormack scheme on the finest grid)
curve (1): single grid method (SG); explicit MacCormack scheme; minimum time step
curve (2): single grid method (SG); explicit MacCormack scheme; local time step
curve (3): direct multigrid method (MG,di) using explicit MacCormack scheme; minimum time step
curve (4): direct multigrid method (MG,di) using explicit MacCormack scheme; local time step
finest grid: 33x33 grid points
V-Cycle: 5 grid levels
ND (number of time steps before every restriction) : 4
NU (number of time step after every interpolation) : 2
N1 (number of time steps on the coarsest grid) : 4

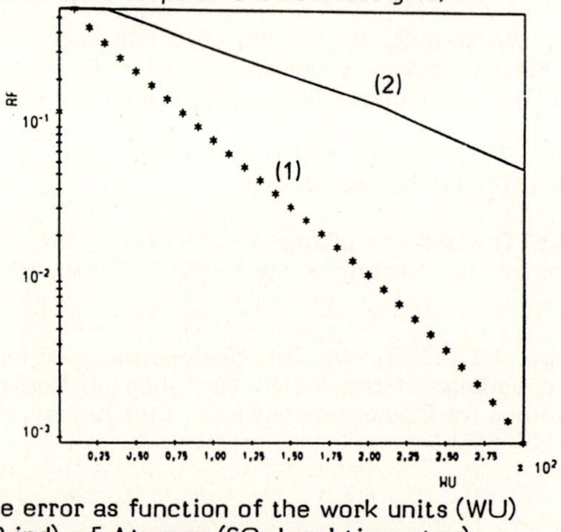

Fig. 2a Relative error as function of the work units (WU)
Δt (MG,ind) = 5 Δt, max (SG, local time step)
curve (1): SG; explicit MacCormack scheme; local time step
curve (2): MG,ind; backward Euler, Gauß-Seidel relaxation
finest grid: 33x33 grid points
V-Cycle: 5 grid levels
ND = 2, NU = 1, N1 = 1
(same notations as before)

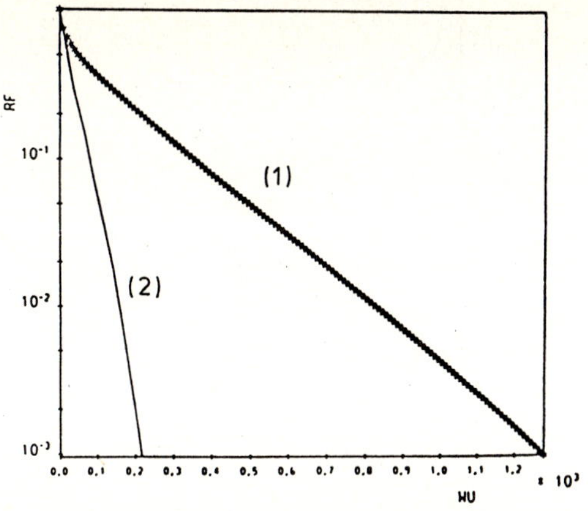

Fig. 2b Relative error as function of the work units (WU)
Δt (MG,ind) = 100 Δt, max (SG, local time step)
curve (1): SG; explicit MacCormack scheme; local time step
curve (2): indirect multigrid method (MG,ind); backward Euler; Gauß-Seidel relaxation
finest grid: 33x33 grid points
V-Cycle: 5 grid levels
ND = 2, NU = 1, N1 = 1
(same notations as in Fig. 1)

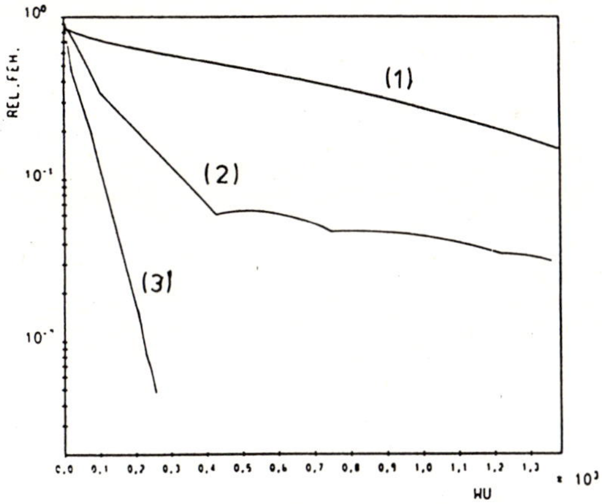

Fig. 3 Relative error as function of the work units (WU),
Re = 10, Ma = 0.5
curve (1): SG; explicit MacCormack scheme; local time step
curve (2): MG,di using explicit MacCormack scheme; local time step
curve (3): MG,ind; backward Euler; Gauß-Seidel relaxation
finest grid: 33x33 grid points
V-Cycle: 5 grid levels
MG,di; ND = 4, NU = 2, N1 = 2
MG,ind; ND = 2, NU = 1, N1 = 1
(same notations as before)

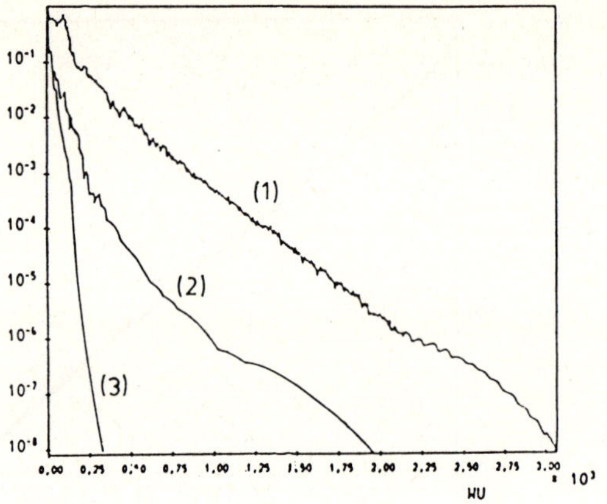

Fig. 4 Maximum residual as function of the work units (WU)
Re = 1000, Ma = 0.5
curve (1): SG; explicit MacCormack scheme; local time step
curve (2): MG,di using explicit MacCormack scheme; local time step
curve (3): MG,ind; backward Euler; collective Switched Evolution/Relaxation (SER) including flux-splitting
finest grid: 17x17 grid points
V-Cycle: 4 grid levels
MG,di: ND = 4, NU = 2, N1 = 2
MG,ind: ND = 2, NU = 1, N1 = 1
(same notations as before)

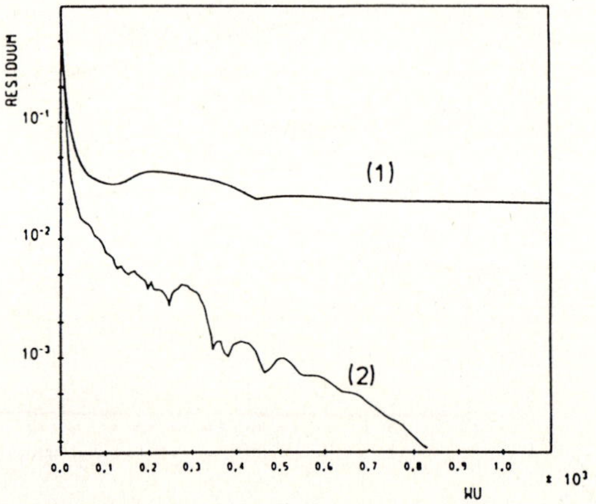

Fig. 5 Maximum residual as function of the work units (WU)
Re = 1000, Ma = 0.8
curve (1): SG; explicit MacCormack scheme; local time step
curve (2): MG,ind; backward Euler; collective Switched Evolution/Relaxation (SER) including flux-splitting
finest grid: 33x33 grid points
V-Cycle: 5 grid levels
ND = 2, NU = 1, N1 = 1
(same notations as before)

A NUMERICAL SIMULATION OF NEWTONIAN AND NON-NEWTONIAN FLOW IN AXISYMMETRIC HYPERBOLIC CONTRACTIONS

P. Schümmer, H.W. Bosch
Institut für Verfahrenstechnik der RWTH Aachen
Turmstr. 46, D - 5100 Aachen, Germany

SUMMARY

A hyperboloid of revolution shall be presented as a geometry for a contraction in which the advantage of a simple mathematical outline - without abrupt corners - is combined to an adapted, orthogonal coordinate system. The numerical process for the determination of the flow of a Newton and of a Maxwell fluid as well as some results are demonstrated.

INTRODUCTION

In the same way as the interest in non-newtonian materials has increased in the last few years, numerical methods for describing the corresponding flows have become more and more important. For the particular problem of the flow through an axisymmetric contraction many papers have been published, nearly all of them dealing with conical nozzles (cf. e.g. [4]) or abrupt constrictions (cf. e.g. [2]). In order to avoid corners, i.e. singularities, we want to present a hyperboloid of revolution as contracted conduit. Here the boundary can be described by a smooth curve. Besides orthogonal coordinate systems exist for which a specific coordinate describes the wall.

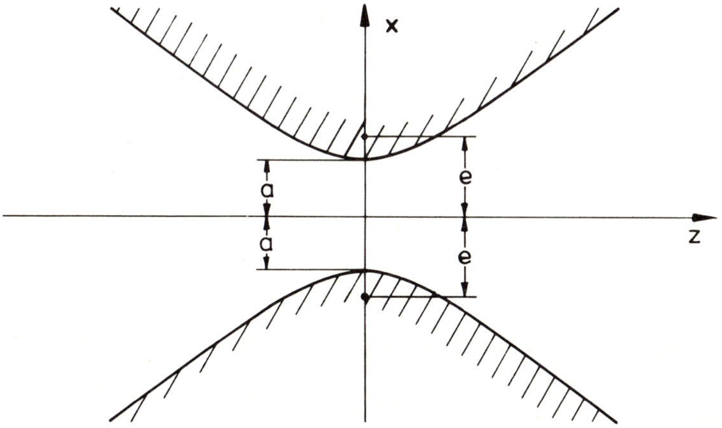

Fig. 1: Hyperboloid of revolution
e: excentricity of the hyperbolas

By variation of the geometric parameters different outlines reaching from an aperture to a pipe as well as conical nozzles can be shaped.
Considering sticking to the wall, symmetry conditions and creeping newtonian flow behaviour [3] in the up- and downstream regions the steady flow of a Newton and of a Maxwell fluid through this geometry shall be discussed in relation to the employment of finite differences.
For reasons of simplicity we apply the Maxwell model for a viscoelastic fluid. Nevertheless the process described depends only slightly on this model and can easily be adapted for a more realistic rheological model.

PROBLEM STATEMENT

ELLIPTIC COORDINATES

The axisymmetric hyperbolic contraction is created by rotation of a hyperbola around the z-axis with the angle $0 \leq \varphi < 2\pi$.
In the plane y=0 confocal orthogonal hyperbolas and ellipses can be defined:

hyperbolas

$$\frac{x^2}{e^2(1-\tau^2)} - \frac{z^2}{e^2\tau^2} = 1 \quad , \quad 0 < \tau < 1 \tag{1}$$

ellipses

$$\frac{x^2}{e^2(1+\sigma^2)} + \frac{z^2}{e^2\sigma^2} = 1 \quad , \quad -\infty < \sigma < \infty \tag{2}$$

which contain the hyperbolas of the wall for $\tau = \tau_0$ with

$$\tau_0 = \sqrt{1 - \frac{a^2}{e^2}} \quad . \tag{3}$$

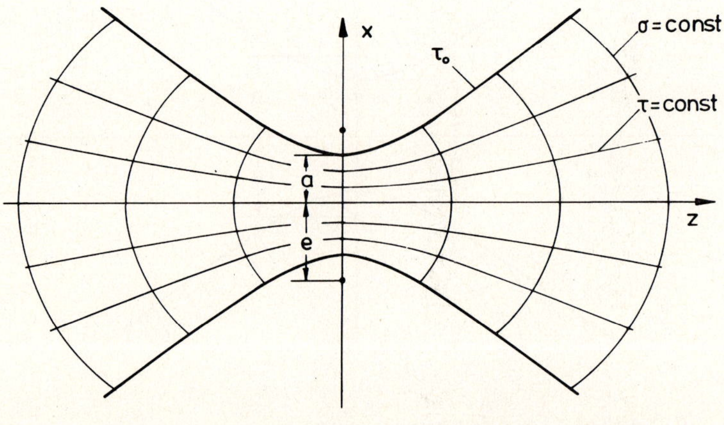

Fig. 2: Hyperboloid of revolution and adapted coordinate system

We take $\sigma \in (-\infty,\infty)$ to differentiate between the in- and outflow region.
σ, τ and the angle φ can be taken as coordinates of a three-dimensional, orthogonal elliptic coordinate system.
If we cover the region $-\Sigma \leqslant \sigma \leqslant \Sigma$, $\tau_0 \leqslant \tau \leqslant 1$ with an equidistant grid, we obtain in the x-z-plane only few lines which are near the axis or the contraction region (cf. the left part of fig. 3). Furthermore some functions, e.g. the shear stress, show indefinite gradients at the axis.

An additional transformation

$$\begin{aligned} \sigma &= \sinh \xi \quad , \quad \xi \in (-\infty,\infty) \\ \tau &= \sin \eta \quad , \quad \eta \in [\eta_0, \pi/2] \end{aligned} \tag{4}$$

- with $\eta_0 = \arcsin \tau_0$ -, yields as transformation of the coordinates (with $\alpha = \cosh \xi$, $\beta = \cos \eta$, $\gamma = \sqrt{\sinh^2\xi + \sin^2\eta}$)

$$\begin{aligned} x &= \alpha\beta \quad \cos \varphi \; , \\ y &= \alpha\beta \quad \sin \varphi \; , \\ z &= \sinh \xi \quad \sin \eta \; . \end{aligned} \tag{5}$$

In the right part of fig. 3 it becomes obvious that an equidistant ξ-η-grid is suited to perform the geometrical division of the hyperboloid. The above mentioned problems regarding the behaviour of some functions near the axis do not occur here.

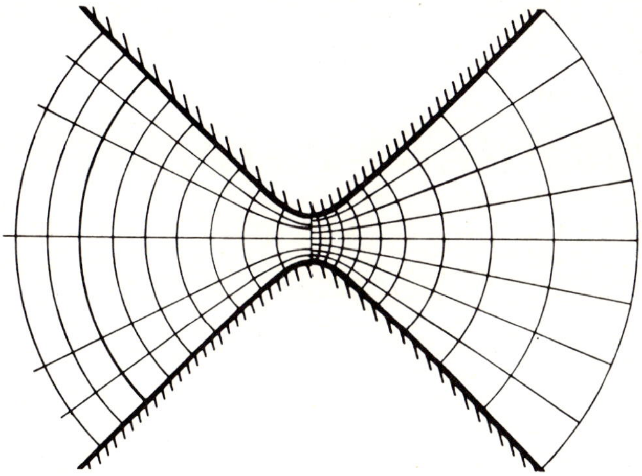

Fig. 3: Mapping of a grid in the σ-τ-plane (left) and of a grid in the ξ-η-plane (right) into the x-z-plane

EQUATIONS

The steady flow of an incompressible, viscoelastic fluid is described by the conservation laws and the equation of state (here: upper convected Maxwell)

$$\nabla \cdot \underline{v} = 0 \quad , \tag{6}$$

$$\rho \, \underline{v} \cdot \nabla \underline{v} = \nabla \cdot \underline{\underline{S}} - \nabla p \quad , \tag{7}$$

$$\underline{\underline{S}} + t_0 \, (\underline{v} \cdot \nabla \underline{\underline{S}} - \nabla v^T \cdot \underline{\underline{S}} - \underline{\underline{S}} \cdot \nabla \underline{v}) = \mu(\nabla \underline{v} + \nabla \underline{v}^T) \quad , \tag{8}$$

as well as by suitable boundary conditions.
$\underline{\underline{S}}$ refers to an anisotropic and p to an isotropic part of the stress tensor.
By the aid of viscosity μ, density ρ and relaxation time t_0 of the fluid, and by means of the volume rate $\dot{V}$ and the excentricity of the hyperbolas e, all appearing quantities are - in the following without special signature - reformulated in a dimensionless way. Now the problem can be described by the asymptotic cone angle η_0, the Reynolds-number

$$Re = \frac{\rho \dot{V}}{e\mu} \tag{9}$$

and the Deborah-number

$$De = \frac{t_0 \cdot \dot{V}}{e^3} \tag{10}$$

whereby one obtains the newtonian fluid for De=0.

By introducing a streamfunction Ψ the equation (6) is identically fulfilled. By application of "∇x" on (7) the isotropic part p can be eliminated (p has to be obtained after having fixed all other values). This gives rise to a differential equation that includes third derivations of Ψ. To reduce the order of differentiation a vorticity function Ω as the $2\alpha\beta$-fold of the φ-component of the vorticity vector is introduced. By this an equation of second order in Ψ is built and the former equation is now of first order in Ψ and Ω, but of second order in $\underline{\underline{S}}_E$.
By splitting the stress tensor into a linear part

$$\underline{\underline{S}}_L = \nabla \underline{v} + \nabla \underline{v}^T \quad , \tag{11}$$

and an extra part $\underline{\underline{S}}_E$ according to

$$\underline{\underline{S}} = \underline{\underline{S}}_L + \underline{\underline{S}}_E \quad , \tag{12}$$

the former equation becomes an equation of second order in Ω, too, and the corresponding differential operator is elliptic. The extra part $\underline{\underline{S}}_E$ is calculated from the modified equation of state,which is obtained by placing (12) into the Maxwell model.
An analysis of the second derivations yielded that the discretized forms of

$$\frac{\partial^2 \Psi}{\partial \eta^2} \quad \text{and} \quad \frac{\sin \eta}{\beta} \cdot \frac{\partial \Psi}{\partial \eta}$$

became small towards the axis, with the same power but with inverse signs. These terms are therefore summed up by

$$\beta \, \frac{\partial}{\partial \eta} \left(\frac{1}{\beta} \frac{\partial \Psi}{\partial \eta} \right) . \tag{13}$$

A similar summation was carried out in the ξ-direction. Thereby the maximum relative error in calculating the creeping newtonian shear stress could be lowered from 27 to 0.5 percent. With these modifications the describing equations are:

$$\Omega = \frac{\alpha}{\gamma^2}\frac{\partial}{\partial\xi}\left(\frac{1}{\alpha}\frac{\partial\Psi}{\partial\xi}\right) + \frac{\beta}{\gamma^2}\frac{\partial}{\partial\eta}\left(\frac{1}{\beta}\frac{\partial\Psi}{\partial\eta}\right) \quad , \tag{14}$$

$$\begin{aligned} &\mathrm{Re}\left\{\gamma^2 L\Omega + \frac{2}{\alpha\beta}\left(\frac{\sigma}{\alpha}\frac{\partial\Psi}{\partial\eta} + \frac{\tau}{\beta}\frac{\partial\Psi}{\partial\xi}\right)\Omega\right\} \\ &-\alpha\frac{\partial}{\partial\xi}\left(\frac{1}{\alpha}\frac{\partial\Omega}{\partial\xi}\right) - \beta\frac{\partial}{\partial\eta}\left(\frac{1}{\beta}\frac{\partial\Omega}{\partial\eta}\right) = G(\underline{\underline{S}}_E) \quad , \end{aligned} \tag{15}$$

$$\begin{aligned} A^{\xi\eta} S_E^{\xi\eta} + \mathrm{De}\, L\, S_E^{\xi\eta} &= C\, S_E^{\xi\xi} + B\, S_E^{\eta\eta} + F^{\xi\eta} \quad , \\ A^{\xi\xi} S_E^{\xi\xi} + \mathrm{De}\, L\, S_E^{\xi\xi} &= 2B\, S_E^{\xi\eta} + F^{\xi\xi} \quad , \\ A^{\eta\eta} S_E^{\eta\eta} + \mathrm{De}\, L\, S_E^{\eta\eta} &= 2C\, S_E^{\xi\eta} + F^{\eta\eta} \quad , \\ A^{\varphi\varphi} S_E^{\varphi\varphi} + \mathrm{De}\, L\, S_E^{\varphi\varphi} &= F^{\varphi\varphi} \quad . \end{aligned} \tag{16}$$

Here L means the differential operator

$$L = \frac{1}{\alpha\beta\gamma^2}\left(-\frac{\partial\Psi}{\partial\eta}\frac{\partial}{\partial\xi} + \frac{\partial\Psi}{\partial\xi}\frac{\partial}{\partial\eta}\right) \quad , \tag{17}$$

and A^{ik}, B, C, F^{ik} are special functions of ξ, η and of the derivatives of Ψ. Correspondingly, G depends on ξ, η, the physical components of $\underline{\underline{S}}_E$ and its derivatives.

BOUNDARY CONDITIONS

Down- and upstream region

It is realistic to assume that the fluid in regions far from the contraction is only exposed to slight deformation rates. We therefore suppose that in these regions a flow will develop which asymptotically is the creeping flow of a newtonian fluid. In this case the streamlines are confocal hyperbolas [3], that means straight lines in the ξ-η-plane.
Instead of prescribing the values of the searched functions at the up- and downstream boundary ($\pm\Xi$), we prefer to use the Neumann condition

$$\frac{\partial\Psi}{\partial\xi} = 0 \quad , \quad \text{e.g.} \quad v^\eta = 0 \quad . \tag{18}$$

With the so obtained Ψ-values the boundary values of all other functions are calculated from the corresponding equations for the creeping flow of a newtonian fluid. The value of Ξ is chosen in a way that the creeping flow of a newtonian fluid is asymptotically obtained near the boundaries. For Re$\leqslant$70 (>70) we took Ξ=8.7 (17.4). Then we observed straight streamlines in an extended region near the up- and downstream boundaries. We can therefore assume that the creeping flow behaviour is sufficiently fulfilled.

Axis of symmetry

Because the axis is a streamline, Ψ could be normalized here to zero. Furthermore the symmetry yields

$$\Omega = S^{\xi\eta} = 0 . \tag{19}$$

As the axis is a characteristic of the material law, ordinary differential equations arise for the stress components:

$$\begin{aligned} \tilde{A}^{\xi\xi} S_E^{\xi\xi} + De\, \tilde{L}\, S_E^{\xi\xi} &= \tilde{F}^{\xi\xi} , \\ \tilde{A}^{\eta\eta} S_E^{\eta\eta} + De\, \tilde{L}\, S_E^{\eta\eta} &= \tilde{F}^{\eta\eta} , \\ S_E^{\varphi\varphi} &= S_E^{\eta\eta} . \end{aligned} \tag{20}$$

Here $\tilde{L}$ means the differential operator

$$\tilde{L} = \frac{1}{\alpha^3} \frac{\partial^2 \Psi}{\partial \eta^2} \frac{\partial}{\partial \xi} \tag{21}$$

and $\tilde{A}^{ik}$, $\tilde{F}^{ik}$ are special functions of ξ, η and of the derivatives of Ψ.

Solid wall

As the fluid sticks to the wall and the volume rate is equal in every cross section we have

$$\Psi = \frac{1}{2\pi} , \tag{22}$$

and therefore

$$S_E^{\xi\eta} = S_E^{\eta\eta} = S_E^{\varphi\varphi} = 0 , \tag{23}$$

$$S_E^{\xi\xi} = \frac{2\, De}{\alpha^2 \beta^2 \gamma^4} \left(\frac{\partial^2 \Psi}{\partial \eta^2}\right)^2 , \tag{24}$$

$$\Omega = \frac{1}{\gamma^2} \frac{\partial^2 \Psi}{\partial \eta^2} . \tag{25}$$

NUMERICAL CONSIDERATIONS

The ξ-η-plane, with $-\Xi \leqslant \xi \leqslant \Xi$, $\eta_0 \leqslant \eta \leqslant \pi/2$, is covered with a rectangular grid, being equidistant in ξ- as well as in η-direction. The boundaries of the grid coincide with the four lines $\xi \pm \Xi$, $\eta = \eta_0$, $\eta = \pi/2$. The field functions are substituted in the usual way, by taking $f_{i,j}$ for the function f at the grid-point i,j: $1 \leq i \leq N$, $1 \leq j \leq M$. In most cases the arising differential quotients can be approximated by the usual differences of second order. In handling the boundary conditions (24),(25) we choose the Jensen formula

$$\frac{\partial^2 \Psi}{\partial \eta^2} \rightarrow \frac{1}{2(\Delta\eta)^2} (-7\ \Psi_{i,1} + 8\ \Psi_{i,2} - \Psi_{i,3}) \tag{26}$$

and for the term (13) we take

$$\frac{\beta_j}{(\Delta\eta)^2}\left(\frac{1}{\beta_{j+0.5}}\Psi_{i,j+1} - \left(\frac{1}{\beta_{j+0.5}} + \frac{1}{\beta_{j-0.5}}\right)\Psi_{i,j} + \frac{1}{\beta_{j-0.5}}\Psi_{i,j-1}\right) \tag{27}$$

(analogously in ξ-direction).
The numerical process was planned as an iteration method in which the creeping newtonian flow or even earlier calculated values are taken as initial condition. In every sweep Ω is determined from (15), where Ψ and $\underline{\underline{S}}_E$ are taken as known values. Proceeding from this Ω-solution, the new values of Ψ can be calculated from (14). The new Ψ serves to correct the boundary values of Ω. Then, the new extra stress values $\underline{\underline{S}}_E$ are calculated from (16). These four equations are written in a form in which only one component occurs on the left side. Therefore, the new values of $\underline{\underline{S}}_E$ can be calculated by inserting the known values of $\underline{\underline{S}}_E$ into the right side and solving the four equations one by one.
For solving the poisson-equation (14) we choose the GS-method. For the vorticity equation (15) we apply a SOR-method. The optimum SOR-parameter is approximately equal to one in the newtonian case for Re<50 and has to be reduced for higher Re-numbers. In the non-newtonian case, a reduction of the parameter is necessary even for small De-numbers in order to achieve a convergent solution. The stress components in the field and at the axis are determined with the upwind-method described in [1].

DISCUSSION OF SOME RESULTS

For our calculations we take a mesh with N=101 and M=9 lines. The iteration was broken off, when after insertion of the solution the difference of the right and left side of (15) was smaller than 10^{-6}. Then the difference of Ψ, Ω or $\underline{\underline{S}}_E$ between two sweeps is of order 10^{-10}, 10^{-9} or 10^{-8}, respectively. These differences could usually be obtained with less than 100 iterations. Only for De-numbers higher than 0.3 more iterations are needed. Because of the increasing number of iterations the attainable De-limit could only inaccurately be determined until now. Nevertheless it is about 1.0 for medium cone angles.

Variation of the Re-number in the newtonian case

For a hyperboloid with a cone angle of 45° we obtained the flow-pattern for Re between 0 and 100.

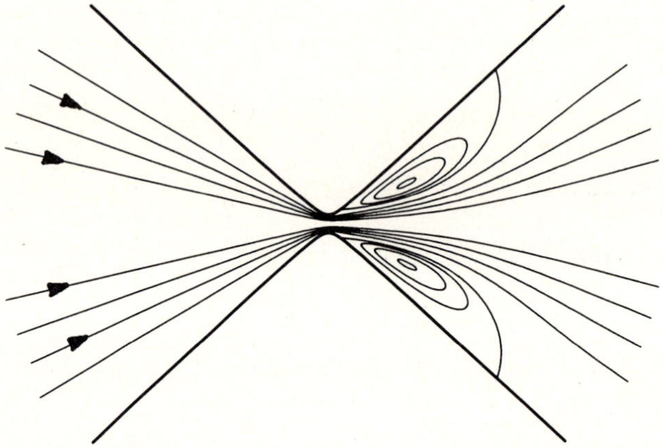

Fig. 4: Typical streamlines, Re=30, η_0=0.785, z ∈ [-24,24]

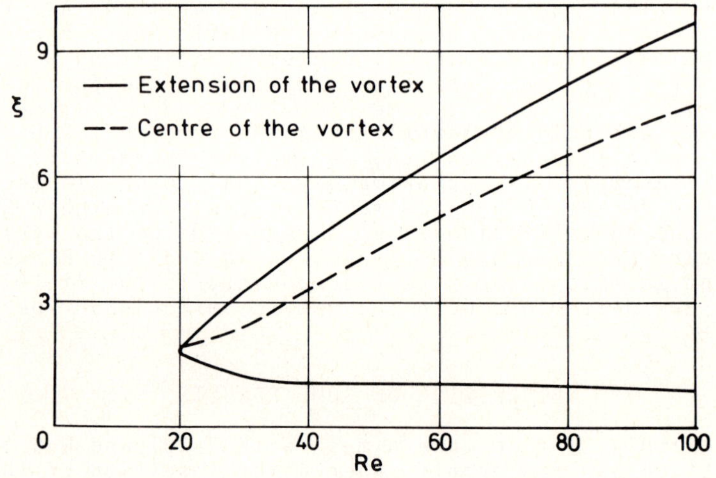

Fig. 5: Vortex extension and centre of the vortex in ξ-direction η_0=0.785

For Re nearly equal to 18 vortices arise in the outflow region near the wall which quickly grow with increasing Re-numbers. Because of the large vortices the velocity profile becomes more and more unsymmetrical concering the contraction.

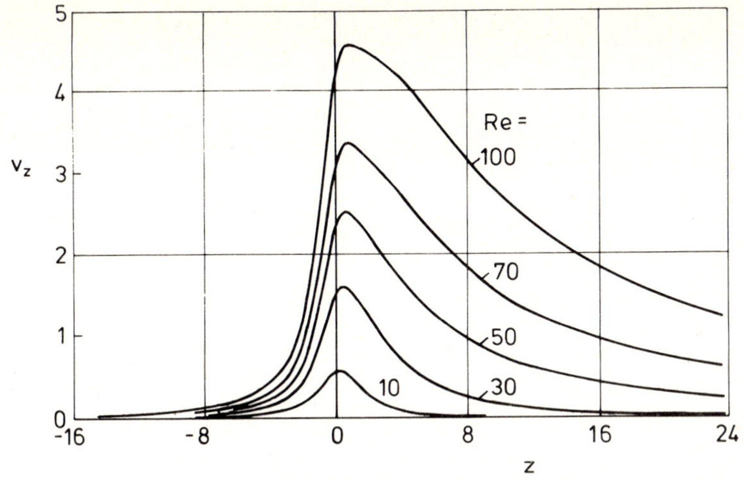

Fig. 6: Velocity in z-direction along the axis , η_0=0.785

Variation of the asymptotic cone angle

For various cone angles between 84^0 and 4^0, we determined the newtonian flow for different Re-values. For smaller or greater cone angles some terms become nearly zero. In order to determine the flowpatterns in these cases specific considerations are necessary.

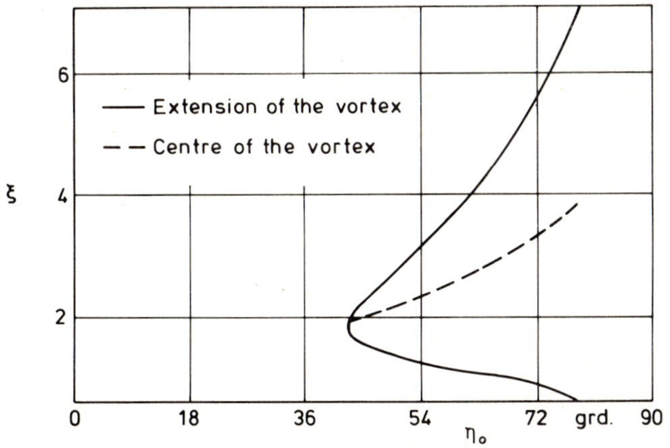

Fig. 7: Vortex extension and centre of the vortex in ξ-direction for Re*= 30 (newtonian fluid)

We formulated the dimensionless equations with the aid of the eccentricity e. When we look towards the geometry of a tube, i.e. a cone angle of 0°, the distance of the hyperbolas of the walls approach zero for fixed e. In order to interpret the results for various cone angles it seems to be easier to fix a and to formulate the dimensionless number by this. The so obtained quantities are marked by "*".

Variation of the De-number

For small Re-numbers and certain cone angles, it was possible to determine the non-newtonian flow up to De-numbers of about 1.0 . This De-limit does not allow a description of strong elastic effects such as "elastic" vortices (cf. e.g. [4] for a cone). Up to now, only some deformations of the streamlines, a reduction of the "inertia" vortices and a slight change in the pressure drop could be observed.

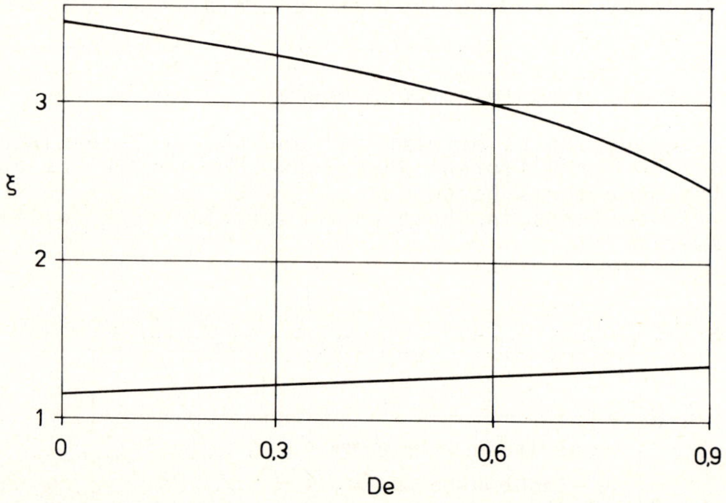

Fig. 8: Vortex extension for Re=30, η_0=0.785 and various De-numbers

As the hyperboloid has a smooth outline we hope to be able to discuss the De-limit to some extend for example with regard to mesh-refinement.

Pressure drop

By integration of the corresponding component of (7) along the axis one obtains the difference Δp of the variable p between the inflow and the outflow region. Since the stress components vanish far in front and behind the narrowing Δp is the measurable pressure drop. In the newtonian case for very small Re-numbers (Re<2) the pressure drop Δp is nearly independent of Re (creeping flow). For non-creeping flow one gets a quadratic dependence of Δp on Re for Re-numbers up to nearly 30.

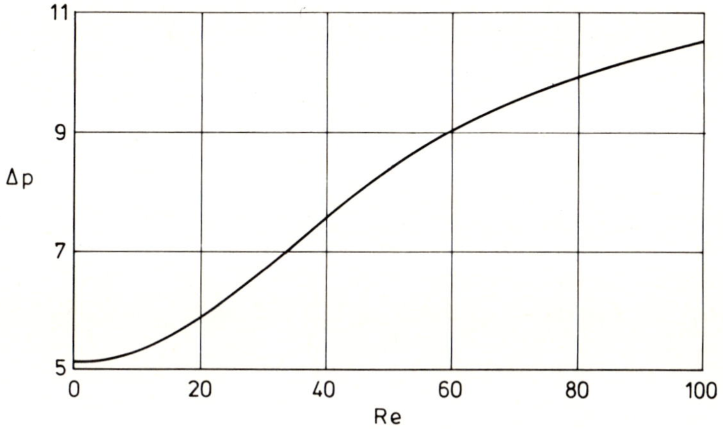

Fig. 9: Pressure drop for a hyperboloid with a cone angle of 45^0 (newtonian fluid)

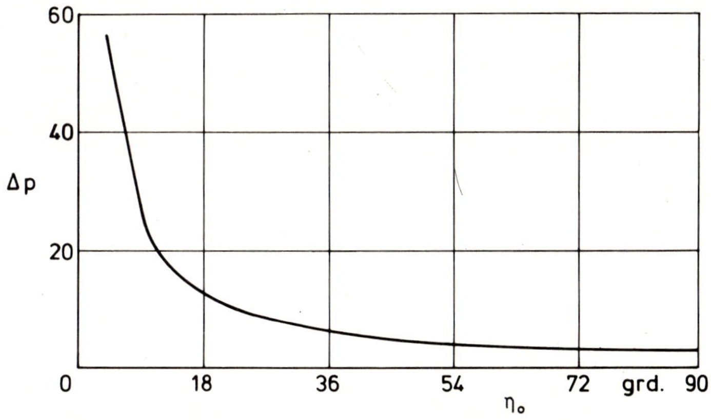

Fig. 10: Pressure drop for various hyperbolas (described by the cone angle η_0) with $Re^* = 30$ (newtonian fluid)

ACKNOWLEDGEMENT

The authors gratefully acknowledge the Deutsche Forschungsgemeinschaft for support to this work.

[1] M.J. Crochet, A.R. Davies, K. Walters
Numerical Simulation of Non-Newtonian Flow
Elsevier, 1984.

[2] A.R. Davies, S.J. Lee, M.F. Webster
J. Non-Newtonian Fluid Mech., 16 (1984), pp. 117-139.

[3] D. Schlegel, P. Schümmer, R.H. Worthoff
Rheol. Acta, 14 (1975), pp. 963-967.

[4] P. Schümmer, Y. Xu
Rheol. Acta, accepted for publication.

CURRENT ACTIVITIES IN BASIC RESEARCH WORK ON PANEL METHODS IN GERMANY

S. Wagner and Ch. Urban

University of the Armed Forces of the
Federal Republic of Germany, Munich
Werner-Heisenberg-Weg 37, D-8014 Neubiberg

SUMMARY

Current activities in basic research work on panel methods in Germany are discussed. This includes higher order methods for subsonic and supersonic flow and indications when higher order methods are obligatory. In addition, the side-edge and leading-edge separation of small- and high aspect-ratio wings is treated including the development and roll-up of the wake. The progress in predicting the aerodynamic characteristics of complex configurations in subsonic and supersonic flow is pointed out by comparison with measurement.

INTRODUCTION

Geometrical complexities, strong aerodynamic interference effects of high performance aircraft and the ever increasing demand in the accuracy of predicting the aerodynamic characteristics of airplanes exert big pressure on the aeronautical engineer to develop proper computational methods in order to meet this challenge. Enormous progress has been made during the last decade both, in the development of new algorithms and in the performance of high speed computers. Thus, the full potential equations, or even the Euler equations, could be solved by applying finite difference or finite element schemes. However, these solutions were restricted to three-dimensional wings or simple wing-body configurations.

For many applications, e.g. analyses of complex configurations in subsonic or supersonic flow, design loop with a large parameter variation, panel methods are still a powerful tool of the design engineer due to their relatively low consumption of computer time and their flexibility in application.

In recent years, new types of singularities were created, higher order methods were developed, and new, more proper formulations of boundary conditions were found. Even the side-edge and leading-edge separation of small and high aspect-ratio wings, including the free development and roll-up of the wake, was modeled by panel methods. Some of these current activities in basic research work on panel methods in Germany will be subject of the following paper. The progress in predicting the aerodynamic characteristics of complex configurations in subsonic and supersonic flow will be pointed out by comparison with measurement.

GENERAL REMARKS

Subsonic flow can be described by the linearized potential equation

$$(1 - M_\infty^2)\frac{\partial^2\Phi}{\partial x^2} + \frac{\partial^2\Phi}{\partial y^2} + \frac{\partial^2\Phi}{\partial z^2} = 0 \tag{1}$$

assuming that the flow can be considered as inviscid, irrotational and stationary and that the approximation of small perturbations can be applied. For general three-dimensional configurations equation (1) cannot be solved analytically. Therefore, in the panel method approach, the differential equation is converted into an integral equation over the body surface by means of Green's Theorem. This integral equation can be solved by a discreting procedure: The configuration is composed of a net of panels to which certain singularity distributions are assigned. Imposing the physical boundary conditions to a discrete number of control points, a system of linear equations is derived which determines the unknown singularity strengths and the induced velocity field, subsequently.

In the last two decades a great variety of panel methods has been developed for subsonic and supersonic flow conditions. In a general classification, they can be divided into first-order panel methods [1] - [3], [5], [26] and higher-order panel methods [4], [14], [31] - [36] referring to the applied order of singularity distributions and geometry representation.
The range of application of panel methods is not necessarily limited to attached flow. Especially for the lifting flow problem, various panel methods [6] - [9] and [21] - [24] have been derived or are under development, indicating that discreting techniques can be applied successfully to vortex separation at sharp wing-edges, high Reynolds-numbers presumed.

Accuracy of potential flow method

In general, first-order panel methods (stepwise constant source/doublet distributions) show a considerable sensitivity of the results to panel spacings and poor accuracy in regions of strong flow gradients. Referring to an application to lifting surfaces, investigations by Hirschel und Fornasier [12], [13] prove, that first-order or low-order panel methods are not apt to fulfill a compatibility condition at the wing trailing-edge, Fig. 1:

$$\Omega = 2\ |\underline{v}_{e,u}|\sin\Psi_u = d\Gamma/dy\ . \tag{2}$$

Stepwise constant or even linearly discontinuous doublet distributions, applied to match lifting effects, lead to an erroneous computation of the v-component (local derivative $\partial\mu/\partial y$) of the velocity vector near the trailing edge. This problem does not arise with higher-order panel methods, e.g. Fornasier [14], that apply quadratic doublet distributions in chordwise and spanwise direction. Additional investigations of an Euler method [15] and an analytical lifting surface method [16] prove the accuracy of these methods in terms of the compatibility condition (1).

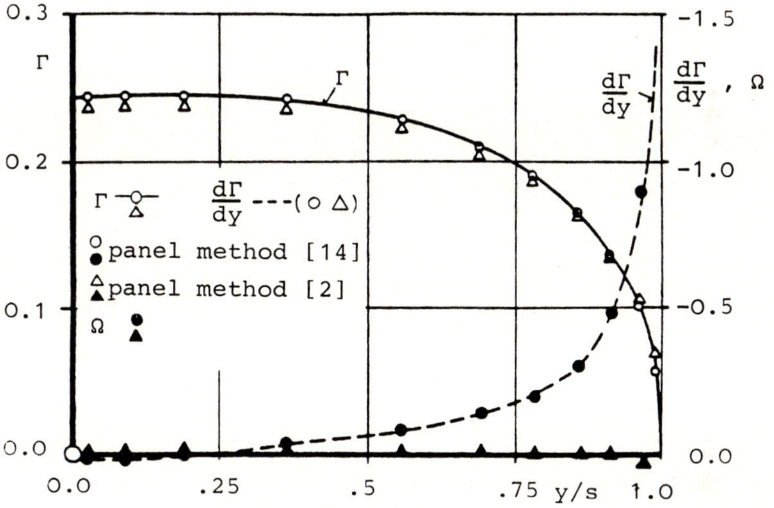

Fig. 1: Circulation distribution Γ, its derivative dΓ/dy, and vorticity content Ω of the wake for a yawed wing, $M_\infty = .25$, $\alpha = 8.2^\circ$, [13]

Although the effects of erroneous v-components on the inviscid pressures are quite negligible, the response of three-dimensional boundary layer analysis to these errors can result in significantly incorrect predictions of skin friction drag and separation characteristics. As a result, inviscid flowfields computed by low-order panel methods represent no adequate basis for boundary layer studies in their present form.

Compressibility effects

Considering pure subsonic flow and assuming small perturbation,

the linearized potential equation (1) can be transformed into the Laplace equation by means of the Göthert-rule. Thus, the analysis of a subsonic flow problem is reduced to incompressible flows about a transformed body shape and the resulting pressure distributions can be recalculated to the original configuration.
Most of the panel methods exploit this similarity rule to cover compressible flow conditions. Another possible way is indicated by Das [17]: The physical processes due to movements of singularities are incorporated into Mach-number dependent influence factors. Within the scope of linearized theory, these factors consider the actual compressible perturbation quantities which are induced by a singularity at a point of the flowfield. Therefore, the kinematic flow condition can be satisfied directly at the original body surface. A successful implementation of the procedure into a first-order panel method [2] is shown by Kohlmeier [18].

Mathematical studies

As indicated, the low-order panel methods cannot compete with the accuracy of higher-order panel codes, which, however, are connected with an increased mathematical effort and numerical expense. In order to gain simple but adequate schemes helpful mathematical investigations are carried out by Hackbusch [19]. They concern multi-level discretizations and series developments for the near-field and far-field influence coefficients.

Selected panel methods

In the following chapters current German developments of panel methods for subsonic and supersonic flow are shortly described: A higher-order panel method [14] is outlined which is intended for complex aircraft configurations and is based on linear lifting surface theory. Yet, in modern aircraft design, interactions between wings and free vortex sheets play an important roll, i.e. for instance the well-known lift-increasing effect of leading-edge vortices [10], [11]. In order to gain a predetermination of such interference effects, various panel methods are in development [21] - [24]. At present, the range of application is restricted to wings only assuming incompressible flow, complete vortex separation and stable leading-edge or side-edge vortices. Informations about the phenomenon of vortex breakdown, which is connected with considerable changes in the aerodynamic wing loads and therefore is of great practical interest, cannot be obtained by these potential flow methods.

HIGHER-ORDER PANEL METHOD HISSS

- LINEARIZED SUBSONIC FLOW PROBLEM -

The HISSS method, whose development is still in progress at MBB Munich, is a higher-order panel method for the solution of linearized steady potential flow about arbitrary three-dimensional configurations at subsonic and supersonic speeds, Fornasier [14].

Geometry and singularity modeling

For subsonic flow conditions, the HISSS method provides a simplified, but in many cases sufficient singularity model which is a straight-forward extension of the first order method [2] to higher-order formulations: The configuration is divided in networks of flat or curved singularity panels, Fig. 2.

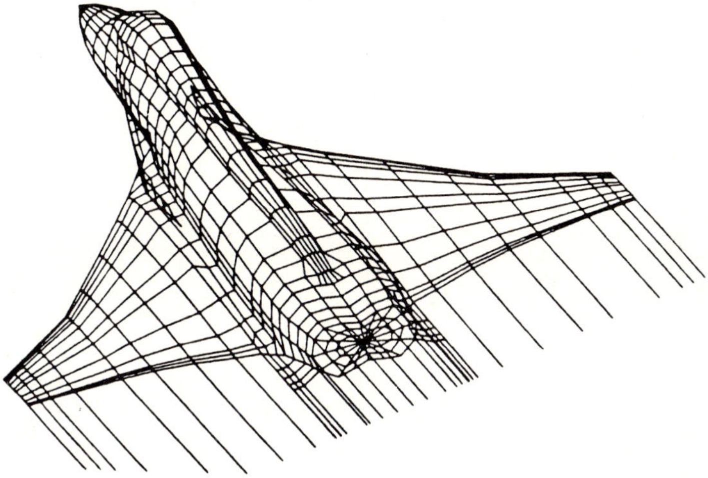

Fig. 2: Wing-body configuration paneling, [14]

Displacement effects are covered by fitting the external body surface with a source network consisting of linear source distributions. Lifting surfaces are represented by an additional doublet network with quadratic doublet distributions at the camber lines of the wing. Extending this doublet network into a fuselage (constant circulation assumed) and in a plane behind the wing provides the wing-body carry-over model and the wake representation, respectively. The doublet networks of lifting wings have a predetermined chordwise distribution

$$\mu(\xi) = \begin{cases} (2 - \xi^{\nu})\ \xi^{\nu} & \text{in the interior} \\ 1 & \text{in the wake} \end{cases} \quad . \qquad (3)$$

Boundary conditions

Neumann-type boundary conditions are applied at the surface panel centers. A source-equality Kutta-condition, Fornasier [20], is enforced at control points located directly at the wing trailing edge and thus eliminating errors by introducing the Kutta-condition at a small, empirically predetermined distance behind the trailing edge, compare [2]. Additional boundary conditions, such as mass flux or velocity vectors, can be specified to simulate inlet or jet effects.

The solution of the resulting set of linear equations succeeds without problems - yet in some cases increased computational storage capacities have to be provided.

Results

The subsonic version of the HISSS method, as described here, has overcome or reduced many deficiences of low order panel methods concerning, for instance, the compatibility condition [13], the Kutta condition and leakage flow. For the treatment of complex aircraft configurations, the general modeling scheme that is obligatory for supersonic flow should be preferred. As a sample result, computations of overall forces and moments of a fighter model are compared with experimental data in Fig. 3.

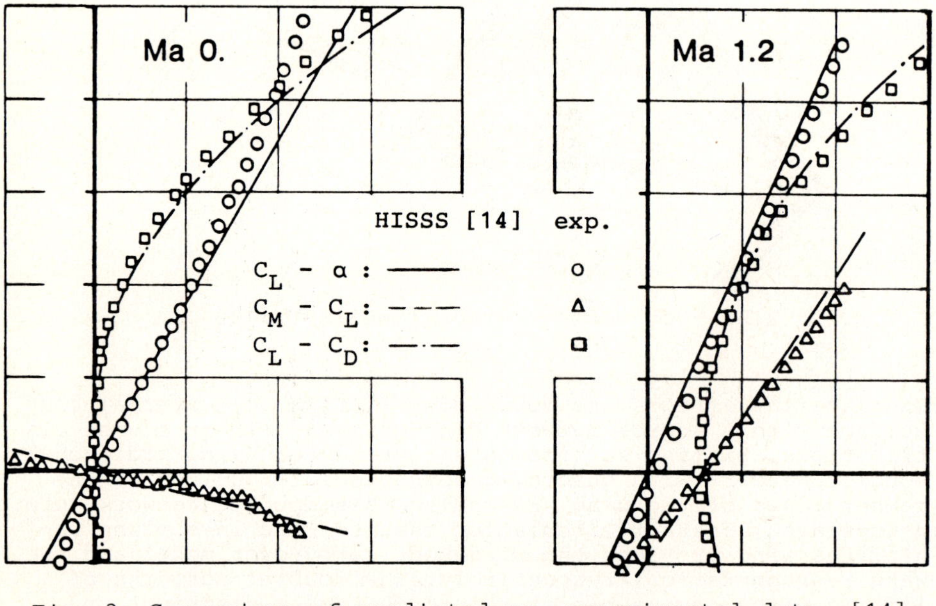

Fig. 3: Comparison of predicted vs. experimental data, [14]

POTENTIAL FLOWS ABOUT THICK LIFTING WINGS WITH SIDE-EDGE VORTEX SEPARATION

Eppler and Jäger [21] are working on a higher-order panel method to calculate incompressible, steady potential flows about thick lifting wings. Free vortex sheets emanating from the sharp trailing and side edges of the wing are incorporated. The main goal of the study is the reliable determination of induced drag.

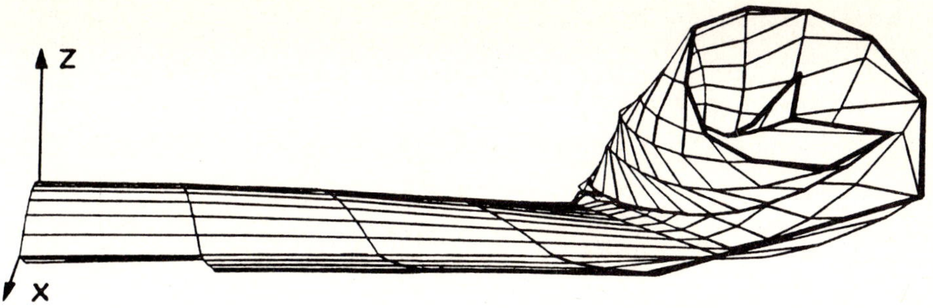

Fig. 4: Rectangular wing, AR=2, with side-edge vortex separation, number of panels 24*4, [21]

Singularity model

The external wing surface is approximated by a net of plane panels having quadratic doublet distributions. The free vortex sheets are replaced by discrete vortex lines appended to the panel corners along the wing edges, Fig. 4. On stripes between two vortex lines constant doublet distribution is assumed.

Procedure of Solution

Neumann-type boundary conditions and the Kutta condition enforced by equation (4) lead to a system of equations which is basically nonlinear due to the a priori unknown shape of the free vortex sheets. The solution is derived by linearizing the problem and applying an iterative procedure: Starting from a prescribed position of the free shear layers the doublet distributions are determined. Subsequently the positions of the vortex lines are corrected relative to streamline directions.

Kutta condition

The Kutta condition is fulfilled by requiring

$$V_x \left[\frac{\partial \mu_u}{\partial x_u} + \frac{\partial \mu_l}{\partial x_l}\right] - V_y \left[\frac{\partial \mu_u}{\partial x_u} - \frac{\partial \mu_l}{\partial x_l}\right] = 0 \qquad (4)$$

along the trailing and side-edges of the wing; in this case x, y, z stand for surface oriented coordinates. Compared with equation (4) results of a simplified version, which assumes $V_y \ll V_x$, exhibits an overestimation of the pressure distribution at the wing tip area and consequently confirm the necessity of a proper representation of spanwise velocity components in panel methods.

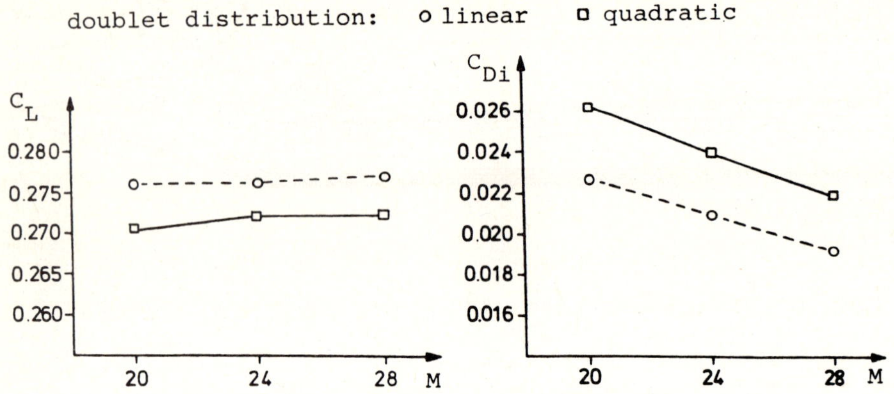

Fig. 5: Lift and induced drag depending on the number M of panels on a chordwise stripe; rectangular wing, AR=2, with side-edge separation, [21]

Induced drag

Investigations of the computed induced drag show that the results strongly depend on the chordwise number of panels, especially on the panel density at the wing nose area, Fig. 5. Eppler and Jäger [21] refer this behaviour to the inconsistent structure of the method applying quadratic doublet distributions but plane panels. In order to achieve a reliable prediction of the induced drag a new method is in development at the present: It is based on curved triangular panels, source/vortex-distributions and combined tangential and normal boundary conditions at the internal and external wing surface.

POTENTIAL FLOWS ABOUT THIN DELTA WINGS WITH LEADING-EDGE VORTEX SEPARATION

The prediction of leading-edge vortex effects which can result in a considerable increase of lift is of great technical interest. Therefore a higher-order panel concept is elaborated by Ballmann, Haag and Wiemer [22]. The method's range of application is, at present, limited to thin and plane delta-shaped wings in incompressible steady potential flows.

General outline

The discontinuity surfaces (wing plus free shear layers) are discretized into panels; biquadratic doublet distributions are applied to these elements. Consideration of the physical boundary conditions (kinematic and dynamic flow condition plus Kutta-condition) then leads to a nonlinear system of equations which is solved iteratively by a Newton procedure. A Taylor-development of this system of equations gives for the kth iteration step

$$\underline{\underline{J}}^k \quad \Delta\underline{x}^k \quad = \quad - \quad \underline{R}^k \qquad \begin{array}{l} \Delta\underline{x} = \text{correction vector} \\ \underline{R} = \text{boundary condition} \\ \underline{\underline{J}} = \text{Jacobi-matrix} \; . \end{array} \qquad (5)$$

Fundamental studies

These studies concern the physical accuracy of the singularity model and the convergence behavior of the iterative process. The main results are as follows: A steady description of the geometry of the discontinuity layers and the doublet distributions has to be guaranteed beyond panel edges, otherwise line vortices are generated between the panel edges which violates the Helmholtz vortex laws. Therefore, in the present method additional boundary conditions are required eliminating the jump in the first derivative of the doublet distributions perpendicular to the panel edges. Investigations of higher-order codes, e.g. [25], indicated that convergence problems can be removed by an accurately calculated Jacobi-matrix and by eliminating forced conditions for the free vortex sheets.
These results are incorporated into a "vortex line adapted" panel method exhibiting a convergence behavior which is significantly independent of the supposed starting solution.

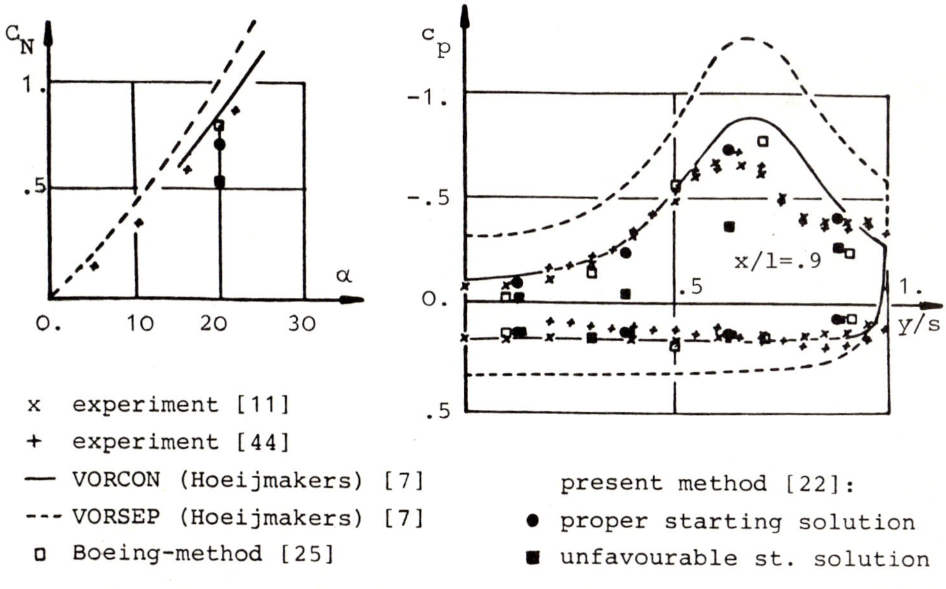

Fig. 6: Overall normal force and pressure distributions of a delta wing, AR=1, with leading-edge separation, [22]

Results

At present, the geometry of the free shear layers emanating from the wing leading edges is completely variable, whereas kinematic restrictions still exist for the feeding sheet represented by a line vortex and for the wake of the wing. This simplification of the physical reality does not disturb the convergence of the numerical procedure due to the accurately calculated Jacobi-matrix. Yet, depending on the starting solution, it can cause considerable differences between computed and experimental data. On the other hand, a properly assumed starting solution leads to a good agreement with measurements, Fig. 6.

In order to achieve a method which is in principle independent of the starting solution, it is planned to improve the simple feeding sheet construction and to eliminate the remaining forced conditions for the free vortex sheet geometry.

A VORTEX-LATTICE METHOD FOR THE CALCULATION OF INTERFERING WINGS WITH VORTEX SEPARATION

As shown in the preceding two chapters application of higher-order panel methods to lifting wings with vortex separation has - at the moment - still to face some numerical difficulties. In order to gain a fast and reliable analysis of interference effects between wings and free vortex sheets a first-order-method is developed by Wagner and Urban [23], [24]. The procedure is designed for incompressible potential flows and single or coupled thin, plane wings with or without leading-edge or side-edge vortex separation.

General outline

The method is based on the application of discretely distributed doublet elements of constant strength (i.e. vortex rings) representing the wing and the free vortex sheets. It applies a time-dependent, stepwise procedure during which - starting with an impulsive motion of the configuration - the development of all free shear layers is observed until steady wing loads are reached. Unsteady flow conditions can be simulated by the variable $\underline{U}_\infty(t)$, low Strouhal-numbers presumed.

After each time step, a system of linear equations for the bound vorticity of the wing is derived by enforcing the kinematic flow condition at the wing control points. All other boundary conditions are implicitly fulfilled by the singularity model which avoids an iterative solution process: The doublet elements emanating from the wing edges at a certain moment retain their actual singularity strengths for all further time steps (dynamic flow condition, Helmholtz vortex laws). After each time increment the new position of the free vortex sheet elements is determined from the resulting stream flow vector at the panel corners (kinematic flow condition). On wing edges with separation the Kutta condition is implemented by requiring $\partial\mu/\partial n = o$ normal to these edges.

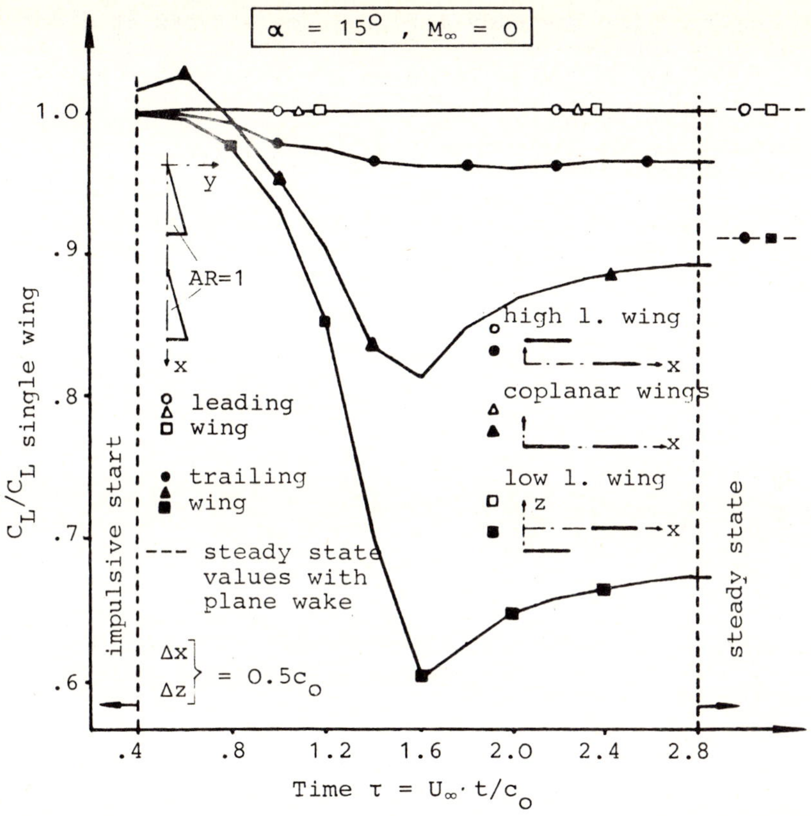

Fig. 7: Time-dependent development of lift-coefficients of a delta-delta-configuration without leading-edge separation, AR=1, α = 15°, [24]

Induced velocities

In general, the Biot-Savart-law is used to calculate vortex-line induced velocities. Yet, applying this law to strongly rolled-up vortex sheets or free shear layers located closely to wing surfaces necessarily leads to unrealistic values of both, amount and direction of the vortex-induced velocities. In order to solve this problem of the near-field, the local derivatives $\nabla\Gamma$ are approximated by a numerical procedure, which, in addition, enforces the flow condition of tangential velocities on the entire wing surface.

Results

The method provides reliable information about the aerodynamic wing loads as well as about the shape and the velocity field of leading-edge or side-edge vortices at favorable computational costs. An example for unsteady interference effects at a delta-delta configuration is shown in Fig. 7: When the starting vortex (or any similar shedded vorticity) of the first wing is carried through the near-field of the following wing, its lift is diminished simultaneously. In steady state, the lift values of the delta wings do not coincide with results of linear lifting surface theory which neglects the realistic shape of the free shear layers but assumes a plane wake behind the wings. The second example, Fig. 8, refers to the steady overall lift of coupled delta wings with leading-edge separation which agrees well with experimental data.

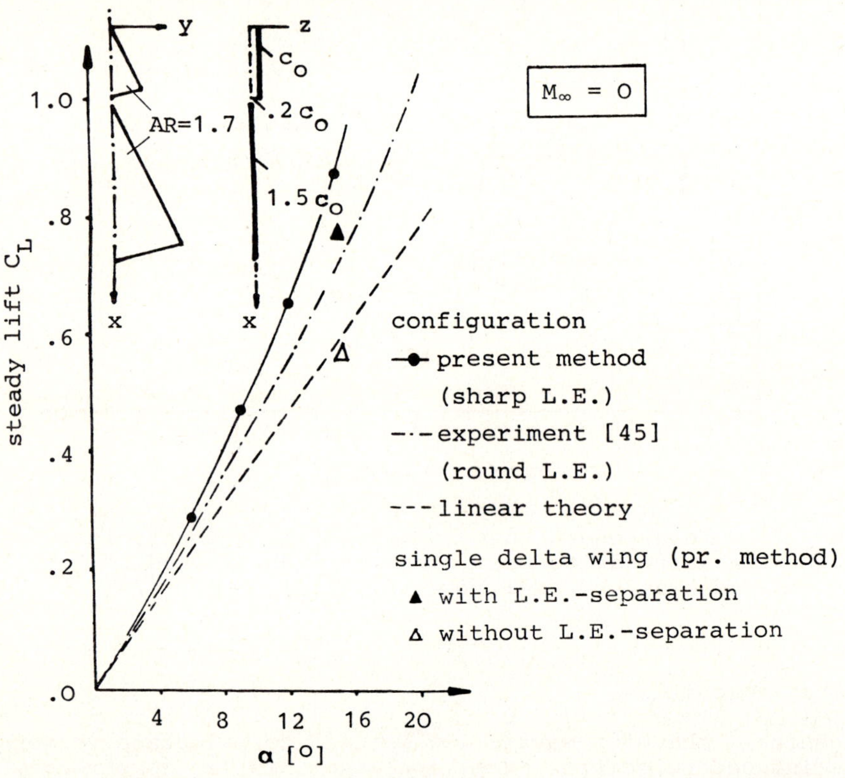

Fig. 8: Steady lift vs. angle of attack for a delta-canard-configuration with leading-edge separation, [24]

PANEL METHODS FOR SUPERSONIC FLOW

General problem

Although the extension of the subsonic surface singularity technique to supersonic flow had been started very early, e.g.

[26], supersonic panel methods had often problems with stability. Only a few years ago, remarkable progress was achieved in the United States, e.g. by Woodward [27] to [30], Ehlers et al. [31] to [34], Carmichael et al. [35], Tinoco et al. [36], and Cenko et al. [37], [38].
Since several codes based on these publications are restricted to U.S. companies only German institutions were forced to develop their own computer programs. For instance, Fornasier [14], [39] to [43] used the published background theory of USSAERO and PAN AIR and developed his own code named HISSS. In addition, some of the shortcomings of these U.S. codes were eliminated in order to improve the accuracy of results and/or to reduce computational effort.

Two essential problems arise when surface singularity methods are extended to the supersonic flow regime [41]. Both of them stem from the hyperbolic character of the linearized differential equation (1) for the velocity potential. The first problem is that singularities commonly applied in subsonic flow (e.g. sources, vortices or doublets) can propagate two kinds of waves in supersonic flow that are named "real" and "virtual" ones in [41]. Real waves exist as compression or expansion waves in reality and propagate in exterior flow around wings and bodies. Virtual waves are the result of numerical disturbances generated by singularity surfaces. These virtual waves can propagate in both, the exterior and the interior space, especially inside a closed surface if this body is represented by source panels only [41]. Repeated reflections and amplifications of such virtual waves can occur. Typical results of such interior wave propagations and reflections are fluctuations in surface pressure distributions [40], as shown in Fig. 9.

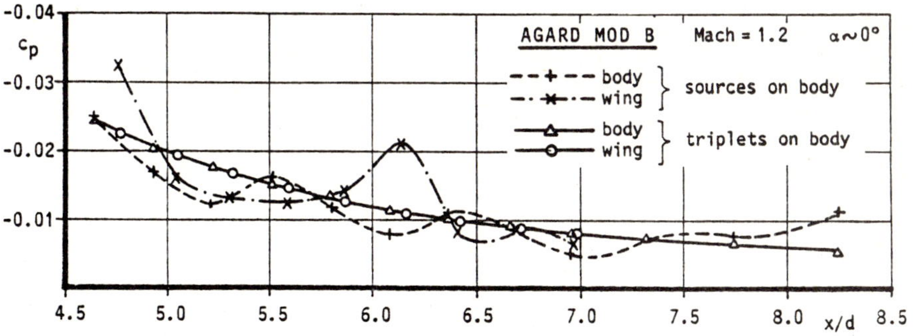

Fig. 9: Pressure distribution comparisons along wing root of AGARD model B, [40]

The second main problem is the peculiarity of the wave propagation mechanism in supersonic flow, i.e., the inducing effect does not go to zero with distance from the generating source. On the contrary, singular terms of the velocities induced by singularity distributions can spuriously influence the induced velocity at a control point that is close to characteristic lines emanating from panel edges. This situation makes results enormously sensitive to control point locations, panel spacing, and free-stream Mach number [41].

HISSS Method

The HISSS method [14] can also be used to calculate supersonic flow over or inside three-dimensional configurations. The most important features are:

- incorporation of a logically independent network concept
- usage of singularity distributions of higher order
- fulfillment of continuity of doublet strength over complete configuration
- accomplishment of continuity of geometry
- choice of a variety of possible specifications of boundary conditions.

Some of the special features that are included in PAN AIR to handle source unique problems of supersonic flow increase computing cost considerably without significantly increasing accuracy in predicting subsonic flow characteristics. Therefore, Fornasier [14] uses a concept that discretisizes the complete configuration into a useful number of subdomains called networks. Each network is a homogeneous set of panels of the same type, number and order of singularity distributions. The user has to specify the geometry of the network and a set of boundary conditions. Following these specifications, the HISSS procedure assigns the location and the total number of control points in such a manner that the singularity distributions of all types of networks are uniquely defined by the boundary conditions Fig. 10.

The network geometry is defined by a rectangular array of grid points. Panel geometry does not have any restrictions with one exception: Panels in supersonic flow should be inclined to the asymptotic velocity vector less than the Mach angle of free stream, at least in the current version of HISSS. Possible implementation of the so-called superinclined panels [33] could render the panel geometry completely independent of Mach number in the future. The singularity distributions provided in HISSS are linearly varying sources and quadratically varying doublets. Thus, the sensitivity to spacing and density of panel distributions is reduced and the doublet distribution can be forced to be continuous all over the configuration. Hence, singular disturbances that could be caused by a discontinuity at panel edges of lower order doublet strength distributions in supersonic flow are avoided from the beginning.
Continuity of doublet strength within a network is automatically fulfilled by a spline fitting procedure. The continuity across

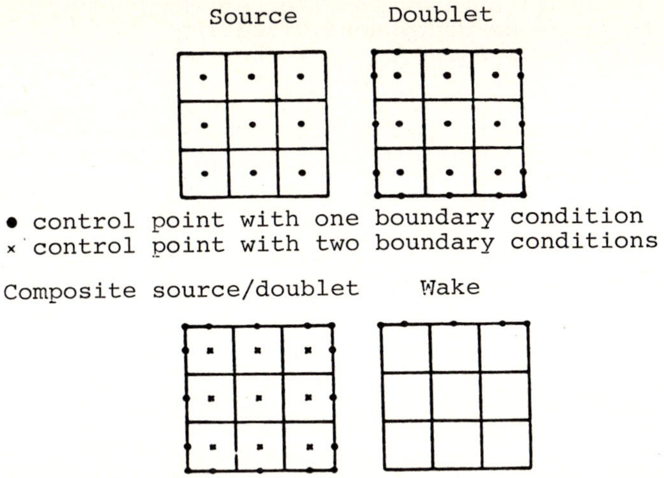

Fig. 10: Control points for different typs of networks,[41]

edges of adjacent networks is provided by special matching conditions [14], see also Fig. 11.

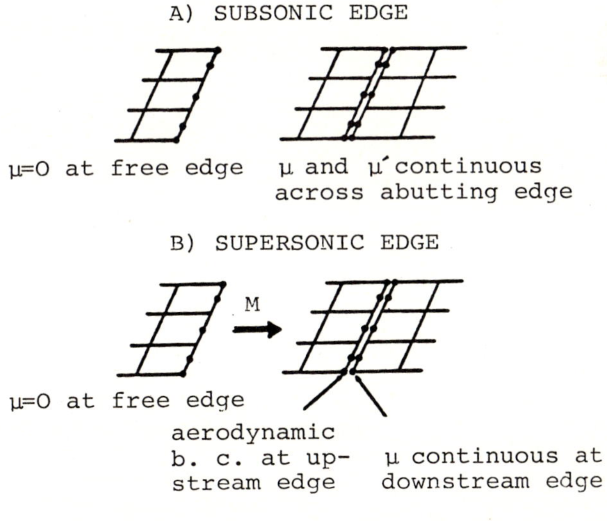

Fig. 11: Doublet matching at network edges, [41]

At subsonic abutments both the doublet strength and its derivative are arranged to be continuous. Treating supersonic abutments the domain of dependence has to be observed. In this case, the doublet distributions of the upstream networks are decoupled from the downstream ones by enforcing a boundary condition equivalent to the one at panel center points. In HISSS, the control points are relocated a small distance upstream of the edge and the aerodynamic influence coefficients are calculated there. The control points of the downstream networks match only the doublet strength.

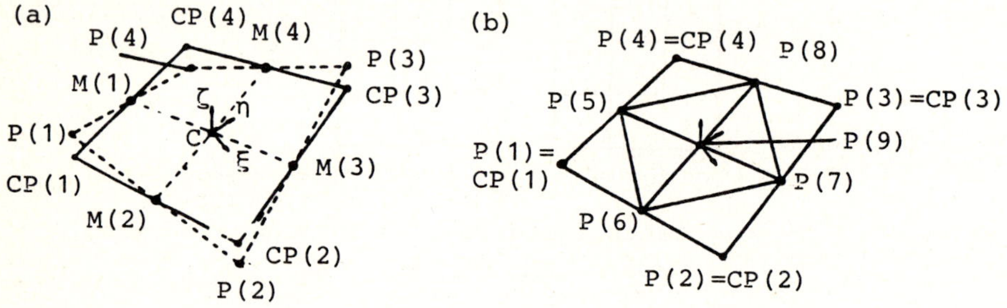

Fig. 12: Flat panel and subpanel geometries, [14]

A similar problem is continuity of geometry. Quadrilateral flat panels that are mostly applied in low order methods cannot generally fit a warped surface without generating gaps, Fig. 11a. Such gaps would lead to jumps in singularity strength and thus to numerical instabilities in supersonic flow. In addition, the spline system used to approximate the quadratic doublet distribution over the panel can provide a global continuous distribution over the network only in a least-square sense. These problems are solved by an idea of Ehlers et al. [8]. Each panel is subdivided into 8 triangular subpanels, Fig. 11b and a separate quadratic distribution is taken in each subpanel. Since this procedure can become very expensive a test based on the hyperbolic distance between the control point and the centroid of the inducing panel is used to decide whether subpaneling is necessary or not.

A large variety of boundary conditions is offered in HISSS. Composite networks are defined when both source and doublet distributions are assigned to a network. In this case, two boundary conditions are fulfilled in a single control point. It is possible to control both, the exterior and the interior flow by applying composite networks (source/doublet) and by specifying mixed external and internal boundary conditions. Thus, the total or perturbation potential inside closed volumes of networks can be forced to be zero and, as a consequence, Mach wave propagation inside any closed volume can be prevented.

It should be mentioned that Woodward [29], [30] developed a new singularity called triplet to overcome this problem of reflected Mach waves inside closed volumes. The triplet singularity is gained by superimposing a constant vorticity distribution on a source panel with constant strength. It was successfully applied to the analysis of isolated bodies with regular paneling. However, when applied in its low order form to complex configurations [41] or to non equally spaced paneling it can cause numerical instability which is again cured by applying a higher order doublet distribution across panel edges [14].

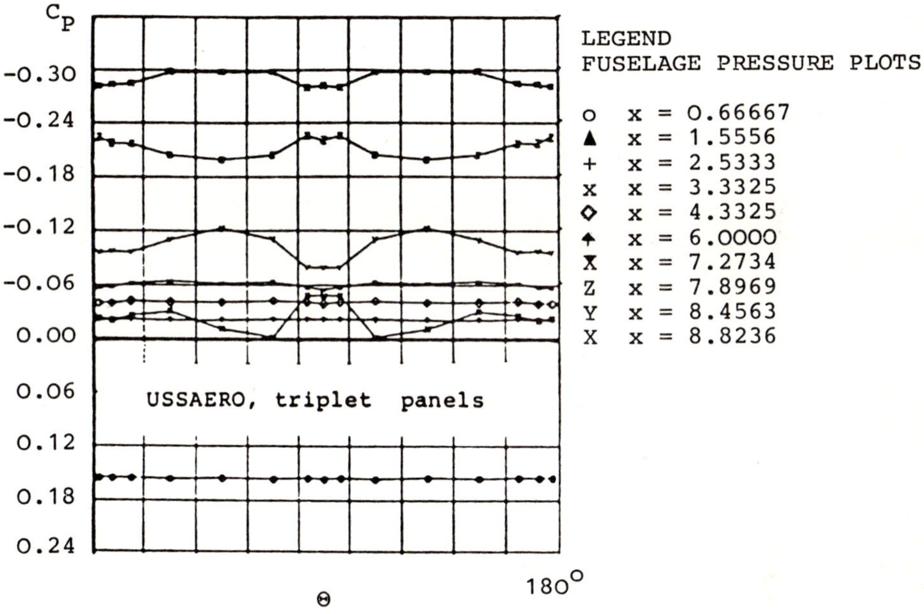

Fig. 13: Cone-cylinder-cone circumferential pressures at $M_\infty = 2.0$, $\alpha = 0^{\circ}$, [41]

Fig. 13 shows that the triplet singularity in its low order form produces **fluctuations** in the circumferential pressures of a cone-cylinder-cone configuration with non equally spaced paneling. Therefore, it is regarded not to be suitable for general surface paneling. Application of the HISSS method [14] to the same paneling and the same configuration cured the problem as demonstrated by Fig. 14. The nine degree of freedom spline used to fit the quadratically varying doublet distribution of HISSS minimizes the pressure fluctuations to a great extend.

Another remarkable feature of HISSS is the flexibility in specifying the boundary conditions. A special specification is the so-called Morino boundary condition of PAN AIR that forces zero normal mass flux. This is achieved by setting the source strengths equal to the negative of the normal component of the freestream velocity and by determining the doublet distributions in such a way that the interior perturbation potential is zero.

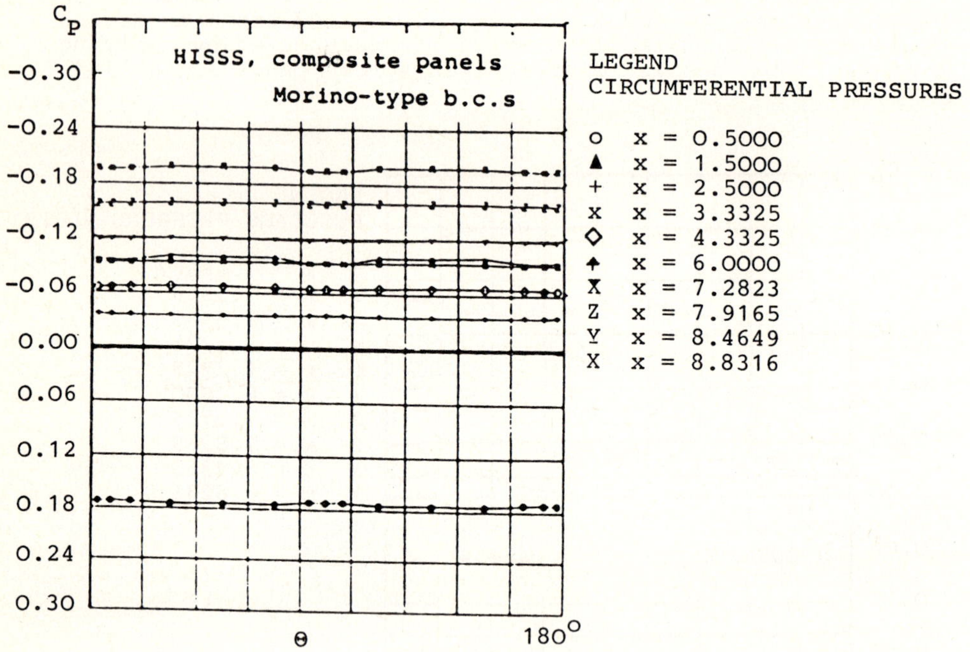

Fig. 14: Cone-cylinder-cone circumferential pressures at $M_\infty = 2.0$, $\alpha = 0^O$, [41]

Thus, the mass-flux through the surface of a closed body is avoided. The advantage of this type of boundary condition is that the source strengths are assigned a priori and only the doublet strengths must be determined. In HISSS, a variation of the Morino boundary condition is used which specifies the mass-flux directly or explicitly, i.e., zero internal perturbation potential is prescribed, but, in contrary to the original formulation, the impermeability condition is enforced by setting the normal component of the external mass-flux zero. It is shown in [41] that the modified Morino boundary condition leads to a better pressure distribution of a cone-cylinder-cone configuration as compared with the original one. In addition, Woodward et al. [41] found that the original Morino formulation could not totally delete external mass-flux in this region. Several other boundary conditions are discussed in [41].

Finally, the calculation of the longitudinal aerodynamic characteristics of a realistic wing-body configuration, Fig. 2, at a Mach number of 1.2 is compared with experiments, Fig. 3, of a quite similar configuration. The agreement between theory and experiment is quite satisfying.

It is intended to use all the experiences gained so far with this analysis to develop a design procedure.

CONCLUSIONS

The current developments of panel methods reveal the potential, the flexibility and the accuracy of discrete singularity concepts.
Panel methods have been presented for the treatment of complex aircraft configurations in subsonic and supersonic flow corresponding to linear lifting surface theory, as well as for the calculation of incompressible interference effects between wings and free vortex sheets. It has been shown that the application of higher-order singularity distributions and geometry representations leads to a considerable improvement of the accuracy of panel methods, particularly for supersonic flow problems. The importance of fulfilling the Kutta-condition directly at wing trailing edges has been emphasized once more.
Concerning wing-vortex interactions the employment of higher-order schemes is at the present still connected with some numerical difficulties and deficiencies: They arise from the demand of a continuous description of wing and free vortex sheet geometries and from kinematic restrictions to the shape of free shear layers.
A first-order panel method combined with spline procedures avoids these problems without remarkable loss of accuracy and admits the calculation of steady and unsteady interference effects of coupled wings with vortex separation.

REFERENCES

[1] HESS, J.G., and SMITH, A.M.O.: "Calculation of Nonlifting Potential Flow about Arbitrary Three-Dimensional Bodies", Journal of Ship Research, 8, No. 2 (1964), pp. 22-24.

[2] KRAUS, W., and SACHER, P.: "Das MBB-Unterschall-Panel-Verfahren - Dreidimensionale Potentialtheorie bei beliebig vorgegebener Mehrkörperanordnung", Messerschmitt-Boelkow-Blohm, Ottobrunn, FRG, MBB Ber. Nr. 672-70 (1970).

[3] HUNT, B., and SEMPLE, W.G.: "Economic Improvements to the Mathematical Model in a Plane/Constant-Strength Panel Method", Paper Presented at Euromech Colloquium 75, Braunschweig, FRG (1975).

[4] JOHNSON, F.T., and RUBBERT, P.E.: "Advanced Panel-Type Influence Coefficient Methods Applied to Subsonic Flows", AIAA 75-50 (1975).

[5] MASKEW, B.: "Prediction of Subsonic Aerodynamic Characteristics: A Case for Low-Order Panel Methods", J. Aircraft, 19, No. 2 (1982), pp. 157-163.

[6] KANDIL, O., CHU, L-C. und TUREAUD, T.: "A Nonlinear Hybrid Vortex Method for Wings at Large Angle of Attack", AIAA Journal, 22, No. 3 (1984).

[7] HOEIJMAKERS, H.W.M. und VAATSTRA, W.: "On the Vortex Flow over Delta and Double-Delta-Wings", AIAA/ASME 3rd Joint Thermophysics, Fluids, Plasma and Heat Transfer Conference, St. Louis (1982).

[8] JOHNSON, F., TINOCO, E., LU, P. und EPTON, M.: "Recent Advances in the Solution of Three-Dimensional Flows over Wings with Leading Edge Vortex Separation", AIAA 79-0282 (1979).

[9] SUMMA, J.M.: "A Numerical Method for the Exact Calculation of Airloads Associated with Impulsively Started Wing", AIAA 77-2 (1977).

[10] WENTZ, W. und MC. MAHON M.: "An Experimental Investigation of the Flow Fields About Delta and Double-Delta Wings at Low Speeds", NASA CR-521 (1966).

[11] HUMMEL, D.:"On the Vortex Formation over a Slender Wing at Large Angles of Indidence", AGARD-CP-247 (1978).

[12] HIRSCHEL, E.H.: "Considerations of the Vorticity Field on Wings", in "Recent Contributions to Fluid Mechanics", (W. Haase, ed.), Springer Verlag, Berlin-Heidelberg-New York (1982), pp. 129-137.

[13] HIRSCHEL, E.H., and FORNASIER, L.: "Flowfield and Vorticity Distribution near Wing Trailing Edges", AIAA 84-421 (1984).

[14] FORNASIER, L.: "HISSS - A Higher-Order Subsonic/Supersonic Singularity Method for Calculating Linearized Potential Flow", AIAA 84-1646 (1984).

[15] RIZZI, A.: "Damped Euler-Equations Method to Compute Transonic Flow Around Wing-Body Combinations", AIAA-Paper 81-0999BR (1981).

[16] WAGNER, S.: "On the Singularity Method of Subsonic Lifting Surface Theory", J. of Aircraft, 6 (1969), pp.549-558.

[17] DAS, A.: "Wellenausbreitung und Lösungsmethoden der Wellengleichung bei bewegten Störquellen", DFVLR-FB 83-07 (1983).

[18] KOHLMEIER, H.H.: "Eine direkte Berechnungsmethode für Störfelder bewegter Körper in kompressiblen Medien bei Unterschall-Machzahl". DFVLR-FB 85-52 (1985).

[19] HACKBUSCH, W., NOWAK, Z.P.: "Multi-Level Discretization and Solution Method for Potential Flow Problems in Three Dimensions", page 71-89 of the present book.

[20] FORNASIER, L.: "A Source-Equality Kutta Condition for Surface Singularity Methods Using Doublet of Predetermined Shape", AIAA Journal, 8 (1984), pp. 1167-1169.

[21] EPPLER, R., JÄGER, H.: "Praktische Berechnung dreidimensionaler Potentialströmung um dicke Flügel unter Berücksichtigung freier Wirbelschichten", Arbeitsbericht Ep 5/9-3, Universität Stuttgart(1985).

[22] BALLMANN, J., HAAG, K., WIEMER, P.: "Vorderkantenwirbel an Deltaflügeln mit gekrümmten Anströmkanten", Forschungsbericht, Rheinisch-Westfälische Technische Hochschule Aachen (1985).

[23] WAGNER, S., URBAN, C.: "Berechnung der nichtlinearen aerodynamischen Kräfte und Momente interferierender Tragflächen", Bericht Nr. 85/1, Universität der Bundeswehr München (1985).

[24] URBAN, C., WAGNER, S.: "Ein Wirbelgitterverfahren zur Berechnung nichtlinearer aerodynamischer Charakteristika von interferierenden Tragflächen", DGLR-Nr. 85-122 (1985).

[25] BRUNE, G.W., WEBER, J.A., JOHNSON, F.T., LU, P., RUBBERT, P.E.: "A Three-Dimensional Solution of Flows over Wings with Leading-Edge Vortex Separation", NASA CR 132709 (1975).

[26] CARMICHAEL, R.L., WOODWARD, F.A.: "An Integrated Approach to the Analysis and Design of Wings and Wing-Body Combinations in Supersonic Flow", NASA TND-3685 (1966).

[27] WOODWARD, F.A.: "An Improved Method for the Aerodynamic Analysis of Wing-Body-Tail Configurations in Subsonic and Supersonic Flow, Part I - Theory and Application", NASA CR-2228 (1973).

[28] WOODWARD, F.A.: "USSAERO Computer Program Development, Versions B and C", NASA CR-3227 (1980).

[29] WOODWARD, F.A., LANDRUM, E.J.: "The Supersonic Triplet - A New Aerodynamic Panel Singularity With Directional Properties", AIAA Journal, 18 (1980), pp. 138-142.

[30] WOODWARD, F.A.: "Development of the Triplet Singularity for the Analysis of Wings and Bodies in Supersonic Flow", NASA CR-3466, (1981).

[31] EHLERS, F.E., JOHNSON, F.T., RUBBERT, P.E.: "A Higher Order Panel Method for Linearized Supersonic Flow", AIAA Paper 76-381 (1976).

[32] EHLERS, F.E., EPTON, M.A., JOHNSON, F.T., MAGNUS, A.E., RUBBERT, P.E.: "An Improved Higher Order Panel Method for Linearized Supersonic Flow", AIAA Paper 78-15 (1978).

[33] EHLERS, F.E., EPTON, M.A., JOHNSON, F.T., MAGNUS, A.E., RUBBERT, P.E. "A Higher Order Panel Method for Linearized Supersonic Flow", NASA CR-3062 (1979).

[34] MAGNUS, A.E., EHLERS, F.E., EPTON, M.A.: "PAN AIR - A Computer Program for Predicting Subsonic or Supersonic Linear Potential Flow About Arbitrary Configurations Using a Higher Order Panel Method", NASA CR-3251, (1980).

[35] CARMICHAEL, R.L., ERICKSON, L.L.: "PAN AIR - A Higher Order Panel Method for Predicting Subsonic or Supersonic Linear Potential Flows About Arbitrary Configurations",

AIAA Paper 81-1255 (1981).

[36] TINOCO, E.N., JOHNSON, F.T., FREEMAN, L.M.: "Application of a Higher Order Panel Method to Realistic Supersonic Configurations", J. Aircraft, 17 (1980), pp. 38-44.

[37] CENKO, A.: "Advances in Supersonic Configuration Design Methods", AIAA Paper 79-0233 (1979).

[38] CENKO, A., TINOCO, E.N., DYER, R.D., DE JONGH, J.: "PAN AIR Applications to Weapons Carriage and Separation", J. Aircraft, 18 (1981), pp. 128-134.

[39] FORNASIER, L.: "Calculation of Supersonic Flow Over Realistic Configurations by an Updated Low-Order Panel Method", AIAA 83-0010 (1983).

[40] FORNASIER, L.: "Treatment of Supersonic Configurations by an Updated Low-Order Panel Method", J. Aircraft, 21, (1984), pp. 227-237.

[41] WOODWARD, F.A., FORNASIER, L.: "Investigation of the Triplet Concept Using a Higher-Order Supersonic Panel Method", ICAS 84-111 (1984).

[42] FORNASIER, L.: "Linearized Potential Flow Analysis of Complex Aircraft Configurations by HISSS, a Higher-Order Panel Method", AIAA - 85-0281 (1985).

[43] FORNASIER, L., RIZZI, A.: "Comparisons of Results from a Panel Method and an Euler Code for a Cranked Delta Wing", AIAA - 85-4091 (1985).

[44] VERHAGEN, N.G.: "Measurement of the Pressure Distribution on a Biconvex Delta Wing with Aspect Ratio 1", Dept. of Aeronautical Engineering, Delft University of Technology, Delft, the Netherlands, unpublished report (1977).

[45] BEHRBOHM, H.: "Basic Low Speed Aerodynamics of the Short Coupled Configuration of Small Aspect Ratio", SAAB TN-60, Sweden (1965).